T0074741

Advanced Computational Techniques for Sustainable Computing

Advanced Computational Techniques for Sustainable Computing is considered a multidisciplinary field encompassing advanced computational techniques across several domains, including Computer Science, Statistical Computation, and Electronics Engineering. The core idea of sustainable computing is to deploy algorithms, models, policies, and protocols to improve energy efficiency and management of resources, enhancing ecological balance, biological sustenance, and other services on societal contexts.

FEATURES

- Offers comprehensive coverage of the most essential topics
- Provides insight on building smart sustainable solutions
- Includes details of applying mining, learning, Internet of Things, and sensor-based techniques for sustainable computing
- Details data extraction from various sources followed with preprocessing of data and how to make effective use of extracted data for application-based research
- Involves practical usage of data analytic language, including R, Python, etc., for improving sustainable services offered by multidisciplinary domains
- Encompasses comparison and analysis of recent technologies and trends
- Includes development of smart models for information gain and effective decision-making with visualization

Readers will become acquainted with the utilization of massive datasets for intelligent mining and processing. This book includes the integration of data-mining techniques for effective decision-making in the social, economic, and global environmental domains to achieve sustainability. The implementation of computational frameworks can be accomplished using open-source software for the building of resource-efficient models. The content of this book demonstrates the usage of data science and the Internet of Things for the advent of smart and realistic solutions for attaining sustainability.

Advanced Computational Techniques for Sustainable Computing

Edited by
Megha Rathi
Adwitiya Sinha

CRC Press is an imprint of the
Taylor & Francis Group, an **informa** business
A CHAPMAN & HALL BOOK

First Edition published 2023
by CRC Press
6000 Broken Sound Parkway NW, Suite 300, Boca Raton, FL 33487-2742

and by CRC Press
4 Park Square, Milton Park, Abingdon, Oxon, OX14 4RN

CRC Press is an imprint of Taylor & Francis Group, LLC

Library of Congress Cataloging-in-Publication Data
Names: Rathi, Megha, editor. | Sinha, Adwitiya, editor.
Title: Advanced computational techniques for sustainable computing /
edited by Megha Rathi, Adwitiya Sinha.
Description: First edition. | Boca Raton : Chapman & Hall/CRC Press, [2022] |
Includes bibliographical references and index. |
Identifiers: LCCN 2021027887 (print) | LCCN 2021027888 (ebook) |
ISBN 9780367495220 (hbk) | ISBN 9780367495282 (pbk) | ISBN 9781003046431 (ebk)
Subjects: LCSH: Sustainable engineering–Data processing. |
Sustainable development–Data processing. | Industries–Data processing.
Classification: LCC TA170 .A35 2022 (print) | LCC TA170 (ebook) | DDC 628–dc23
LC record available at https://lccn.loc.gov/2021027887
LC ebook record available at https://lccn.loc.gov/2021027888

ISBN: 978-0-367-49522-0 (hbk)
ISBN: 978-0-367-49528-2 (pbk)
ISBN: 978-1-003-04643-1 (ebk)

DOI: 10.1201/9781003046431

Typeset in Palatino
by codeMantra

Contents

Preface

Sustainable computing is considered a multidisciplinary field with primary emphasis on applications of advanced computational techniques from cross disciplines, including Information Technology, Statistics, Electronics, and Computer Science, to improve and manage sustainable evolution. This encompasses the development of intelligent models, frameworks, and applications for effective decision-making dealing with the overall administration and management of the resources for providing solutions to critical and real-world problems. The core idea of sustainable computing is to explore the variety of aspects that may have societal, economic, or ecological impacts. It includes a large spectrum of applications that require addressing several open challenges associated with real-world problem domains by applying a computational paradigm, which involves new architectures and technological approaches. Sustainability allows framing policies and developing strategies to reduce overall power consumption, recycle waste products, minimize resource and energy requirements, enhance the quality of ecological, manage sports, and improve biological and social services. It also aims at reducing wastage of natural reserves, thereby improving the efficacy of utilization, monitoring, management, and critical surveillance of primary resources.

Developing sustainable ecofriendly solutions for real-world problems requires a wide range of disciplines, such as statistics, optimization, decision support systems, data science, artificial intelligence, machine learning, and big social mining. It helps in building novel algorithms, models, computational frameworks, and techniques for real-time as well as high-performance computing. Sustainable computing is a holistic approach that extends from modeling power-aware systems to societal development and combating terrorism, and from efficient usage of solar power to performing sustainable resource management. The notion of sustainable computing estimates the impact and benefit of leading-edge technology and state-of-the-art techniques.

Our book is about the development and application of smart computational techniques for offering smart solutions to achieve sustainability. The readers would get acquainted with the utilization of massive datasets for intelligent mining and processing. It includes the integration of data-mining techniques for effective decision-making in the social, economic, and global environmental domains to achieve sustainability. The implementation of computational frameworks can be accomplished using open-source software for the building of resource-efficient models. The content of the book demonstrates the usage of data science and the Internet of Things for the advent of smart and realistic solutions for attaining sustainability.

Editors

Dr. Megha Rathi has obtained her Ph.D. in Computer Science from Banasthali University and is presently working as an Assistant Professor (Senior Grade) in the Department of Computer Science, Jaypee Institute of Information Technology (JIIT), Noida, Sector 62, Uttar Pradesh, India. She has 10 years of teaching experience and has worked on a research project at the National Informatics Centre (NIC), Delhi. She has experience in software development and worked as a Project Associate at the Indian Institute of Technology (IIT), Delhi. She has organized several special sessions at international conferences and also delivered invited talk. Her research areas include sustainable computing, data mining, data science, health analytics, and machine learning.

Dr. Adwitiya Sinha has received her Ph.D. in Computer Science from School of Computer & Systems Sciences, Jawaharlal Nehru University (JNU), New Delhi, India in 2015. She has obtained Masters of Technology in Computer Science and Technology in 2010 from JNU. She was awarded with First Rank certificate in M.Tech. batch 2008. She received Senior Research Fellowship (SRF) from the Council of Scientific & Industrial Research (CSIR), New Delhi and also awarded research scholarship from the University Grants Commission (UGC) for her research in wireless sensor networks. She has published more than 80 papers in international journals, international conferences, book chapters, and books. She has delivered several invited talks in lecture series on Networks & Graphs, Wireless Sensor Networks, and Performance Analysis of Computing Systems, organized by the Consortium for Educational Communication (CEC), UGC, New Delhi in the form of EDUSAT live lectures. Presently, she is working as an Associate Professor in Jaypee Institute of Information Technology, Uttar Pradesh, India. She is a Senior Member of IEEE. Her research area includes complex network analysis, online social media, data science, large-scale graphs, and performance analysis of wireless sensor and actuator networks.

Contributors

Jigyasa Agarwal
Department of Computer Science & Engineering and Information Technology
Jaypee Institute of Information Technology
Noida, India

Mitushi Agarwal
Department of Computer Science & Engineering and Information Technology
Jaypee Institute of Information Technology
Noida, India

Ayushi Aggarwal
Department of Computer Science & Engineering and Information Technology
Jaypee Institute of Information Technology
Noida, India

Ujjwal Alreja
Department of Computer Science & Engineering and Information Technology
Jaypee Institute of Information Technology
Noida, India

Abhinna Arjun
Department of Computer Science & Engineering and Information Technology
Jaypee Institute of Information Technology
Noida, India

Nitya Arora
Department of Computer Science & Engineering and Information Technology
Jaypee Institute of Information Technology
Noida, India

Palak Arora
Department of Computer Science & Engineering and Information Technology
Jaypee Institute of Information Technology
Noida, India

Jagriti Bhandari
Department of Computer Science & Engineering and Information Technology
Jaypee Institute of Information Technology
Noida, India

Suparna Biswas
Department of Computer Science and Engineering
Maulana Abul Kalam Azad University of Technology, West Bengal
Kolkata, India

Ekam Singh Chahal
Jaypee Institute of Information Technology
Noida, India

Satish Chandra
Department of Computer Science & Engineering and Information Technology
Jaypee Institute of Information Technology
Noida, India

Paras Chaudhary
Department of Computer Science & Engineering and Information Technology
Jaypee Institute of Information Technology
Noida, India

Anirban Dutta
Department of Computer Science & Engineering and Information Technology
Jaypee Institute of Information Technology
Noida, India

Muskan Garg
Department of Computer Science & Engineering and Information Technology
Jaypee Institute of Information Technology
Noida, India

Sherry Garg
Department of Computer Science & Engineering and Information Technology
Jaypee Institute of Information Technology
Noida, India

Amogh Sanjeev Gupta
Department of Computer Science & Engineering and Information Technology
Jaypee Institute of Information Technology
Noida, India

Anish Gupta
Department of Information Technology
Apex Institute of Technology
Chandigarh University
Ghaziabad, India

Ayush Gupta
Department of Computer Science & Engineering and Information Technology
Jaypee Institute of Information Technology
Noida, India

Chandna Gupta
Department of Computer Science & Engineering and Information Technology
Jaypee Institute of Information Technology
Noida, India

Chetna Gupta
Department of Computer Science & Engineering and Information Technology
Jaypee Institute of Information Technology
Noida, India

Ritik Gupta
Department of Computer Science & Engineering and Information Technology
Jaypee Institute of Information Technology
Noida, India

Ankur Haritosh
Department of Computer Science & Engineering and Information Technology
Jaypee Institute of Information Technology
Noida, India

Somya Jain
Department of Computer Science & Engineering and Information Technology
Jaypee Institute of Information Technology
Noida, India

Madhuri Jha
School of Computational and Integrative Sciences (SCIS)
Jawaharlal Nehru University
New Delhi, India

Dhananjay Jindal
Department of Computer Science & Engineering and Information Technology
Jaypee Institute of Information Technology
Noida, India

Parul Jindal
Department of Computer Science & Engineering and Information Technology
Jaypee Institute of Information Technology
Noida, India

Shishir Khandelwal
Department of Computer Science & Engineering and Information Technology
Jaypee Institute of Information Technology
Noida, India

Rajalakshmi Krishnamurthi
Department of Computer Science & Engineering and Information Technology
Jaypee Institute of Information Technology
Noida, India

Aditya Lahiri
Department of Computer Science & Engineering and Information Technology
Jaypee Institute of Information Technology
Noida, India

Aradhya Mathur
Department of Computer Science & Engineering and Information Technology
Jaypee Institute of Information Technology
Noida, India

Prapti Miglani
Department of Computer Science & Engineering and Information Technology
Jaypee Institute of Information Technology
Noida, India

Anushka Mittal
Department of Computer Science & Engineering and Information Technology
Jaypee Institute of Information Technology
Noida, India

Priyanka Parashar
Department of Computer Science & Engineering and Information Technology
Jaypee Institute of Information Technology
Noida, India

Priyadarshini
Department of Computer Science & Engineering and Information Technology
Jaypee Institute of Information Technology
Noida, India

Ratik Puri
Department of Computer Science & Engineering and Information Technology
Jaypee Institute of Information Technology
Noida, India

Ayush Raj
Department of Computer Science & Engineering and Information Technology
Jaypee Institute of Information Technology
Noida, India

Raja Raubins
Department of Computer Science & Engineering and Information Technology
Jaypee Institute of Information Technology
Noida, India

Rishabh
Department of Computer Science & Engineering and Information Technology
Jaypee Institute of Information Technology
Noida, India

Suruchi Sabherwal
Department of Computer Science and Engineering
J.S.S. Academy of Technical Education, Noida
Noida, India

Vishrut Sacheti
Department of Computer Science & Engineering and Information Technology
Jaypee Institute of Information Technology
Noida, India

Satyam Saini
Department of Computer Science & Engineering and Information Technology
Jaypee Institute of Information Technology
Noida, India

Rachit Shukla
Department of Computer Science & Engineering and Information Technology
Jaypee Institute of Information Technology
Noida, India

Gajendra Pratap Singh
School of Computational and Integrative Sciences
Jawaharlal Nehru University
New Delhi, India

Rizul Singh
Department of Computer Science & Engineering and Information Technology
Jaypee Institute of Information Technology
Noida, India

Sakshi Singh
Department of Computer Science & Engineering and Information Technology
Jaypee Institute of Information Technology
Noida, India

Shresth Singh
Department of Computer Science & Engineering and Information Technology
Jaypee Institute of Information Technology
Noida, India

Tanya Srivastava
Department of Computer Science & Engineering and Information Technology
Jaypee Institute of Information Technology
Noida, India

Manan Thakral
Department of Computer Science & Engineering and Information Technology
Jaypee Institute of Information Technology
Noida, India

Dipanwita Thakur
Department of Computer Science
Banasthali Vidyapith
Rajasthan, India

Mayank Deepak Thar
Department of Computer Science & Engineering and Information Technology
Jaypee Institute of Information Technology
Noida, India

Mayank Tiwari
Department of Computer Science
G L Bajaj Institute of Technology and Management
Greater Noida, India

Omkareshwar Tripathi
Department of Electrical and Electronics
G L Bajaj Institute of Technology and Management
Greater Noida, India

Somya Tripathi
Department of Computer Science and Engineering
J.S.S. Academy of Technical Education, Noida
Noida, India

Vidushi Tripathi
Jaypee Institute of Information Technology
Noida, India

Kartik Tyagi
Jaypee Institute of Information Technology
Noida, India

Japsehaj Singh Wahi
Department of Computer Science
Jaypee Institute of Information Technology
Noida, India

Neeraj Yadav
Department of Computer Science and Engineering
J.S.S. Academy of Technical Education, Noida
Noida, India

1

Sustainable Computing—An Overview

Paras Chaudhary, Adwitiya Sinha, and Somya Jain
Jaypee Institute of Information Technology

CONTENTS

DOI: 10.1201/9781003046431-1

1.1 What Is Sustainability?

Sustainability in a crude sense is the capability of a given set-up to exist and to continue to produce on a long-term basis. However, in the modern context, sustainability has been often defined as the bridging of multiple disciplines with a singular aim of determining the manner in which the resource requirements of the present should be fulfilled without adversely affecting the generations to come in the future (Gewin, 2008). This bridging further paves the way for *Homo sapiens* to do what they have been doing since their very first appearance on the face of the Earth 2.5 million years ago, striving toward better standards of living but this time with more consciousness and a higher sense of responsibility. This incorporation of a sustainability mindset into the general course of development can be termed sustainable development.

1.2 Sustainable Development: Motivations and Obstacles

The term *sustainable development* when searched on Google generates more than 350 million hits (Borowy, 2013). This omnipresent nature of the term lies in sharp contrast with the diffused understanding of the profoundness and eminence of the term in real-life developments. Historically, there have been several attempts to define and describe the term and its functioning, respectively, starting from the mid-20th century. Some of these attempts being those of Kenneth Boulding, Luis Sanchez, International Institute for Environment and Development, International Union for Conservation of Nature, Lester Brown, World Bank, and the World Commission on Environment and Development (WCED) of the United Nations (UN) in the years 1966, 1974, 1979, 1980, 1981, 1983, and 1986 in a report 3 years after its conceptualization, respectively. Despite the aesthetics and simplicity of the definition of sustainable development given by WCED that defines it as fulfilling the present demands without compromising the future generations' ability to fulfill theirs, it is the attempt of Luiz Sanchez at describing eco-development that truly captured the balancing nature of sustainable development by calling it a developmental approach of some ecosystem or locality, in the discussion, that can harmonize ecological and economic aspects to assure optimal use of both the natural and human resources of the area to best meet the aspirations and the needs sustainably of maximum people. Therefore, one can conclude that a successful sustainability strategy reconciles tensions between the following.

1.2.1 Present versus Future Generations

One might argue that the millions of years of human evolution is the direct evidence of the fact that we as a race have mastered the art of living sustainably. This mastery although was never complete and the skills prove insufficient locally, so many times societies have declined, relocated, and/or collapsed because they failed to plan enough. However, now in a globalized world, even local calamities have the potential to threaten the survival and well-being of the whole world. This global era can be defined as *Anthropocene* and in times like these extreme policy changes are needed to urgently stop the present generation's maximizing of benefits for itself that will leave

Present Future

Requirements
Greed
SUSTAINABILITY

FIGURE 1.1
Visual representation of how sustainability is a balancing act between present and future.

nothing but an overwhelming heap of problems, due to resource exhaustion, for the global future generations (Figure 1.1).

1.2.2 Economic versus Environmental Perspectives

Originally, the human economy was always rooted in a context of nature and environment where it is self-evident that the growth is limited. But in the late 19th century, the fragmentation and specialization of the subjects resulted in the elimination of nature from economic theory. Now, economics emphasized mathematical and theoretical modeling methods and ignored contextual and empirical data. Ecological factors were now regarded as "externalities" that were beyond the general interests of economists and hence made the idea of endless growth possible. This separation got so ingrained throughout the 20th century that ecologists and economists found it tough to communicate and agree on issues of importance. And with the human proclivity for more monetary power, majority institutions have been moving with a singular aim of maximizing revenue ignoring the possibility of crashing ecological systems important to human survival. This ignorance toward environmental collapse can also be attributed to human inability to imagine or realistically gauge extremely improbable phenomena (the *black swan* phenomenon), the tendency to underestimate risks related to a factor that also provides benefits, and to ignore obvious signs of danger when other people do so too (bystander apathy). Therefore, a workable concept and policy of sustainable development would overcome these human biases to balance economic and environmental perspectives out (Figure 1.2).

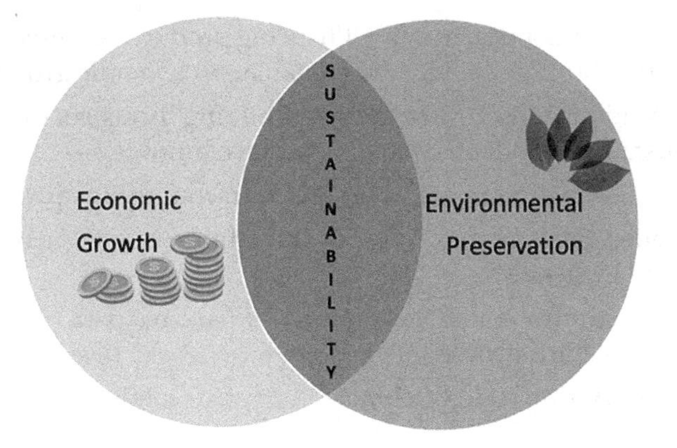

FIGURE 1.2

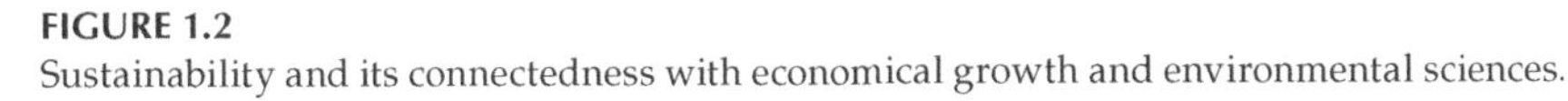

Sustainability and its connectedness with economical growth and environmental sciences.

1.3 Goals to Strive toward Sustainable Development

To implement the theoretical idea of sustainable development and drive policy change in the right direction, it has been broken into three subpillars: social, economic, and environmental. In the 1990s, this idea was promoted by Mohan Munasinghe, a renowned economist and scientist, and was then adopted by United Nations Member States in the Johannesburg Declaration on Sustainable Development's fifth paragraph in September 2002 (DESA, 2014). With the proceedings of world international conferences such as World Summit on Sustainable Development, Rio Declaration on Environment and Development, World Summit for Social Development, Beijing Platform for Action, United Nations Conference on Sustainable Development, and Programme of Action of the International Conference on Population and Development; the states of UN jointly announced 169 associated targets and 17 Sustainable Development Goals (SDGs) that came into effect on January 1 and has been guiding decisions since then and will continue to do so till 2030 (Assembly, 2016). With several leaders from all around the world pledging common action and endeavor across such a universal and broad policy agenda, it became the torchbearer of the idea of sustainable development for the complete benefit of all, for both today's and the future generation. The following are the agenda's 17 agreed-upon SDGs along with their brief descriptions of the UN summit for the adoption of the post-2015 development agenda as per the draft outcome document:

- No Poverty—Eradicating poverty in all possible forms everywhere
- Zero Hunger—Eradicating hunger while promoting sustainable agriculture
- Good Health and Well-Being—Promoting well-being while ensuring healthy lives for all at all ages
- Quality Education—Ensuring quality education that is equitable and conclusive for all
- Gender Equality—Empowering all women and girls and achieving gender equality
- Clean Water and Sanitation—Managing sustainable water and sanitation for all
- Affordable and Clean Energy—Making available reliable, affordable, and modern energy for everyone
- Decent Work and Economic Growth—Ensuring productive employment for everyone by promoting both sustained and sustainable economic growth
- Industry, Innovation, and Infrastructure—Building infrastructure that is resilient to promote sustainable industrialization to drive innovation
- Reduced Inequalities—Reducing inter- and intracountry inequalities
- Sustainable Cities and Communities—Making human settlements like cities safe, sustainable, and inclusive
- Responsible Consumption and Production—Ensuring patterns in consumption and production that are sustainable
- Climate Action—Acting to fight climate change urgently to reduce its long-term impacts

- Life below Water—Conserving the water bodies such as seas and oceans while sustainably using marine resources
- Life on Land—Restoring and promoting sustainable use of ecosystems that are terrestrial by combating desertification, and halting land degradation and biodiversity loss
- Peace, Justice, and Strong Institutions—Providing easy access to justice for everyone by building accountable and effective institutions at all levels
- Partnerships for the Goals—Strengthening the ways to implement and revitalize the Global Partnerships to collaborate to achieve sustainable development (Figure 1.3)

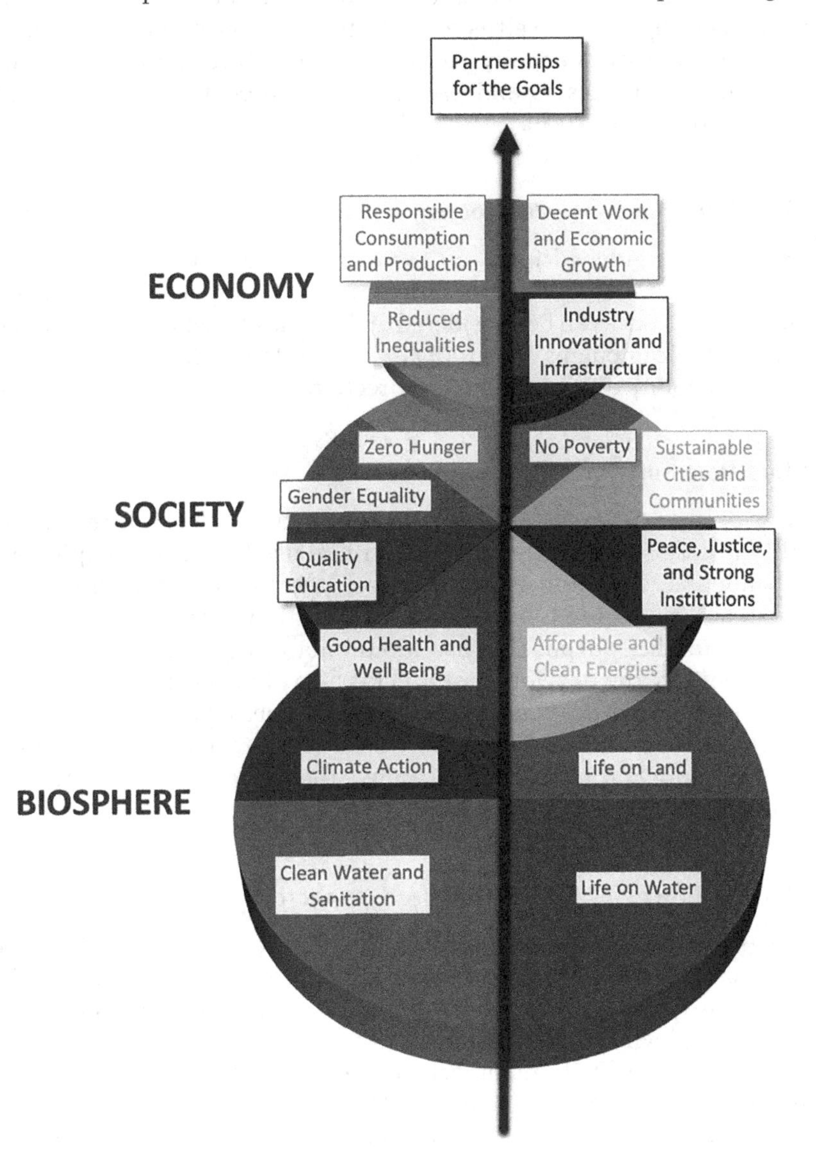

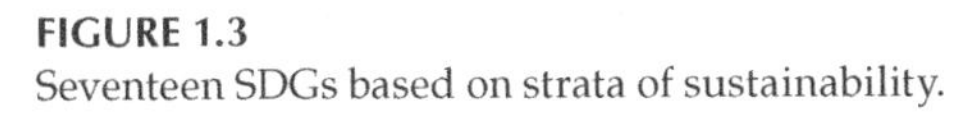

FIGURE 1.3
Seventeen SDGs based on strata of sustainability.

1.4 Sustainability and Computing

In an age that is governed by computing and when the success and failure of almost all human ventures hinge so much on the use of contemporary technologies, it was of prime importance to come up with a concept that could govern and also lay down the foundations of what could be called good and responsible practices in the field of computing. The term *sustainable computing* has been utilized famously to shift the political idea of sustainability to the context of computational systems, which includes both materialistic components known as hardware as well as informational components called software and their respective development as well as consumption processes. On both the levels, that is, hardware and software, sustainability can be said to have three different areas of focus: (a) product, (b) production processes, and (c) consumption processes, giving us the following six focus areas (Mocigemba, 2006).

1.4.1 Product from Hardware Perspective

This can include a heightened focus on the durability and longevity of products and components, for instance, the provision of upgradeable components by physical product manufacturers. Such issues can be considered an integral element of the sustainable development concept; the campaigns for products to have longer life cycles serve as a great example of sustainability discourses in this direction.

1.4.2 Product from Software Perspective

A focus on a product from the software's point of view elicits a certain amount of responsibility in terms of how the product will be used after its completion. For it to be called sustainable, it should either be developed to aid the solving of any of the general sustainability-related issues (like climate action, clean water, etc.) or even if it is developed for another purpose, its use should not go against the concept of sustainability. Accessibility of the developed product is also of prime importance; the technology that empowers only a particular sect of society goes against the global concept of sustainability; and developers of such software should focus on making the products inclusive so that more and more people can use them irrespective of their differences (physical and mental).

1.4.3 Production Processes from Hardware Perspective

The few things that can be tweaked in the orthodox ways of manufacturing computer hardware to make them more pro-sustainable are the conducting of a thorough life cycle assessment and at the same time the enforcing of proper codetermination rights and rights related to employment. For example, the production of certain hardware is highly resource-intensive, especially the manufacturing of processors and processors like Intel486 required 16–19 tons of unprocessed material (2/3rds of that required for an average-sized vehicle throughout its lifecycle). Furthermore, electronics waste can be extremely hazardous and tough to dispose of; hence, it is also the responsibility of the manufacturer to use environmental designs.

1.4.4 Production Processes from Software Perspective

A sustainable production process for software enforces enhanced participation so that the accumulated intellectual effort overrides the need for higher resource allocation and eventual wastage. Therefore, concepts like FreeSoftware and OpenSource are highly relevant for sustainable development discourses. On a similar premise, it has also been seen that while code recycling is a good practice, it is up for debate by the researchers if software patents are pro- or anti-sustainability as they are in economic interest but go against equal opportunities both of which are aspects of sustainability.

1.4.5 Consumption Processes from Hardware Perspective

The electronic waste-recycling requires to be carried out responsibly, as the electronic materials often emit harmful chemicals and radiations. Hence, in order to achieve sustainability, the recycling process should be ensured through maximum reusability, followed by safe discard procedures. It should be noted how the potential energy that could be saved from any optimization that leads to a hardware system utilizing lesser energy for a task is more than compensated by increased demand and eventual use of such products, called the rebound effect (Hilty et al., 2006), and hence to reduce energy consumption one has to employ proper sufficiency strategies. Also whenever possible existing machines should be used, for example, allowing routing or printing to be done on older machines while processing can be done on newer ones, setting up arrangements like these for a certain amount is the business model of companies that call themselves sustainable computing providers.

1.4.6 Consumption Processes from Software Perspective

The consumption processes associated with the software are the most versatile as they have what is famously known as the potential for social transforming, and this majorly is in the collaborative use of the software. Computer software has the power of unifying people from all demographics and geographical locations toward a single objective, and when an objective like this is chosen to be a social one, one that promotes sustainability, major feats in the favor of humanity and the world that it inhabits can be achieved. Furthermore, social network analysis techniques as a tool can be used to operationalize and model the social quality of such software products and thereafter even test the extent to which it was achieved (Figure 1.4).

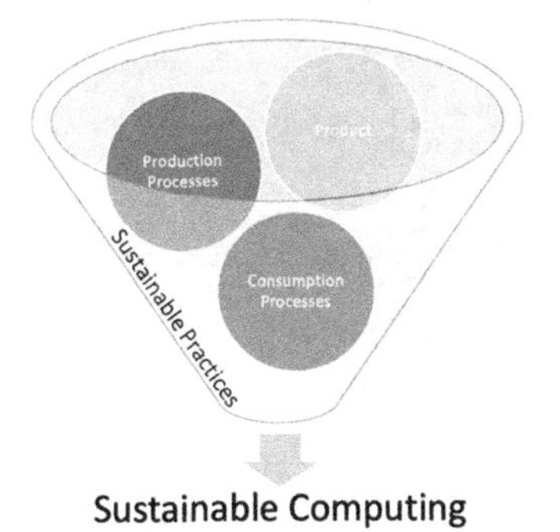

FIGURE 1.4
Combining effect of hardware/software product, its production processes, and its consumption processes.

1.5 Computing Paradigms for Individual Sustainable Development Goals

This section discusses some of the most bleeding-edge technologies that have been researched and developed in a manner to fit the sustainability concept following at least one of the six focus areas discussed above. The section is divided into 17 subheadings, each of which references one of the all 17 SDGs laid down by the UN (Assembly, 2016), to elucidate how significant technical work has been done in all of these subfields (Figure 1.5).

1.5.1 No Poverty

It comes as no surprise that before poverty can be tackled through political policies and initiatives, what is of prime importance is to understand its nature and thoroughly understand its distribution and severity in different geopolitical areas. There has been significant progress in doing the cluster analysis of poverty in South Asian countries where people living in low-income households are many: these results have also been verified through empirical analysis on government records. K-means was applied for the task on Indonesian provincial data (Sano and Nindito, 2016): other machine learning (ML) models like those based on k-Nearest Neighbors, Decision Tree, and Naïve Bayes were tested to classify the bottom 40% of households in Malaysia (Sani et al., 2018), and even Hidden Markov Models have been used for the same in Pasay City (Panaligan et al., 2018). Furthermore, energy poverty has been tried to be mitigated by coming up with automated demand-side management energy-scheduling techniques based on income designed as a mixed-integer linear programming problem (Longe and Ouahada, 2018). Moreover, several graph-based convolutional network approaches have been developed for performing better modeling of the multi-view nature of social networks, thereby outperforming the failed traditional methods (Khan and Blumenstock, 2019).

1.5.2 Zero Hunger

In an age where novel logistical methods are being devised every day to solve problems of world hunger, food delivery systems are more active than ever; it is highly critical for

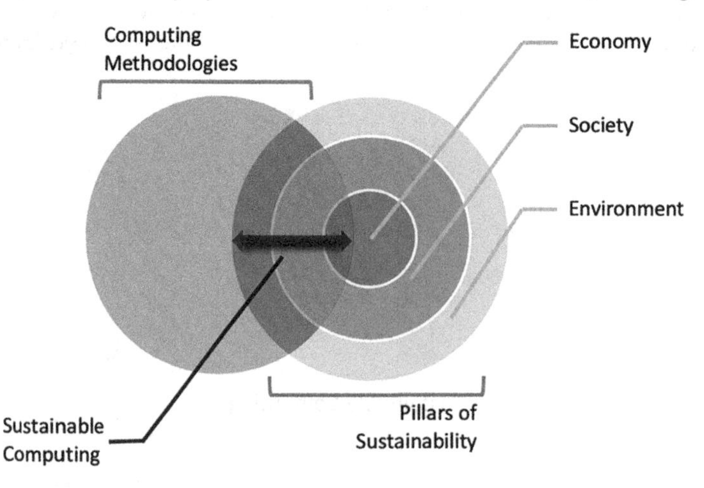

FIGURE 1.5
Three pillars of sustainability based on sustainable computing technologies.

researchers to come up with algorithms that either utilize or aid these systems that are trying to solve hunger through food rescue and delivery. In food rescue that were envy-free allocations, both pickup and delivery were formulated as an NP-hard problem and a Benders' decomposition-based cutting plane algorithm was presented, which used a new heuristics algorithm that utilizes local and greedy search (Rey et al., 2018), while an integrated linear programming model with heuristic based on Tabu Search was also tested to work well for the same task (Nair et al. 2018) and so was Sequential Resource Allocation modeling for nonprofit organizations operating in this field (Yang, 2018). Even a portal that achieved the same incompleteness was implemented, called eFeed-Hungers.com (Sharma et al., 2018). Significant advances have been made in more creative solutions to world hunger too like the conception of a matching and ML algorithms-based recommendation system for the nourishment of children (Banerjee and Nigar, 2019), use of remote sensing and artificial intelligence to identify food insecure zones (Priyadarshini et al., 2018), and for the Indian context of food grain supply chain devising a multiperiod inventory transportation model (Mogale et al., 2017).

1.5.3 Good Health and Well-Being

Similar to the importance of solving scheduling and routing problems in Section 1.4.2, it is of prime importance to solve them for timely access to Home Health Care systems but with different parameters as the priorities are different in this scenario (Rathi et. al., 2019, 2020). The problem has been tried to be solved using a hybrid genetic algorithm that considers the demand to be fuzzy (Shi et al., 2017) and another that uses a memetic algorithm for the same (Decerle et al., 2018). The biggest and the most sensitive element of the healthcare systems is the patient data or the electronic health records; oftentimes, these data can be huge and might serve as the starting point for big data analysis. Therefore, it is of prime importance that these data are consistent as well as complete. This problem of identification of patients with high data-completeness has been done by algorithms that were externally validated for comparative effectiveness research by Lin et al. (2020) and health big data-based global maximum frequent item-sets are effectively mined through an incremental mining algorithm (He et al. 2017). Such records after having cutting-edge ML algorithms applied to them have resulted in successful identification of Autism Spectrum Patient Cohort (Lingren et al., 2016), bleeding events in admitted patients (Moriarty et al., 2017), acute kidney injury (Mohamadlou et al., 2018), and colonoscopy indication (Burnett-Hartman et al., 2019).

1.5.4 Quality Education

Education like all other fields has been deeply impacted by the advent of technology, and based on the consumption of education industry in the last 40 years, it can be segregated into the following distinct stages, namely the growth of computer-based instructions from 1976 to 1986, standalone multimedia-based learning from 1987 to 1996, collaborative learning using networked computers from 1997 to 2006, and digital age's online learning from 2007 to 2016 and beyond (Zawacki-Richter, 2018). Apart from these mainstream adoptions of computers in learning, several creative approaches to the teaching methodology have been researched, like flipped classrooms in which video lectures are provided to see at home and authentic problem-solving work (previously assigned as homework) to be done in a monitored setting of the school hours (Fulford and Paek, 2017). Furthermore, a lot of studies have worked toward suggesting the effectiveness of smart classes

(Abdel-Basset et al., 2019; Hinostroza et al., 2016) and finding sustainable ways to achieve the same (Ray and Dhaheri, 2017).

1.5.5 Gender Equality

In the world, which is offloading rapidly its menial/clerical tasks to automated systems and slowly its cognitive tasks to artificially intelligent systems, it is of prime importance that all such systems be fair and unbiased toward both their recognition of gender and their treatment of it post-recognition (i.e., if required). It has been well documented that since ML and AI systems are trained primarily by observing, any sort of under-representation of women could undo years of advancements in gender equality to perpetuate modern gender ideologies that disadvantage women (Leavy, 2018). Also, it was noted after studying three commercial gender classification systems that dark-skinned females were the most misclassified group and the error rates could be as high as 34.7% (Buolamwini and Gebru, 2018). Some of the other highly relevant researches in this field have been detecting gender stereotypes in online education, through supervised learning (Silva, 2019), and suggesting intervention techniques, through a decision tree algorithm (Sonza and Tumibay, 2020).

1.5.6 Clean Water and Sanitation

Because of its omnipresence in nature and its immense importance in keeping almost all living beings alive, water has been heavily studied in computer literature in the past decades. Since the advent of molecular simulations in the 1960s, a cohort of potential models have been researched and then tested by computer simulations (Guillot, 2002). Apart from simulations, computer systems have primarily been used to achieve two sustainable goals, namely monitoring water flow/level in clean/consumable water to avoid shortage/wastage (Sood et al., 2013; Karray et al., 2016; Johari et al., 2011) and predicting the levels of a naturally occurring water body at a given point in time. Prediction tasks have been primarily achieved using neural networks (Arbain and Wibowo, 2012) and at times using artificial neural networks to first predict attributes, like precipitation, in an area and then feeding the predicted attributes along with the other parameters to finally predict the water level (Bustami et al., 2002).

1.5.7 Affordable/Clean Energy

An affordable and reliable availability of energy is heavily reliant on the fact that the energy is not being created in abundance, and hence needs to be judiciously utilised without wastage. Wastage leads to lower availability of resources that in turn increases the costs of being able to afford it, and hence, it is of prime importance that in the age of technology, we exhaustively employ the information and communication solutions to reduce the overall consumption of energy of an area (Kramers et al., 2014). Apart from a general reduction in consumption, the world has also seen a shift from heavy reliance on fossil fuels to cleaner sources of energy like hydrogen fuel cells, solar energy, etc. But this shift has neither been absolute nor binary; the approaches that have worked the best are those of a hybrid nature. Therefore, the task at hand was to find ways of optimizing the tradeoffs to leverage electric power, which is affordable while being able to minimize carbon emissions (Belgana et al., 2014). Particle swarm optimization has been used on Hybrid Renewable Energy Sources to bring down the Levelised Cost of Energy (Amer et al., 2013), and Artificial Bee Colony optimization has been demonstrated on hybrid PV-Biomass energy systems (Singh and Kaushik, 2016). And despite the lower practicability of a stand-alone PV system, there exists a Maximum PowerPoint Tracking algorithm for it (Gomathy et al., 2012).

1.5.8 Decent Work/Economic Growth

Since the theoretical separation of economic growth from its environmental limits, various mathematical models have been proposed to chart historical trends and then use them to make predictions about future conditions. However, a lot of modern computational algorithms have worked toward bringing the context back to economics and hence come up with models that inform the human race toward a better, safer, and more sustainable future. For example, in the context of a digital economy, for making decisions on the feasibility of cooperation for enterprises building machines, a generalized algorithm has been researched (Kobzev et al. 2019); collaborations like these can heavily reduce wastage in the industry as a whole that too while increasing the efficiencies. One of the biggest problems that mar the economy ubiquitously is that of Economic Load Dispatch, which is the momentary determination of the optimal energy output, to meet the system demand, at the least cost, subject to constraints: this has been tried to be solved using exchange market algorithm (Ghorbani and Babaei, 2016), and energy management systems that utilize coyote algorithm for a heightened fuel economy have also been looked into (Fathy et al., 2019). Some interesting studies have also used ML techniques such as Long Short-Term Memory to draw out interesting correlations between the Economy and Armed Conflict in India (Hao et al., 2020). Lastly, economic growth has also been studied using different algorithms (Buletova et al., 2019; Chiarolla, 2019).

1.5.9 Industry, Innovation, and Infrastructure

The digitization age brought in major advancements to all industries worldwide and laid the groundwork for what is now known as the fourth industrial revolution, and an industry that adheres to these requirements of increased reliance on cyber-physical systems, IoT devices, and intelligent systems can be called smart-industry or Industry 4.0. Algorithms have been devised that advise on how to arrange automated production systems to maximize efficiency and make an Industry "4.0" compliant in general (Gurjanov et al., 2018; Ivanov et al., 2016) and also for specific industries too, like instruments making industry (Zakoldaev et al., 2018). There have also been multiple instances across several industries where novel computing techniques have improved their conditions. For instance, Sandmine (Babajani et al., 2019), Molding (Oztemel and Selam, 2017), Woodworking (Petrov and Pelevin, 2017), Insurance (Moradi and Hosseinkhani, 2017), Automobile Parts (Wu et al., 2018), Capital Goods (Chansombat et al., 2019), Additive (Adnan, 2018), etc. provide better industrial processing through optimal scheduling and planning for addressing industry-specific technical problems.

1.5.10 Reduced Inequalities

As discussed in previous sections, there has developed a heavy reliance on almost all industries on cognitive systems that employ ML techniques to automate some of their processes and even for insights. The systems are also trying to model the industry's consumer behavior that can be a very delicate task to both execute and execute without falling prey to the social evil of inequality. ML models are generally trained to minimize average loss, which results in representational disparity, and this disparity can develop because minority groups (say non-native speakers) contribute less (little less or almost absent) to the training process. Therefore, it is important to develop optimizations that mitigate such issues in all future commercial and noncommercial ML projects. A proven way to do this would be to minimize the worst-case risk overall distributions close to the empirical distribution (Hashimoto et al., 2018). Other studies have researched fairness in reinforcement learning (Jabbari et al., 2017) and equality of opportunity in supervised learning (Hardt et al., 2016). The implications of fairness in decision making in intelligent systems on underlying populations' social equity have been discussed formally too (Mouzannar et al., 2019).

1.5.11 Sustainable Cities/Communities

Sustainable cities and the concept of smart cities developed in tandem with each other can be briefly defined as being those cities where information technology is mixed with architecture, infrastructure, and even its inhabitants to an extent to address environmental, economic, and social problems that too in such a manner that when investments in traditional (transport), human-social capital, and modern (ICT) communication infrastructure are made, they in turn fuel a high quality of life that stems from sustainable economic growth, while wisely managing natural resources by the mode of participatory governance (Höjer and Wangel, 2015). Some projects that work to making a city strive toward being called "smart" is one in which social network analysis is being utilized to streamline the city's traffic movement by first recognizing traffic hotspots and even areas that could become one in the future (Jain and Sinha, 2019). Furthermore, any city that employs one or more of the technologies described under one of these 17 SDG headings is working toward being a sustainable city.

1.5.12 Responsible Consumption/Production

Responsible production and consumption are the cornerstones of sustainability; without first monitoring and then adopting responsible practices, such practices can also be adopted before a lot of damage is done instead of waiting for things to get out of hand if reliable and accurate prediction methods are in place. The discourses of consumption and production are often about energy and its sources. Through the years, petroleum energy consumption and production have been correlated with a lot of other national metrics; a study employed a genetic algorithm to estimate it using vehicle ownership and GDP (Ozturk et al., 2004), while other methods of the forecast being the use of neural networks trained by flower pollination method for OPEC petroleum consumption (Chiroma et al., 2016). Electrical energy consumption has been predicted using genetic algorithms (Azadeh and Tarverdian, 2007), that too for an hourly basis for buildings that are net-zero energy (Garshasbi et al., 2016). In addition, good production planning practices can be employed to minimize energy consumption, while increasing efficiency of both single-machine (Yildirim and Mouzon, 2011) and complete shop floors (Liu et al., 2016).

1.5.13 Climate Action

Formerly, there was a heavy reliance on hypothesis-driven statistics for the collection of climate data, while future climate projections were based on computational models that were based on physics. However, a collection of new datasets that provide a more data-centric approach are possible. Especially, complex networks are particularly well suited for both predictive modeling and descriptive analysis tasks (Steinhaeuser et al., 2011), while supervised learning techniques do well in prediction tasks. Nonconvex climate problems have been tried to be solved using global optimization methods, such as evolutionary computation methods, adaptive stochastic methods, and deterministic/hybrid techniques (Moles et al., 2004). Although a lot of research is also being diverted toward Greenhouse Gas (GHG) emissions, which are a big determinant of the global temperatures, particle swarm optimization has been used to predict GHG temperatures (Coelho et al., 2005); and even in water supply mitigation techniques in case of climate change, GHG emissions are one of the multiple objectives that have been taken into consideration in the evolutionary algorithm framework (Paton et al., 2014).

1.5.14 Life below Water

To preserve the aquatic ecology the first and foremost task would be to see it and understand it better: for this task of underwater exploration, a lot of wired thruster-based bots have been used, but a biomimetic soft robotic fish, developed by a team of researchers at MIT, which is remotely operated, equipped with cameras and can do agile maneuvers, looks promising in its abilities to noninvasively explore life underwater by integrating more naturally with the environment (Katzschmann et al., 2018). The surveillance video recorded by such underwater exploratory robots needs to be transmitted successfully for which it is suggested to send a selected group of Autonomous Underwater Vehicles a scalable coded video through the underwater acoustic channel (Rahmati and Pompili, 2019). Maximum entropy modeling of ecological niches of aquatic life and its distribution of population has been successful (King et al., 2017), while life histories along with distribution have been able to map submerged aquatic vegetation when used in tandem with multispectral satellite remote sensing (Luo et al., 2017). Dissolved oxygen, one of the most integral life-giving chemicals in the waters, can also be satisfactorily predicted using gene expression theory (Mehdipour et al., 2017).

1.5.15 Life on Land

Humans control several domestic species and their lifestyles for companionship, conservation, production, and research. With computer vision enabled systems, it becomes easier to track the behavioural traits and movement of the domestic herds. But manually assessing a video of animal behavior is both extremely taxing and prone to human error; hence, 3D computer vision systems have been developed that can further recognize the postures, trajectory, and pattern of movements in animals (Barnard et al., 2016) to solve occlusions that might occur when tracking multiple animals: efficient algorithms have been developed (Rodriguez et al., 2017), and in this way, great inferences can be made from video streams of animals. Even though environmental sensors have always been used to keep track of animals, there has been progress in the direction of sustainability in this particular use case as optimizations that could have the energy consumption of such setups (compared to square-grid) have been researched (Piña-Covarrubias et al., 2019).

1.5.16 Peace/Justice/Strong Institutions

Through the years, the way justice has been granted has changed quite a bit, and now with the presence of computers in almost all settings, it would be impossible to claim that there have not been ways that have been researched that drastically change the orthodox courtroom to make the access to timely justice more feasible. One such research has been assisting access to justice through computers, by Israeli scientists, via modeling of formal jurisprudence. This would allow people self-assess their legal rights in a specific situation through an interview-like procedure that would be easier to access as it would be made available through the Internet (Bar-Sinai et al., 2019). Justice management system prototypes have also been developed and deployed in universities to test their feasibility (Conrad et al., 2019). Clinical workers have also been excessively using computer-mediated intervention techniques to further their goal of providing an effective social justice framework (Dennis, 2018).

1.5.17 Partnerships for the Goals

The Internet has been one of the most pivotal elements that led to geographical boundaries becoming diffuse enough to aid the world reach the levels of globalization. Remote collaborations across continents would have not been possible if the world was not connected to the Internet, and this Internet is a byproduct of the world's advancements in the field of computing and networking. Today, this same Internet can and is being used by people very well to partner across borders irrespective of the nations that they belong to work toward a more sustainable world by accomplishing SDGs together. For example, there are documented accounts of collaboration between American and Swedish students, aided by computers, for gaining a Global Criminal Justice perspective (Gallo et al., 2018). Also, virtual partnerships are engaging students from India and Australia using communication mediated by computers in E-service learning (Harris, 2017).

1.6 Conclusion and Future Scope of Research

Therefore, it can be seen that there has been significant progress in all the 17 SDGs where sustainable computing is concerned significant. However, still certain aspects need to be extended. Some of the shortcomings arise due to the conflicting nature of computing in general and those of the SDGs; for instance, one of the working assumptions of sustainable computing is the computational overhead. Though it is worth the goal that it is solving, this assumption holds weak in front of the goal of responsible consumption and production, particularly in the context of energy, or affordable and clean energy, when we start employing computation heavy methods that take up a lot of electrical energy to automate and achieve tasks that were being done equally well before without all the energy overheads. Another reason where special care and looking into is required is the ubiquitous fast-paced cognitive offloading onto intelligent systems. However, artificial intelligence has its flaws and biases at this stage of development, and hence its use in highly sensitive applications like in criminal law (Huq, 2018) could result in poor decision, which may lead to increased inequalities (Casacuberta, Guersenzvaig, 2019). Hence, we can conclude that a computing task is sustainable not only when it strives to solve a problem, but also is well thought out and has empirically shown the merits of using it to achieve a goal (SDG).

References

Abdel-Basset, M., Manogaran, G., Mohamed, M. and Rushdy, E. (2019) Internet of Things in smart education environment: Supportive framework in the decision-making process, *Concurrency and Computation: Practice and Experience*, 31(10), p. e4515.

Adnan, F. A., Romlay, F. R. M. and Shafiq, M. (2018) Real-time slicing algorithm for Stereolithography (STL) CAD model applied in additive manufacturing industry, In IOP Conference Series: Materials Science and Engineering, Vol. 342, No. 1, p. 012016. IOP Publishing.

Amer, M., Namaane, A. and M'sirdi, N. K. (2013) Optimization of hybrid renewable energy systems (HRES) using PSO for cost reduction, *Energy Procedia*, 42, pp. 318–327.

Arbain, S. H. and Wibowo, A. (2012) Neural networks based nonlinear time series regression for water level forecasting of Dungun River, *Journal of Computer Science*, 8(9), p. 1506.

Azadeh, A. and Tarverdian, S. (2007) Integration of genetic algorithm, computer simulation and design of experiments for forecasting electrical energy consumption, *Energy Policy*, 35(10), pp. 5229–5241.

Babajani, R., Abbasi, M., Taher Azar, A., Bastan, M., Yazdanparast, R. and Hamid, M. (2019) Integrated safety and economic factors in a sand mine industry: A multivariate algorithm, *International Journal of Computer Applications in Technology*, 60(4), pp. 351–359.

Banerjee, A. and Nigar, N. (2019) Nourishment Recommendation Framework for Children Using Machine Learning and Matching Algorithm. In 2019 International Conference on Computer Communication and Informatics (ICCCI), pp. 1–6. IEEE.

Bar-Sinai, M., Tadjer, M. and Vilozni, M. (2019) Computer Assisted Access to Justice via Formal Jurisprudence Modeling. arXiv preprint arXiv:1910.13518.

Barnard, S., Calderara, S., Pistocchi, S., Cucchiara, R., Podaliri-Vulpiani, M., Messori, S. and Ferri, N. (2016) Quick, accurate, smart: 3D computer vision technology helps assessing confined animals' behavior, *PloS One*, 11(7).

Belgana, A., Rimal, B. P. and Maier, M. (2014) Open energy market strategies in microgrids: A Stackelberg game approach based on a hybrid multiobjective evolutionary algorithm, *IEEE Transactions on Smart Grid*, 6(3), pp. 1243–1252.

Borowy, I. (2013) Defining sustainable development for our common future: A history of the World Commission on Environment and Development (Brundtland Commission). Routledge.

Buletova, N. E., Zlochevsky, I. A. and Stepanova, E. V. (2019) Intersectorial Structure of National Economy: Algorithm for Studying Industrialization Rate. In 2nd International Scientific conference on New Industrialization: Global, national, regional dimension (SICNI 2018). Atlantis Press.

Buolamwini, J. and Gebru, T. (2018) Gender shades: Intersectional accuracy disparities in commercial gender classification. In Conference on fairness, accountability and transparency, pp. 77–91.

Burnett-Hartman, A. N., Kamineni, A., Corley, D. A., Singal, A. G., Halm, E. A., Rutter, C. M. and Doria-Rose, V. P. (2019) Colonoscopy indication algorithm performance across diverse health care systems in the PROSPR consortium, *eGEMs*, 7(1).

Bustami, R., Bessaih, N., Bong, C. and Suhaili, S. (2007) Artificial neural network for precipitation and water level predictions of bedup river, *IAENG International Journal of Computer Science*, 34(2).

Casacuberta, D. and Guersenzvaig, A. (2019) Using Dreyfus' legacy to understand justice in algorithm-based processes, *AI & Society*, 34(2), pp. 313–319.

Chansombat, S., Musikapun, P., Pongcharoen, P. and Hicks, C. (2019) A Hybrid Discrete Bat Algorithm with Krill Herd-based advanced planning and scheduling tool for the capital goods industry, *International Journal of Production Research*, 57(21), pp. 6705–6726.

Chiarolla, M. B. (2019) An algorithm for equilibrium in a dynamic stochastic monetary economy, *International Journal of Contemporary Mathematical Sciences*, 14(4), pp. 225–235.
Chiroma, H., Khan, A., Abubakar, A. I., Saadi, Y., Hamza, M. F., Shuib, L. and Herawan, T. (2016) A new approach for forecasting OPEC petroleum consumption based on neural network train by using flower pollination algorithm, *Applied Soft Computing*, 48, 50–58.
Coelho, J. P., de Moura Oliveira, P. B. and Cunha, J. B. (2005) Greenhouse air temperature predictive control using the particle swarm optimisation algorithm, *Computers and Electronics in Agriculture*, 49(3), pp. 330–344.
Conrad, M., Moses, A. and Ngige, J. (2019) A Computer-Aided Justice Management System: A prototype for Universities in Uganda.
Decerle, J., Grunder, O., El Hassani, A. H. and Barakat, O. (2018) A memetic algorithm for a home health care routing and scheduling problem, *Operations Research for Health Care*, 16, pp. 59–71.
Dennis, K. S. (2018) Clinical Social Workers' Use of Computer-mediated Intervention and Social Justice (Doctoral dissertation, Fordham University).
Fathy, A., Al-Dhaifallah, M. and Rezk, H. (2019) Recent coyote algorithm-based energy management strategy for enhancing fuel economy of hybrid FC/Battery/SC system, *IEEE Access*, 7, pp. 179409–179419.
Fulford, C. P. and Paek, S. (2017) Maximizing quality class time using computers for a flipped classroom approach. In 2017 40th International Convention on Information and Communication Technology, Electronics and Microelectronics (MIPRO), pp. 649–654. IEEE.
Gallo, C., Fowlin, J. and Lilja, M. (2018) Gaining a global criminal justice perspective: a computer-supported collaboration between students in Sweden and the United States, *Journal of Criminal Justice Education*, 29(4), pp. 531–550.
Garshasbi, S., Kurnitski, J. and Mohammadi, Y. (2016) A hybrid Genetic Algorithm and Monte Carlo simulation approach to predict hourly energy consumption and generation by a cluster of Net Zero Energy Buildings, *Applied Energy*, 179, pp. 626–637.
General Assembly. (2016) Resolution adopted by the General Assembly on 19 September 2016. A/RES/71/1, 3 October 2016 (The New York Declaration).
Gewin, V. (2008) Sustenance for sustainability, *Nature*, 455(7217), pp. 1276–1276.
Ghorbani, N. and Babaei, E. (2016) Exchange market algorithm for economic load dispatch, *International Journal of Electrical Power & Energy Systems*, 75, pp. 19–27.
Gomathy, S. S. T. S., Saravanan, S. and Thangavel, S. (2012) Design and implementation of maximum power point tracking (MPPT) algorithm for a standalone PV system, *International Journal of Scientific & Engineering Research*, 3(3), pp. 1–7.
Guillot, B. (2002) A reappraisal of what we have learnt during three decades of computer simulations on water, *Journal of molecular liquids*, 101(1–3), pp. 219–260.
Gurjanov, A. V., Zakoldaev, D. A., Shukalov, A. V. and Zharinov, I. O. (2018) Algorithm for designing smart factory Industry 4.0. In IOP Conference Series: Materials Science and Engineering, Vol. 327, No. 2, p. 022111. IOP Publishing.
Hao, M., Fu, J., Jiang, D., Ding, F. and Chen, S. (2020) Simulating the Linkages Between Economy and Armed Conflict in India With a Long Short-Term Memory Algorithm. Risk Analysis.
Hardt, M., Price, E. and Srebro, N. (2016) Equality of opportunity in supervised learning. In Advances in neural information processing systems, pp. 3315–3323.
Harris, U. S. (2017) Virtual partnerships: engaging students in e-service learning using computer-mediated communication, *Asia Pacific Media Educator*, 27(1), pp. 103–117.
Hashimoto, T. B., Srivastava, M., Namkoong, H. and Liang, P. (2018) Fairness without demographics in repeated loss minimization. arXiv preprint arXiv:1806.08010.
He, B., Pei, J. and Zhang, H. (2017) An incremental mining algorithm for global maximum frequent itemsets based on health big data. In 2017 IEEE 2nd Information Technology, Networking, Electronic and Automation Control Conference (ITNEC), pp. 1631–1635. IEEE.
Hilty, L. M., Köhler, A., Von Schéele, F., Zah, R. and Ruddy, T. (2006) Rebound effects of progress in information technology, *Poiesis & Praxis*, 4(1), pp. 19–38.

Hinostroza, J. E., Ibieta, A. I., Claro, M. and Labbé, C. (2016) Characterisation of teachers' use of computers and Internet inside and outside the classroom: The need to focus on the quality. *Education and Information Technologies*, 21(6), pp. 1595–1610.

Höjer, M. and Wangel, J. (2015) Smart sustainable cities: definition and challenges. In ICT innovations for sustainability, pp. 333–349. Springer, Cham.

Huq, A. Z. (2018) Racial equity in algorithmic criminal justice, *Duke LJ*, 68, 1043.

Ivanov, D., Dolgui, A., Sokolov, B., Werner, F. and Ivanova, M. (2016) A dynamic model and an algorithm for short-term supply chain scheduling in the smart factory industry 4.0, *International Journal of Production Research*, 54(2), pp. 386–402.

Jabbari, S., Joseph, M., Kearns, M., Morgenstern, J. and Roth, A. (2017) Fairness in reinforcement learning. In Proceedings of the 34th International Conference on Machine Learning-Volume 70, pp. 1617–1626. JMLR. org.

Jain, S. and Sinha, A. (2019) Social network sustainability for transport planning with complex interconnections, Sustainable Computing: Informatics and Systems, 24, 100351.

Johari, A., Wahab, M. H. A., Latif, N. S. A., Ayob, M. E., Ayob, M. I., Ayob, M. A. and Mohd, M. N. H. (2011) Tank water level monitoring system using GSM network, *International Journal of Computer Science and Information Technologies*, 2(3), pp. 1114–1115.

Karray, F., Garcia-Ortiz, A., Jmal, M. W., Obeid, A. M. and Abid, M. (2016) Earnpipe: A testbed for smart water pipeline monitoring using wireless sensor network, *Procedia Computer Science*, 96, pp. 285–294.

Katzschmann, R. K., DelPreto, J., MacCurdy, R. and Rus, D. (2018) Exploration of underwater life with an acoustically controlled soft robotic fish, *Science Robotics*, 3(16), p. eaar3449.

Khan, M. R. and Blumenstock, J. E. (2019) Multi-GCN: Graph convolutional networks for multi-view networks, with applications to global poverty. In Proceedings of the AAAI Conference on Artificial Intelligence, Vol. 33, pp. 606–613.

King, R. J., Batista-Navarro, R., Nicolas, M., Hilomen, V. and Solano, G. (2017) Ecological niche modelling tool for aquatic life population distribution using maximum entropy model. In 2017 8th International Conference on Information, Intelligence, Systems & Applications (IISA) pp. 1–6. IEEE.

Kobzev, V., Skorobogatov, A. and Izmaylov, M. (2019) The generalized algorithm of making decisions on practicability of cooperation for machine building enterprises in the context of digital economy. In IOP Conference Series: Materials Science and Engineering, Vol. 497, No. 1, p. 012008. IOP Publishing.

Kramers, A., Höjer, M., Lövehagen, N. and Wangel, J. (2014) Smart sustainable cities–Exploring ICT solutions for reduced energy use in cities, *Environmental Modelling & Software*, 56, pp. 52–62.

Leavy, S. (2018) Gender bias in artificial intelligence: The need for diversity and gender theory in machine learning. In Proceedings of the 1st International Workshop on Gender Equality in Software Engineering, pp. 14–16.

Lin, K. J., Rosenthal, G. E., Murphy, S. N., Mandl, K. D., Jin, Y., Glynn, R. J. and Schneeweiss, S. (2020) External validation of an algorithm to identify patients with high data-completeness in electronic health records for comparative effectiveness research, *Clinical Epidemiology*, 12, p. 133.

Lingren, T., Chen, P., Bochenek, J., Doshi-Velez, F., Manning-Courtney, P., Bickel, J. and Barbaresi, W. (2016) Electronic health record based algorithm to identify patients with autism spectrum disorder, *PloS One*, 11(7).

Liu, Y., Dong, H., Lohse, N. and Petrovic, S. (2016) A multi-objective genetic algorithm for optimisation of energy consumption and shop floor production performance, *International Journal of Production Economics*, 179, pp. 259–272.

Longe, O. M. and Ouahada, K. (2018) Mitigating household energy poverty through energy expenditure affordability algorithm in a smart grid, *Energies*, 11(4), p. 947.

Luo, J., Duan, H., Ma, R., Jin, X., Li, F., Hu, W. and Huang, W. (2017) Mapping species of submerged aquatic vegetation with multi-seasonal satellite images and considering life history information, *International Journal of Applied Earth Observation and Geoinformation*, 57, pp. 154–165.

Mehdipour, V., Memarianfard, M. and Homayounfar, F. (2017) Application of Gene Expression Programming to water dissolved oxygen concentration prediction, *International Journal of Human Cap. Urban Management*, 2(1), pp. 1–10.

Mocigemba, D. (2006) Sustainable computing, *Poiesis & Praxis*, 4(3), pp. 163–184.

Mogale, D. G., Dolgui, A., Kandhway, R., Kumar, S. K. and Tiwari, M. K. (2017) A multi-period inventory transportation model for tactical planning of food grain supply chain, *Computers & Industrial Engineering*, 110, pp. 379–394.

Mohamadlou, H., Lynn-Palevsky, A., Barton, C., Chettipally, U., Shieh, L., Calvert, J. and Das, R. (2018) Prediction of acute kidney injury with a machine learning algorithm using electronic health record data, *Canadian Journal of Kidney Health and Disease*, 5, p. 2054358118776326.

Moles, C. G., Banga, J. R. and Keller, K. (2004) Solving nonconvex climate control problems: pitfalls and algorithm performances, *Applied Soft Computing*, 5(1), pp. 35–44.

Moradi, H. and Hosseinkhani, J. (2017) A Data Mining Approach for Analysis of Customer Behavior in order to Improve Policies in Insurance Industry based on Combination of Particle Swarm Optimization and k-Means Algorithm.

Moriarty, J. P., Daniels, P. R., Manning, D. M., O'Meara, J. G., Ou, N. N., Berg, T. M. and Naessens, J. M. (2017) Going beyond administrative data: retrospective evaluation of an algorithm using the electronic health record to help identify bleeding events among hospitalized medical patients on Warfarin, *American Journal of Medical Quality*, 32(4), pp. 391–396.

Mouzannar, H., Ohannessian, M. I. and Srebro, N. (2019) From fair decision making to social equality. In Proceedings of the Conference on Fairness, Accountability, and Transparency, pp. 359–368.

Nair, D. J., Grzybowska, H., Fu, Y. and Dixit, V. V. (2018) Scheduling and routing models for food rescue and delivery operations, *Socio-Economic Planning Sciences*, 63, pp. 18–32.

Oztemel, E. and Selam, A. A. (2017) Bees Algorithm for multi-mode, resource-constrained project scheduling in molding industry, *Computers & Industrial Engineering*, 112, pp. 187–196.

Ozturk, H. K., Ceylan, H., Hepbasli, A. and Utlu, Z. (2004) Estimating petroleum exergy production and consumption using vehicle ownership and GDP based on genetic algorithm approach, *Renewable and Sustainable Energy Reviews*, 8(3), pp. 289–302.

Panaligan, M. C., Yao, J. A. and Arcilla, R. (2018) Clustering Poverty Data from Community Based Monitoring System of Pasay City Using Hidden Markov Models.

Paton, F. L., Maier, H. R. and Dandy, G. C. (2014) Including adaptation and mitigation responses to climate change in a multiobjective evolutionary algorithm framework for urban water supply systems incorporating GHG emissions, *Water Resources Research*, 50(8), pp. 6285–6304.

Petrov, A. and Pelevin, V. (2017) Image binarization algorithm using GPU for woodworking industry applications. In AIP Conference Proceedings, Vol. 1886, No. 1, p. 020104. AIP Publishing LLC.

Piña-Covarrubias, E., Hill, A. P., Prince, P., Snaddon, J. L., Rogers, A. and Doncaster, C. P. (2019) Optimization of sensor deployment for acoustic detection and localization in terrestrial environments, *Remote Sensing in Ecology and Conservation*, 5(2), pp. 180–192.

Priyadarshini, N. K., Kumar, M. and Kumaraswamy, K. (2018) Identification of food insecure zones using remote sensing and artificial intelligence techniques, *The International Archives of the Photogrammetry, Remote Sensing and Spatial Information Sciences*, 42(5), pp. 659–664.

Rahmati, M. and Pompili, D. (2019) UW-SVC: Scalable Video Coding Transmission for In-network Underwater Imagery Analysis. arXiv preprint arXiv:1910.08844.

Rathi, M., Grover, V. and Kheterpal, T. (2020) Dr. Query: A predictive mobile-based healthcare tool-for querying drug, *International Journal of Swarm Intelligence Research (IJSIR)*, 11(1), pp. 44–64.

Rathi, M., Jain, N., Bist, P. and Agrawal, T. (2020). Smart HealthCare Model: An End-to-end Framework for Disease Prediction and Recommendation of Drugs and Hospitals. In High Performance Vision Intelligence, pp. 245–264. Springer, Singapore.

Rathi, M. and Pareek, V. (2019) Mobile based healthcare tool an integrated disease prediction & recommendation system, *International Journal of Knowledge and Systems Science (IJKSS)*, 10(1), pp. 38–62.

Ray, S. and Al Dhaheri, A. (2017) Using Single Board Computers in University Education: A Case Study. In World Conference on Information Systems and Technologies, pp. 371–377. Springer, Cham.

Rey, D., Almi'ani, K. and Nair, D. J. (2018) Exact and heuristic algorithms for finding envy-free allocations in food rescue pickup and delivery logistics, *Transportation Research Part E: Logistics and Transportation Review*, 112, pp. 19–46.

Rodriguez, A., Zhang, H., Klaminder, J., Brodin, T., and Andersson, M. (2017) ToxId: An efficient algorithm to solve occlusions when tracking multiple animals, *Scientific Reports*, 7(1), 1–8.

Sani, N. S., Rahman, M. A., Bakar, A. A., Sahran, S. and Sarim, H. M. (2018) Machine learning approach for bottom 40 percent households (B40) poverty classification, *International Journal Advance Science Engineering Information Technology*, 8(4–2), p. 1698.

Sano, A. V. D. and Nindito, H. (2016) Application of K-means algorithm for cluster analysis on poverty of provinces in Indonesia, *ComTech: Computer, Mathematics and Engineering Applications*, 7(2), pp. 141–150.

Sharma, S., Shandilya, R., Tim, U. S. and Wong, J. (2018) eFeed-Hungers. com: Mitigating global hunger crisis using next generation technologies, *Telematics and Informatics*, 35(2), pp. 446–456.

Shi, Y., Boudouh, T. and Grunder, O. (2017) A hybrid genetic algorithm for a home health care routing problem with time window and fuzzy demand, *Expert Systems with Applications*, 72, pp. 160–176.

Silva, J. D. A. (2019) A supervised learning approach to detect gender stereotype in online educational technologies.

Singh, S. and Kaushik, S. C. (2016) Optimal sizing of grid integrated hybrid PV-biomass energy system using artificial bee colony algorithm, *IET Renewable Power Generation*, 10(5), pp. 642–650.

Sonza, R. L. and Tumibay, G. M. (2020) Decision tree algorithm in identifying specific interventions for gender and development issues, *Journal of Computer and Communications*, 8(2), pp. 17–26.

Sood, R., Kaur, M. and Lenka, H. (2013) Design and development of automatic water flow meter. *International Journal of Computer Science, Engineering and Applications*, 3(3), p. 49.

Steinhaeuser, K., Chawla, N. V. and Ganguly, A. R. (2011) Complex networks as a unified framework for descriptive analysis and predictive modeling in climate science, *Statistical Analysis and Data Mining: The ASA Data Science Journal*, 4(5), pp. 497–511.

United Nations Department of Economic and Social Affairs. (2014) Prototype global sustainable development report. United Nations Department of Economic and Social Affairs, Division for Sustainable Development (UN DESA), New York, NY.

Wu, Q., Wang, X., He, Y. D., Xuan, J. and He, W. D. (2018) A robust hybrid heuristic algorithm to solve multi-plant milk-run pickup problem with uncertain demand in automobile parts industry, *Advances in Production Engineering & Management*, 13(2), 169–178.

Yang, Q. (2018) Optimal Allocation Algorithm for Sequential Resource Allocation in the Context of Food Banks Operations.

Yildirim, M. B. and Mouzon, G. (2011) Single-machine sustainable production planning to minimize total energy consumption and total completion time using a multiple objective genetic algorithm, *IEEE Transactions on Engineering Management*, 59(4), pp. 585–597.

Zakoldaev, D. A., Shukalov, A. V., Zharinov, I. O. and Zharinov, O. O. (2018) Algorithm of choosing type of mechanical assembly production of instrument making enterprises of Industry 4.0. In Journal of Physics: Conference Series, Vol. 1015, No. 5, p. 052033. IOP Publishing.

Zawacki-Richter, O. and Latchem, C. (2018) Exploring four decades of research in Computers & Education, *Computers & Education*, 122, pp. 136–152.

2

Ambient Air Quality Analysis and Prediction Using Air Quality Index and Machine Learning Models—The Case Study of Delhi

Megha Rathi, Muskan Garg, Japsehaj Singh Wahi, and Mayank Deepak Thar
Jaypee Institute of Information Technology

CONTENTS

2.1 Introduction

Air pollution has become one of the primary problems humans have been facing over the years. According to WHO, air pollution is the fifth largest killer in India killing around 1.5 million people every year. Due to the increase in vehicular emissions, construction activities, and thermal power stations, air pollution in India has become the biggest threat to human life. The outdoor air quality has successively worsened due to factors such as increase in population, increasing vehicular pollution, etc., and hence it has become necessary to analyze the air quality level and its impact on human health [Ocak & Turalioglu (2010)].

This study considers the ambient air of the capital city of India, Delhi. The analyses were conducted in four stations in Delhi: Anand Vihar, East Delhi; Punjabi Bagh, West Delhi; Mandir Marg, Central Delhi; and RK Puram, South Delhi. Data have been collected in 4 hours from Central Pollution Control Board (CPCB). According to our previous research paper, we predicted the concentrations of major air pollutants, that is, PM2.5, PM10, NO_2, SO_2, and CO_2, and the overall Air Quality Index (AQI) for four stations in Delhi using different machine learning algorithms [Wahi et al. (2019)]. Great advancements have been made in the prediction of air pollution in the last decade, but still the prediction is not

DOI: 10.1201/9781003046431-2

accurate. In the proposed research, we are using deep learning for the prediction of air pollutants and AQI. Deep learning is just another name for artificial neural networks (ANN) but in a more refined and easier way. They are inspired by a biological neural network and are used to approximate a function that depends on vast numbers of inputs, which are generally unknown. In deep learning we have used the H2O package in R.

Ambient air pollution impacts health adversely and is one of the major causes of harmful diseases and deaths globally. The health impacts range from several harmful diseases such as asthma, chronic heart disease, respiratory infection, and cardiovascular disease to increase the chances of premature deaths.

The symptoms of these diseases are most common as they are because of the adverse effects of major air pollutants. The main aim of our project is to predict the possible diseases one can have in the four areas of Delhi according to the predicted concentration of pollutants in that area.

To predict the disease according to symptoms, a dataset was prepared in which the most common system and the pollutant-specific symptoms were used to create a binary table indicating whether the symptom was present in this particular disease or not (Table 2.1). A set of diseases that were mostly due to air pollution were used and a symptom table was created for them. R Shiny and Android have been used as the user interface for showing the predicted pollutants using the mobile application interface.

TABLE 2.1

Pollutant Information

Pollutant	Sources	Effects
Carbon monoxide: A gas that comes from the burning of fossil fuels, mostly in cars. It cannot be seen or smelt.	Carbon monoxide is released when engines burn fossil fuels. Emissions are higher when engines are not tuned properly, and when fuel is not completely burned. Cars emit a lot of the carbon monoxide found outdoors. Furnaces and heaters in the home can emit high concentrations of carbon monoxide, too, if they are not properly maintained.	Carbon monoxide makes it hard for body parts to get the oxygen they need to run correctly. Exposure to carbon monoxide makes people feel dizzy and tired and gives them headaches. In high concentrations it is fatal. Elderly people with heart disease are hospitalized more often when they are exposed to higher amounts of carbon monoxide.
Nitrogen dioxide: A reddish-brown gas that comes from the burning of fossil fuels. It has a strong smell at high levels.	Nitrogen dioxide mostly comes from power plants and cars. Nitrogen dioxide is formed in two ways: when nitrogen in the fuel is burned, or when nitrogen in the air reacts with oxygen at very high temperatures. Nitrogen dioxide can also react in the atmosphere to form ozone, acid rain, and particles.	High levels of nitrogen dioxide exposure can give people coughs and can make them feel short of breath. People who are exposed to nitrogen dioxide for a long time have a higher chance of getting respiratory infections. Nitrogen dioxide reacts in the atmosphere to form acid rain, which can harm plants and animals.
Particulate matter: A solid or liquid matter that is suspended in the air. To remain in the air, particles usually must be less than 0.1 mm wide and can be as small as 0.00005 mm.	Particulate matter can be divided into two types: coarse particles and fine particles. Coarse particles are formed from sources like road dust, sea spray, and construction. Fine particles are formed when fuel is burned in automobiles and power plants.	Particulate matter that is small enough can enter the lungs and cause health problems. Some of these problems include more frequent asthma attacks, respiratory problems, and premature death.

(Continued)

TABLE 2.1 (*Continued*)

Pollutant Information

Pollutant	Sources	Effects
Sulfur dioxide: A corrosive gas that cannot be seen or smelled at low levels but can have a "rotten egg" smell at high levels.	Sulfur dioxide mostly comes from the burning of coal or oil in power plants. It also comes from factories that make chemicals, paper, or fuel. Like nitrogen dioxide, sulfur dioxide reacts in the atmosphere to form acid rain and particles.	Sulfur dioxide exposure can affect people who have asthma or emphysema by making it more difficult for them to breathe. It can also irritate people's eyes, noses, and throats. Sulfur dioxide can harm trees and crops, damage buildings, and make it harder for people to see long distances.

Source: Jonathan Levy, Harvard School of Public Health. Based on the information provided by the Environmental Protection Agency.

2.2 Literature Survey

Much work has been done to relate the fields of air pollution and how it adversely affects the health of humans. The work done by the author [Bui et al. (2018)] for the prediction of values of the parameters involved in pollution in South Korea shows how deep machine learning technologies have solved the problem of inaccuracies that were often part and parcel of standard techniques. These techniques show strong potential of leveraging most information from given data on meteorological parameters. The correct and accurate prediction helps in relating it to the disease dataset and finding important correlations such as an increase of 10 g/m^3 in PM2.5 concentrations can lead to an elevated 4–8% of cardiopulmonary and lung cancer mortality. In the study, the authors [Xiao et al., (2019)] have provided mathematical models so that we can simulate the effects of air pollution at various concentrations. In another survey, various models have been compared with various deep learning models between themselves and also with artificial algorithms and a guide for implementing deep learning models on the given air pollution data [Ayturan et al. (2018)]. Another significant research work has shown the importance of considering both spaces and time in the various deep learning model to have better accuracy for achieving precise prediction values as much as possible [Soh et al. (2018)]. The basics of deep learning, its development, and how to use it in conjunction with big data to extract maximum knowledge from the same was given very comprehensively in the study [Nguyen et al. (2019)]. Proposed research work has shown an implementation by using PM10 and PM2.5 data using Long Short-Term Memory (LSTM). H2O, a deep learning framework, helps us directly in the prediction as it has a package called airpred as a part of its library that makes the prediction work easier [Kim et al. (2019)]. From the study by Sabath et al. (2018), we understood the basics of deep learning by using it to derive knowledge from electricity load. It also entailed the help of the H20 framework for implementing deep learning.

Various diseases have their causal parent in air pollution. In yet another novel contribution [Guarnieri & Balmes (2014)], the authors have shown the effect air pollution has on asthma and which are the various pollutants that cause it. The authors Spieza et al. (2014) conducted a study that had several patients and tested the onset of pulmonary embolism in these patients. Patients living in areas where the concentration of several particular pollutants is high showed a higher risk in these patients for developing that disease. In their study, Shea et al. (2008) have shown exactly the high convergence that exists between accurate air pollution prediction and disease onset. Accurate data of various pollutants were taken and hospital visits of people with an upper respiratory infection. Simulations are done with different concentrations of diseases to find what other effects they can have on the disease. Another research work [Link & Dockery (2010)] has shown a relationship between cardiac arrhythmia and air pollution. They say that people with a history of cardiac events should keep a watchful eye on the AQI data as they are at a much higher risk. Important conclusions are drawn in the study by Chang et al. (2015), which has shown us that with the increase in the amount of carbon monoxide and nitrogen dioxide the risk of osteoporosis increases. Similar studies are there that show that bone densities also decrease in these situations. Livers are also the target of these foul conditions, and this was shown in the proposed work [Kim et al. (2014)]. The pollutants can increase the toxicity that is associated with the liver and can also lead to inflammation. Work done in the study by Che et al. (2017) has shown the instance of surgery with the given pollutant. Patients with dementia were taken for surgery, but the onset of dementia and its negative reaction to the pollutants negatively impacted the surgery. This becomes a very serious issue.

In yet another novel work, an association was established to prove that certain pollutants, such as CO and O_3, are brought about by the factors such as speed of the wind and traffic congestion, and proven to be hazardous for health [Raaschou-Nielsen et al. (2016)]. A research was conducted to find the relation between the amount of particulate pollution and daily mortality rates [Dockery et al. (1993)]. They found some correlation between the two irrespective of different demographic factors such as gender, age, and BMI. Factors such as occupational exposure and smoking status were also found to be correlated. Another significant research was conducted in the high-population density region where the major contributors to pollution levels were the industrial, commercial, and residential factors [Gupta (2008)]. A correlation value was to be found between the contributing factors and the concentration of different pollutants such as that of PM10, NO_2, SO_2, etc. The data about atmospheric factors of wind speed and direction, rainfall, and humidity were found out from IMD. A study of a similar kind during the colder months found that the pollutant level at these months was larger in magnitude irrespective of data-gathering technique and timeline.

A Bayesian classifier was used in a study [Corani & Scanagatta (2016)] to estimate the chances of a pollutant level surpassing a certain predefined value. The study also found that multiple predictions are required for dependent variables. To correct this, they used a multilabel classifier. The authors Xi et al. (2015) researched more than 70 cities in China and tried to find a model that best fits each city. By conducting experiments, they found that the best output is obtained when we use a different group of feature selection and model selection. The ANN model is used in another research work to predict the data [Ahmad et al. (2014)]. They also found out that during the colder months the pollution level is relatively higher and highlighted the reasons for the same. They also stressed that techniques like ANN will be very important for similar future studies. ANN falls in the bracket of an advanced machine learning algorithm. In another significant research work [Gujral & Sinha (2021)], the relationship between air pollutants and COVID-19 was determined using a hybrid dynamic learning model.

The basic idea is that if we have data points from air-monitoring stations, then we can calculate the relation between the variables by making use of mathematical techniques like regression analysis. We also have statistical models to judge the nearness between the predicted and the real-world measurements in exact conditions. The factors that affect pollutant levels are inherently regarded in the air data used to make and improve the current model. Another beneficial feature of using the model is that it is low cost and requires less material usage. Several meteorological factors like humidity, precipitation, speed of the wind, and temperature affects the pollution level in air. A lot of papers that were mentioned above from 1990 to 2008 have addressed this topic. From then on, the interest shifted and the availability of technology allowed us to use the available information to predict the major pollutants from the data and the cause-and-effect relation identified by various studies to predict the concentration of these pollutants. Later with the development of deep learning the accuracy and precision of the predicted data improved and are now good enough to use and make major policy decisions. Air pollution's relation to the health impacts it has is age old and many papers have included this in their study. Many correlations were found out between a given disease and the air pollutant that caused it. Another important feature that many papers have studied is the ability to simulate the effects by changing a fixed amount of a pollutant concentration. This also helps us to figure out what the future effect of doing a certain activity that creates a certain pollutant will be.

2.3 Materials and Methodology

This section provides a detailed description of all the techniques and methods used.

2.3.1 Study Area

We have analyzed the capital city of Delhi. According to the WHO, Delhi has been regarded as the most polluted city in India. It has a population of 18.8 million (2017) and air pollution kills around 80 lives every day. It has an area of 1485 square kilometers. The four stations from where we have collected the dataset are Mandir Marg, RK Puram, Anand Vihar, and Punjabi Bagh.

2.3.2 Dataset Description

Air quality data of major air pollutants and meteorological factors from January 2016 to March 2018 was collected from CPCB [CPCB (1995)]. The data were collected in 4 hours from the four major areas in Delhi. Major air pollutants were PM10, PM2.5, NO_2, SO_2, and CO, and meteorological factors were AT, BP, RH, WS, WD, and SR. For disease prediction, we selected 21 major diseases, namely Lymphoma, Asthma, Bronchitis, Pneumonia, Influenza, Pneumocystis Carinii Pneumonia, Spasm Bronchitis, Dementia, Delusion, Hypertension, Thrombus, Upper Respiratory infection, Tachycardia Sinus, Migraine Disorder, Hyperbilirubinemia, Delirium, Pneumothorax, Cardiomyopathy, Chronic Obstructive Airway Disease, Neutropenia, and Embolism Pulmonary and 18 major symptoms, namely productive cough, wheezing, chest tightness, neurotic pain, sore throat, clammy skin, rhonchus, flushing, hypoxemia, hyper cambia, snoring, distress respiration, fever, vomiting, yellow septum, stuffy nose, abdominal pain, and sinus rhythm, which are caused mainly by air pollutants. We have made a binary table indicating whether the symptom was present in this particular disease or not (Table 2.2).

2.3.3 Flowgraph

Pictorial representation of the entire proposed work is presented in Figure 2.1.

TABLE 2.2
Dataset Description

Pollutant	Symptoms	Disease
Ozone, PM10, PM2.5, NO_2, SO_2	Wheezing, cough, shortness of breath, nonproductive cough, chest tightness, pleuritic pain, productive cough	Asthma
PM10, NO_2, SO_2	Clammy skin, night sweat, flushing, choke, constipation	Emphysema pulmonary
PM10, NO_2, SO_2	Shortness of breath, chest pain, yellow sputum, wheezing, productive cough, distress respiratory	Embolism pulmonary
Ozone, PM10, PM2.5, NO_2, SO_2	Shortness of breath, cough, wheezing, chest tightness, respiratory distress	Chronic obstructive airway disease
PM2.5	Fever, lethargy, agitation, cough, pain, rhonchus	Dementia
PM10, PM2.5	Shortness of breath, productive cough, sore throat, rhonchus, fever, vomiting	Spasm bronchial
PM2.5, PM10, NO_2, SO_2	Fever, fatigue, difficulty passing urine	Lymphoma
NO_2, PM2.5	Yellow sputum, chill, headache, nonproductive cough, neck stiffness, diarrhea	Pneumocystis carinii pneumonia
PM10, NO_2, SO_2	Chest pain, palpitation, sweating increased, increased weight	Paroxysmal dyspnea
PM2.5	Fever, vertigo, lightheadedness, out of breath	Osteoporosis
PM2.5, O_3	Drowsiness, chest pain, shortness of breath, chest discomfort, sweat	Ischemia
PM2.5, PM10	Palpitation, yellow sputum	Cardiomyopathy

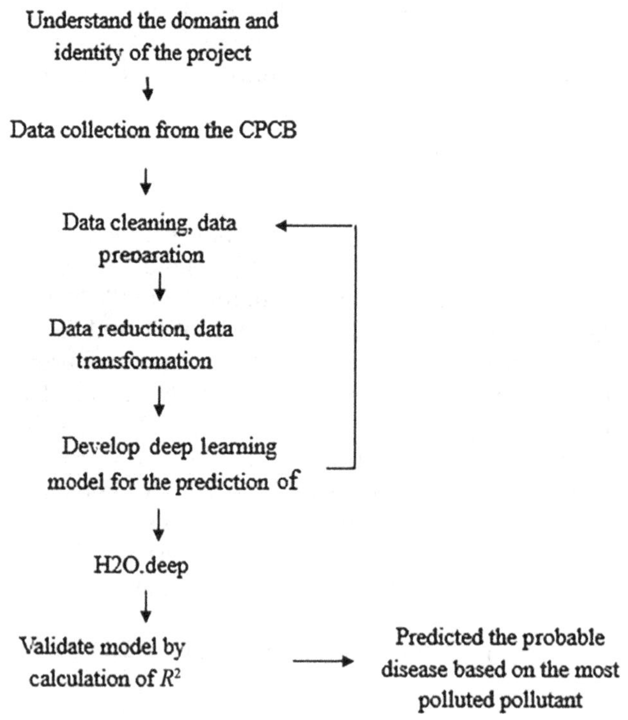

FIGURE 2.1
System architecture of the proposed system.

2.3.4 Data Preprocessing

We matched the meteorological data and air pollutant data through correlation to apply a deep learning technique using the H2O library in R. The data had various missing values for various pollutants and meteorological factors. We replaced the missing values with the mean of the previous day's and the same day's values taken at a 4-hour gap.

To predict the disease, we created a dataset of the most probable symptoms of the various diseases caused by air pollution. We then created a binary dataset mapping the symptoms to their respective diseases.

2.4 Results and Analysis

Statistical software R programming language (R Development Core Team) [Team, R. C. (2013)] and its package openair [Carslaw et al. (2012)] were used to carry out statistical data analysis on pollutants of air to predict AQI and symptoms of disease to predict the disease caused by air pollution. A deep learning technique using the H2O library was used to predict AQI and its health impact. ANOVA [Faraway (2002)] was used to predict the disease due to air pollution.

AQI

Several methods and equations are used for determining the AQI (R Development Core Team) in different countries. We used the basic equation in which AQI can be calculated from the concentration of different pollutants using the following formula:

AQI = [concentration(PM10)/standard(PM10) + concentration(PM2.5)/standard(PM2.5) + concentration(SO_2)/ standard(SO_2) + concentration(NO_2)/ standard(NO_2) + concentration (CO)/ standard(CO)] / 5

Table 2.3 summarizes the various categories of IND-AQI (National Air Quality Index) as approved by the CPCB, New Delhi, 2017. This IND-AQI has six categories.

Table 2.4 summarizes the various categories of IND-AQI as approved by the CPCB, New Delhi, 2017. This IND-AQI has six categories.

H2O

TABLE 2.3

Various Categories of IND-AQI (National Air Quality Index, CPCB, 2017)

AQI Values	Remark	Levels of Health Concern
0–50	Good	Good
51–100	Satisfactory	Moderate
101–200	Moderately polluted	Unhealthy for sensitive groups
201–300	Poor	Unhealthy
301–400	Very poor	Very unhealthy
400–500	Severe	Hazardous

TABLE 2.4

The Means, Maximum, Minimum, and Standard Deviation of Meteorological Parameters and Pollutant Concentration between Years 2016 and 2018

Pollutants and Meteorological Parameters	Mean	Max Value	Min Value	Standard Deviation	Number of Observations
PM_{10} concentration (μg/m³)	306.1749	556.1256	0	205.1429	19,207
$PM_{2.5}$ concentration (μg/m³)	103.6727	535.9634	0	113.1685	19,207
NO_2 concentration (μg/m³)	26.72098	478.2679	0	48.15624	19,207
SO_2 concentration (μg/m³)	21.72098	482.5512	0	18.06924	19,207
CO concentration (mg/m³)	0.876983	502.0934	0	1.54758	19,207
Temperature (°C)	23.52865	46.8500	0	8.264977	19,207
Relative Humidity (%)	52.63869	93.1700	0	18.54115	19,207
Wind Speed (m/s)	0.8996	144.9012	–5.93	2.253728	19,207
Wind Direction	205.591	330.70	0	58.49976	19,207

Note: Negative sign in wind speed indicates that the wind is blowing from north to south (north wind).

Deep learning with neural networks is arguably one of the most rapidly growing applications of machine learning and AI today. H2O's deep learning is based on a multilayer feed-forward ANN that is trained with stochastic gradient descent using back-propagation [Ciaburro & Venkateswaran (2017)]. The network can contain a large number of hidden layers consisting of neurons with tanh, rectifier, and maxout activation functions. A feed-forward ANN model, also known as deep neural network (DNN) or multilayer perceptron (MLP), is the most common type of DNN and the only type that is supported natively in H2O-3.h2o.init is the function that initializes the h2o and makes the required network connections with h2o server online and helps us start an h2o process on our computer. We have then input our data and divided that into training and testing data with a few values kept aside for the prediction data. As h2o does not accept these data, we have to convert them into an h2o frame. This is a handle to the data and all future returns will also be in this format. The next step is to run the h2o.deeplearning function, which will apply the deep learning algorithm and give us back a model that has been created using the training data with validation applied on the testing data and another set of validation done on the full data. A list of different forms of errors was returned to us, and these errors were the criteria on the basis we judged the goodness of the model. Finally, the h2o.predict function was used to judge any future value that we may want to predict.

Disease Prediction

According to the most polluted pollutant list, associated symptoms are generated. These symptoms are provided to the user in the form of a check box using the android app, and according to his/her choices, disease is predicted using a decision tree (Figure 2.2).

The chosen dataset was trained using H2O deep learning on daily data chosen from January 2016 to March 2018. The forecasted values of pollutants were measured with the actual ones. The respective values of R and R^2 are summarized in Table 2.5. The deep learning model is in the acceptable range, thus the model is suited for the AQI prediction and can be used for real-time analysis. The disease has been predicted based on the most polluted pollutants in that area.

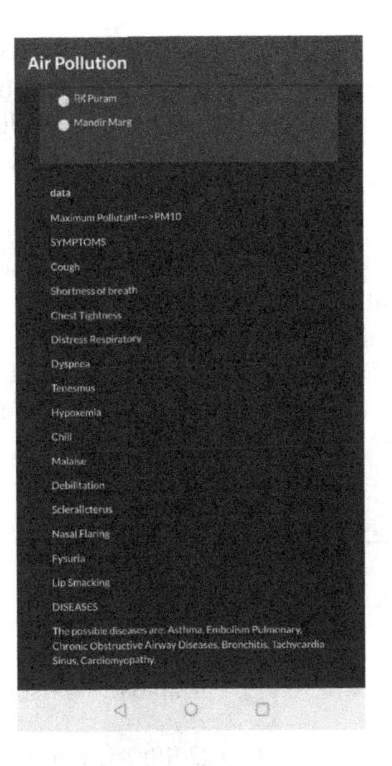

FIGURE 2.2
Screenshot of the app depicting the symptoms and possible diseases.

TABLE 2.5
Model Statistics

Model	R	R^2
H2O. Deep Learning	0.81	0.6561

2.5 Novelty

In our study, we analyzed air quality and predicted the same. The novelty lies at the point where we have compared the three models, namely regression analysis, random forest, and ANN based on the value of R^2. The forecasted value of pollutants has been analyzed by all three methods. Based on the comparison and hence finding out which model is best is then used for predicting the AQI. Moreover, associated health impacts concerning the level of AQI are indicated.

2.6 Conclusion

In this study, the relationship between different air pollutants and meteorological parameters was analyzed using three models, i.e., Regression Analysis, Random Forest, and ANN. The maximum concentration of pollutants was observed during the peak hours in the morning as well as during the night mainly in the colder months (December, January).

Increased vehicular emissions are the main contributors to air pollution. Meteorological parameters also play a significant role in determining air quality.

A statistical model was developed to predict the air pollutant concentrations based on meteorological parameters and the previous day's pollutant concentration. The correlation between the previous day's PM10, NO_2, and SO_2 concentrations and actual PM10, NO_2, and SO_2 concentrations were found as 0.77, 0.59, and 0.61, respectively. Temperature, humidity, and wind speed showed a good correlation between pollutants like PM2.5 and NO_2. Hence, it can be concluded that daily pollutant concentrations are not only influenced by different meteorological parameters but also by the values of the previous day's pollutant concentration as well. The neural network models for the forecasting of the air pollution level are sufficiently more effective than the regression technique and random forest.

References

Ahmad, S. S., Nawaz, M., Wahid, A., Malik, B. and Urooj, R. (2014) Scaling of multivariate ground flora data of gatwala forest park (GFP), Faisalabad Pakistan, *World Applied Sciences Journal*, 29(12), pp. 1492–1496.

Ayturan, Y. A., Ayturan, Z. C. and Altun, H. O. (2018) Air pollution modeling with deep learning: a review, *International Journal of Environmental Pollution and Environmental Modelling*, 1(3), pp. 58–62.

Bui, M., Adjiman, C. S., Bardow, A., Anthony, E. J., Boston, A., Brown, S. and Hallett, J. P. (2018) Carbon capture and storage (CCS): The way forward, *Energy & Environmental Science*, 11(5), pp. 1062–1176.

Carslaw, D. C. and Ropkins, K. (2012) Openair—an R package for air quality data analysis, *Environmental Modelling & Software*, 27, pp. 52–61.

Chang, L., Xu, J., Tie, X. and Wu, J. (2016) Impact of the 2015 El Nino event on winter air quality in China, *Scientific Reports*, 6(1), pp. 1–6.

Che, L., Li, Y. and Gan, C. (2017) Effect of short-term exposure to ambient air particulate matter on the incidence of delirium in a surgical population, *Scientific Reports*, 7(1), pp. 1–7.

Ciaburro, G. and Venkateswaran, B. (2017) Neural Networks with R: Smart models using CNN, RNN, deep learning, and artificial intelligence principles. Packt Publishing Ltd.

Corani, G. and Scanagatta, M. (2016) Air pollution prediction via multi-label classification, *Environmental Modeling & Software*, 80, pp. 259–264.

CPCB, D. P. S. and Intra-portal, C. P. C. B. (1995) Central Pollution Control Board. Government of India.

Dockery, D. W., Pope, C. A., Xu, X., Spengler, J. D., Ware, J. H., Fay, M. E. and Speizer, F. E. (1993) An association between air pollution and mortality in six US cities, *New England Journal of Medicine*, 329(24), pp. 1753–1759.

Faraway, J. J. (2002) Practical Regression and ANOVA using R, Vol. 168. Bath: University of Bath.

Guarnieri, M. and Balmes, J. R. (2014) Outdoor air pollution and asthma, *The Lancet*, 383(9928), pp. 1581–1592.

Gujral, H., Sinha, H. (2021) "Association between exposure to airborne pollutants & COVID-19 in Los Angeles, United States with ensemble-based dynamic emission model," Environmental Research, Elsevier, volume 194, no. 110704, pp. 1–12.

Gupta, U. (2008) Valuation of urban air pollution: A case study of Kanpur City in India, *Environmental and Resource Economics*, 41(3), pp. 315–326.

Kim, J. W., Park, S., Lim, C. W., Lee, K. and Kim, B. (2014) The role of air pollutants in initiating liver disease, *Toxicological Research*, 30(2), pp. 65–70.

Kim, S., Kang, S., Ryu, K. R. and Song, G. (2019) Real-time occupancy prediction in a large exhibition hall using a deep learning approach. *Energy and Buildings*, 199, pp. 216–222.

Levy, J. I., Buonocore, J. J. and Von Stackelberg, K. (2010) Evaluation of the public health impacts of traffic congestion: a health risk assessment, *Environmental Health*, 9(1), p. 65.

Link, M. S. and Dockery, D. W. (2010) Air pollution and the triggering of cardiac arrhythmias, *Current Opinion in Cardiology*, 25(1), p. 16.

Nguyen, Q. (2019) On connected sublevel sets in deep learning. arXiv preprint arXiv:1901.07417.

Ocak, S. and Turalioglu, F. S. (2010) Relationship between air pollutants and some meteorological parameters in Erzurum, Turkey. In Global Warming, pp. 485–499. Springer, Boston, MA.

R Core Team. (2013) R: A language and environment for statistical computing.

Raaschou-Nielsen, O., Beelen, R., Wang, M., Hoek, G., Andersen, Z. J., Hoffmann, B. and Nieuwenhuijsen, M. (2016) Particulate matter air pollution components and risk for lung cancer, *Environment International*, 87, pp. 66–73.

Sabath, M. B., Di, Q., Braun, D., Schwartz, J., Dominici, F. and Choirat, C. (2018) Airpred: A Flexible R Package Implementing Methods for Predicting Air Pollution. In 2018 IEEE 5th International Conference on Data Science and Advanced Analytics (DSAA), pp. 577–583. IEEE.

Shea, K. M., Truckner, R. T., Weber, R. W. and Peden, D. B. (2008) Climate change and allergic disease, *Journal of Allergy and Clinical Immunology*, 122(3), pp. 443–453.

Soh, P. W., Chang, J. W. and Huang, J. W. (2018) Adaptive deep learning-based air quality prediction model using the most relevant spatial-temporal relations, *IEEE Access*, 6, pp. 38186–38199.

Spiezia, L., Campello, E., Bon, M., Maggiolo, S., Pelizzaro, E. and Simioni, P. (2014), Short-term exposure to high levels of air pollution as a risk factor for acute isolated pulmonary embolism, *Thrombosis Research*, 134(2), pp. 259–263.

Wahi, J. S., Thar, M. D., Garg, M., Goyal, C. and Rathi, M. (2019) Analysis of air quality and impacts on human health. In Smart Healthcare Systems, pp. 109–123. Chapman and Hall/CRC.

Xi, X., Wei, Z., Xiaoguang, R., Yijie, W., Xinxin, B., Wenjun, Y. and Jin, D. (2015) A comprehensive evaluation of air pollution prediction improvement by a machine learning method. In 2015 IEEE International Conference on Service Operations And Logistics, And Informatics (SOLI), pp. 176–181. IEEE.

Xiao, D., Fang, F., Zheng, J., Pain, C. C. and Navon, I. M. (2019) Machine learning-based rapid response tools for regional air pollution modeling, *Atmospheric Environment*, 199, pp. 463–473.

3

Assessing Land Cover and Drought Prediction for Sustainable Agriculture

Anirban Dutta, Anushka Mittal, Shishir Khandelwal, and Adwitiya Sinha
Jaypee Institute of Information Technology

CONTENTS

3.1 Introduction

Properly collecting relevant raw data and using the data to carry out predictions and analysis will provide insights about the ongoing trends and suggest appropriate measures. The research has been divided into two case studies: drought prediction and assessing land cover. The first case study aims to extract the raw data of drought-affected areas of certain districts of Maharashtra and further empower the farmers to become aware of future climatic changes. Extracting accurate data and implementing correct algorithms can help formulate solutions to lessen the effects of droughts in drought-prone states. The second case study is based on satellite images taken from the Amazon Rainforest dataset and the UCI Machine Learning Repository sites. Specific ways have been implemented to study the satellite images and the tags corresponding to them. Cluster analysis, t-SNE embedding, and Normalized Difference Vegetation Index (NDVI) have helped to analyze the images. Convolution Neural Networks (CNNs) are employed for the classification of the images.

DOI: 10.1201/9781003046431-3

3.1.1 Contribution to Sustainable Development

Factors such as concern for food security, ever-increasing population, and climate change have used innovative approaches such as artificial intelligence and machine learning to improve and protect the crop yield. Such practices can help farmers learn about the techniques they should employ. Carrying out detailed predictions and analyses would help the farmers in making better decisions.

3.2 Related Work

Various methods have been implemented to predict droughts using stochastic linear models and artificial neural networks (ANNs). Specific extraction learning models have been applied to satellite images as well. Some of them are discussed in this section.

Ramos-Giraldo et al. presented the automated system for predicting drought using machine learning algorithms incorporated with computer vision. A real-time drought prediction system has a significant advantage in preventing loss in crop yield because of changing climatic conditions. The paper provides the response of corn and soybean crops during drought and are documented, and the drought status of crop plants is detected using machine learning. In another work, Kaur et al. proposed a framework for evaluating and predicting drought using fog computing and drought cloud computing. There has been a significant reduction in the execution time of the model by reducing the size of data using Principal Component Analysis (PCA). Neural networks are used to assess drought severity and can effectively target different climatic regions and time ranges.

Deo et al. highlighted an ANN model that shows good forecasting of average precipitation. The report proves that for drought prediction models, predictor variables such as historical observations of rainfall levels, evaporation rates, and temperature variables are suitable. The authors have optimized specific neurons and activation functions in their proposed models. According to another relevant study, the concerned authors have illustrated the forecasting future droughts in the Eastern Part of West Bengal. The authors Mishra et al. used stochastic linear models such as Autoregressive Integrated Moving Average (ARIMA) and Seasonal Autoregressive Integrated Moving Average (SARIMA), and the results were compared with real-time data. The models work accurately for predicting the probability of droughts for the coming 1–2 months. These models have shown the potential to forecast reliably toward the goal of drought forecasting. M. Tejaswini. et al. (2019) have performed land cover classification and change detection using high-resolution satellite images. The images were taken over the years 2013 and 2016 with Landsat-7 and are of the Guntur region. VGG19 architecture was employed for change detection and land classification.

Standardized precipitation index has been used for monitoring relative wetness and dryness, and the standardized precipitation evapotranspiration index has been used as the basis of the machine learning model by the authors S. Poornima et al. The article shows that ARIMA and Long Short-Term Memory (LSTM) are implemented on a large dataset. The LSTM model has better accuracy for predictions on a longer timescale; meanwhile, the ARIMA model has better accuracy for predictions on a smaller timescale. Traditional work typically applies the classification method only on a small number of samples giving unsatisfied classification results. The primary focus of the authors Zhao et al. has been on Landsat image classification's performance enhancement using CNNs. PSPNet, a model based on the original CNN, has been applied to learn from the training set's spectral features. Resnet50 was used to fine-tune the model, giving better results with an overall accuracy of 83%. The results of the study prove the efficiency of CNN in large-scale land cover mapping.

The authors Agana et al. investigated the dry spell expectation issue utilizing profound learning calculations. A belief network was proposed comprising Boltzmann Machines for dry spell expectations for long durations using slacked estimations of standardized streamflow index (SSI) as information sources. The proposed model performs better than conventional techniques using Root Mean Square Error (RMSE) and Mean Absolute Error. Further, the researchers Marmanis et al. applied CNN models for classifying the land cover area into the corresponding labels. Models were implemented on UC Merced dataset. Moreover, they also introduced an efficient algorithm to handle large dimensional images. In yet another study, Castelluccio et al. applied deep learning models to classify satellite images. The data were applied to CaffeNet and GoogLeNet with varying learning parameters. Techniques were applied to the data to remove overfitting and reduce compile time. The models showed significant improvement and applicability.

In Section 3.3, we conduct multiple case studies with artificial intelligence-based models and compare the results with existing counterparts, hence achieving suitable results for predicting droughts and analyzing land usage.

3.3 Case Study 1: Drought Prediction for Sustainable Agriculture

It has been observed that every 3 years, India faces the issue of drought. Drought is a concern for India's rainfed areas that account for significant food grain production and host the largest number of farmers. It impacts the poorest people the most, and it has been noted that a farmer takes 4–5 years to recover from a drought depending on severity. Using machine learning, researchers have found ways to predict droughts and help the farmers to plan accordingly. The most critical parameters that need to be studied for drought prediction are pressure, rainfall, and temperature.

3.3.1 Dataset

The features of the dataset are scrapped from various government sites. The pressure and temperature data have been extracted from 2001 to 2018 from UC Merced, 2020; and the rainfall data are extracted from Rainfall Dataset, 2021, using the Python library Beautiful Soup. Pressure, temperature, and rainfall have been extracted for regions shown in Figure 3.1 and are integrated further for predicting drought labels.

3.3.2 Methodology

The workflow of the drought prediction case study is shown in Figure 3.2. Several classifiers like SVM, K-NN, Decision Tree, and Naive Bayes are used for drought prediction.

3.3.3 Implementation

After successfully scraping data, binning is applied to the raw data using Python tools, which helps to convert the continuous data to categorical. This method is implemented on parameters like pressure, temperature, and rainfall. After binning, the drought labels

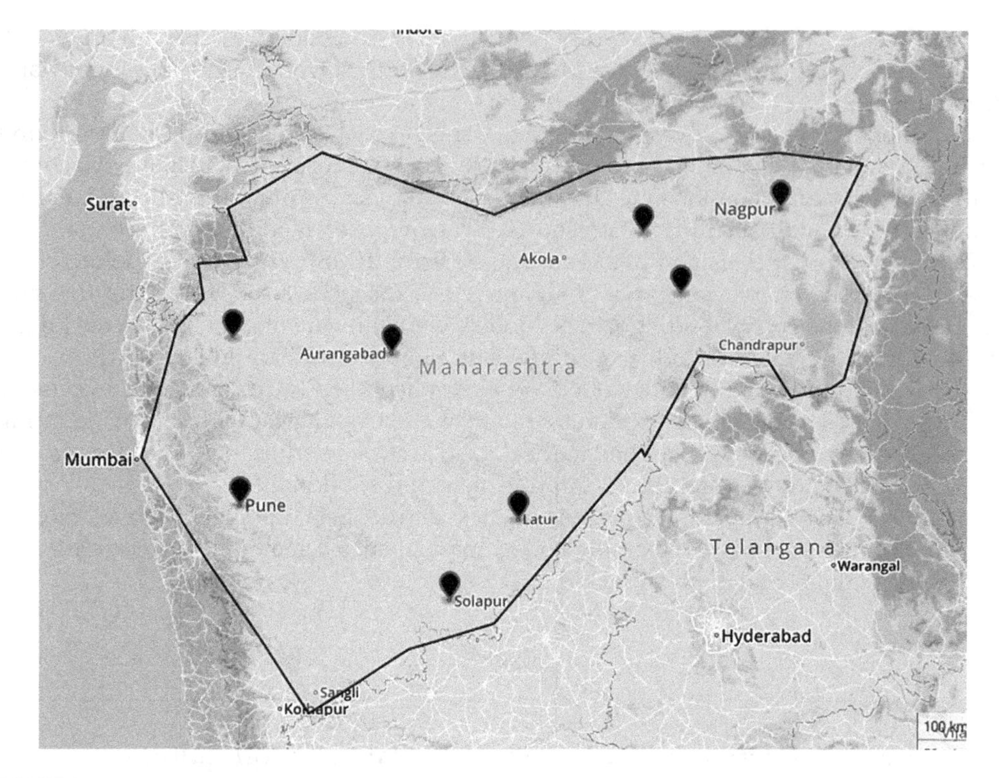

FIGURE 3.1
Drought-affected districts of Maharashtra.

are calculated and analyzed for the selected districts from 2001 to 2018. All the parameters are combined to form a dataset that is to be provided to the classifiers. After the analyses of the dataset, it is observed that the dataset is imbalanced. Hence, under- and oversampling techniques are implemented using the Python tool SMOTE to remove the imbalance (Simonyan and Zisserman, 2014). The final drought prediction dataset is fed into classifiers, and parameters like accuracy, precision, recall, specificity, and sensitivity are calculated.

Figure 3.3 depicts the comparative study between classifiers and their performances. It is observed that the Decision Tree Classifier has provided a maximum accuracy of 95%. Naive Bayes has failed to classify the drought level correctly compared to other classifiers with an accuracy of 69%. Other essential performance metrics are shown in Figure 3.4.

Decision Tree Classifier provides an accuracy of 95% with a specificity of 95% and a sensitivity of 87%.

The ROC curve in Figure 3.5 depicts that the classifier detects more class 0 labels than class 1 labels. The ROC curve of class 0 is more toward the True Positive Rate and away from False Positive Rate. Moreover, the calculated specificity (0.95) is higher than the sensitivity (0.87), proving our analysis to be accurate.

The precision–recall curve as shown in Figure 3.6 is used while handling binary classifications. Because of the minor imbalance between the classes, it is inferred from the graph that the area under the curve of class 0 is greater than the area under class 1. Therefore, the precision–recall of class 0 is higher than that of class 1.

Figure 3.7 depicts the count of True Positive (316), False Positive (26), False Negative (20), and True Negative (350) evaluated by the Decision Tree Classifier on the drought prediction dataset.

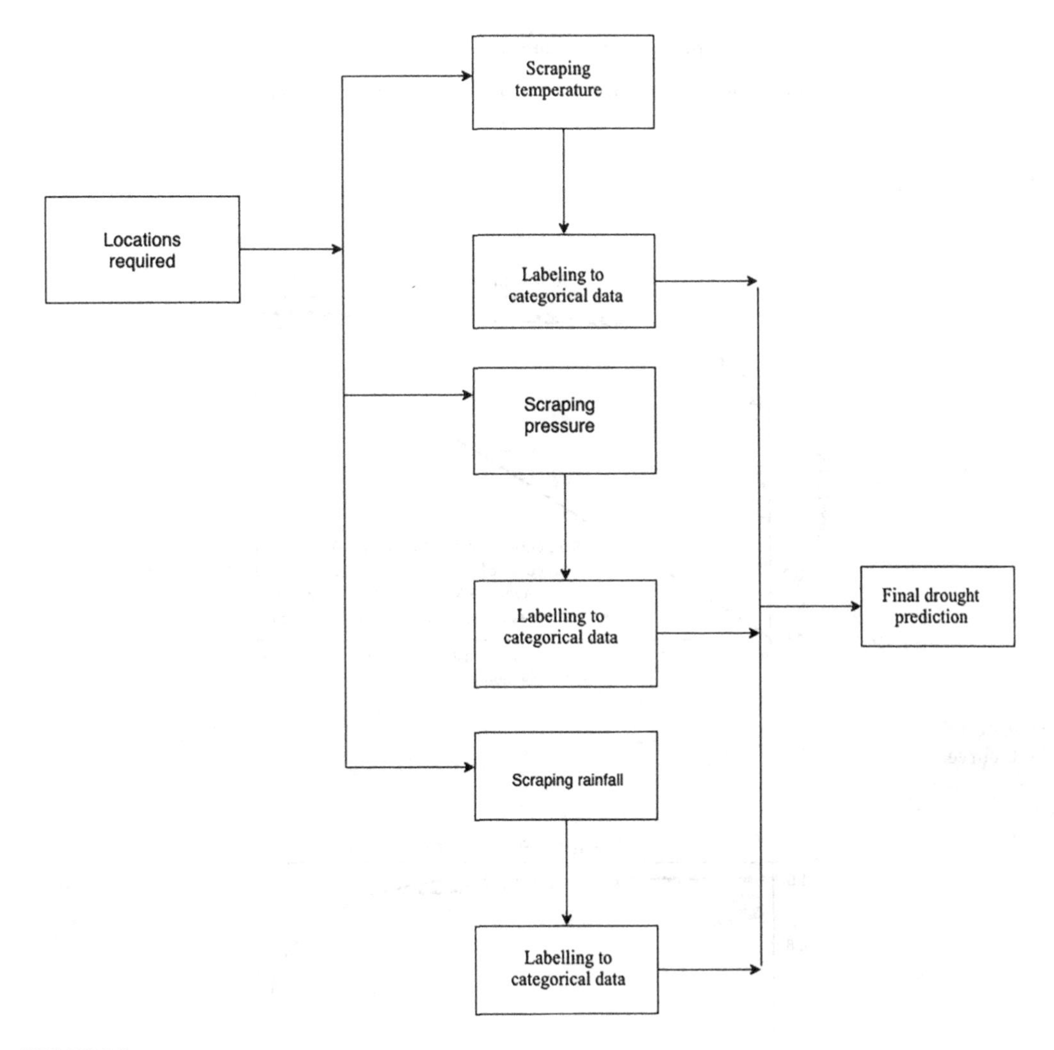

FIGURE 3.2
Flowchart for the drought prediction.

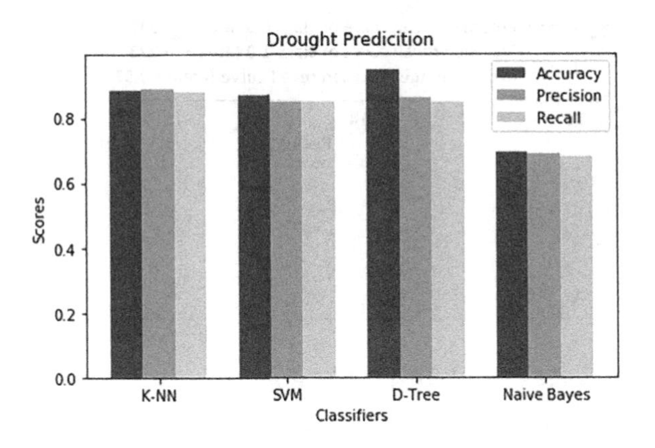

FIGURE 3.3
Classification results of drought-predicted regions.

	precision	recall	f1-score	support
class 0	0.96	0.93	0.95	342
class 1	0.94	0.97	0.95	370

FIGURE 3.4
Classification report.

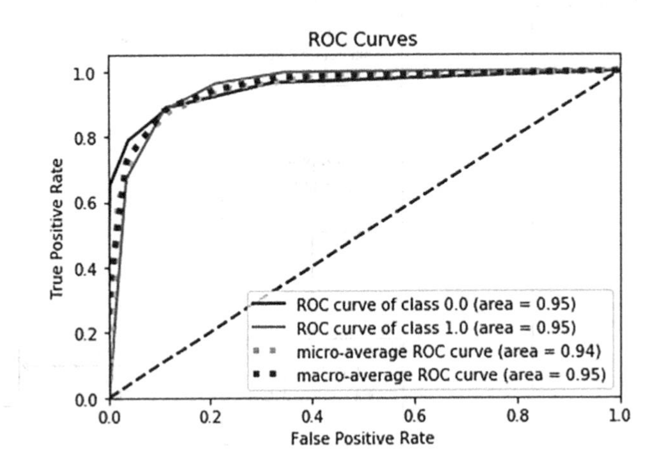

FIGURE 3.5
ROC curve.

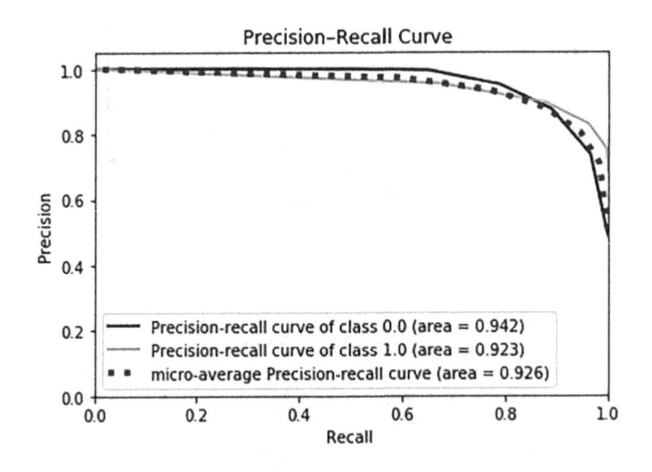

FIGURE 3.6
Precision–recall curve.

3.4 Case Study 2: Assessing Land Cover

Recently, satellite images have been widely used by researchers for various purposes. Using the images, we can smartly classify the images and extract meaningful information from them. For such purposes, CNN models, t-SNE, and clustering are employed.

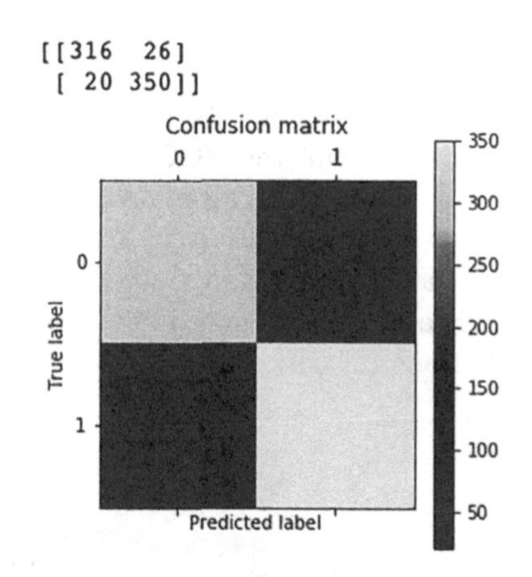

FIGURE 3.7
Confusion matrix for Decision Tree Classifier.

3.4.1 UC Merced

The UC Merced dataset consists of manually extracted images from the USGS National Map Urban Area Imagery collection (Land Use Dataset, 2020).

3.4.1.1 Dataset

The images are classified into 21 labels in the dataset and have been downloaded from the UCI Machine Learning Repository. The images from the dataset are shown in Figure 3.8.

The dataset consists of 2100 images, with 100 images belonging to each of the 21 classes. Moreover, each image measures 256×256 pixels. The output labels in the dataset include classes like agriculture, airplane, baseball, beach, river, forest, and runway.

3.4.1.2 Methodology

The images are preprocessed, cropped, and resized before being fed to classification models. As an equal number of images were present for each class, no data augmentation was required. The images were cropped and resized so that the models could learn the features in a better manner. Various CNN models such as VGG16, VGG19, Xception, and InceptionV3 were employed for classification (Chollet, 2017).

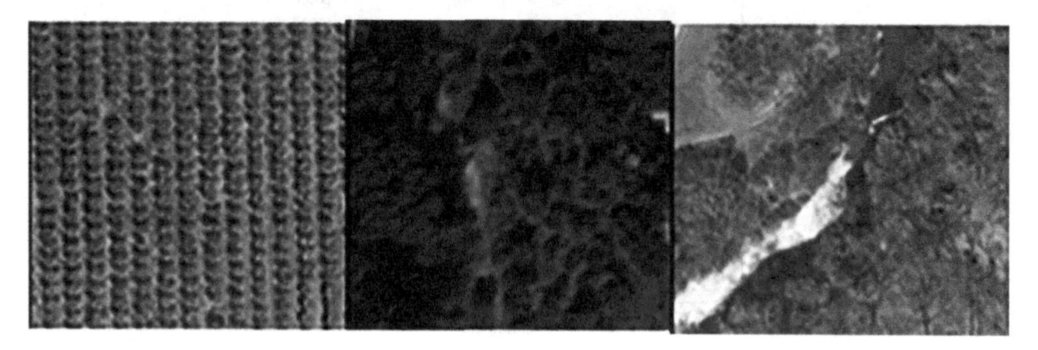

FIGURE 3.8
Images in the UC Merced dataset.

3.4.1.3 Implementation

The graph in Figure 3.9 depicts the comparative study between the different CNN models and their performance. It is observed that the VGG19 model provides maximum accuracy of 94.16% with a precision of 0.92 and a recall of 0.91. The models such as VGG16 and InceptionV3 show equally good performance accuracy of 93.5% and 92.5%, respectively.

Figure 3.10 provides the confusion matrix as evaluated by the VGG19 model. We have the actual labels along the y-axis and the predicted labels along the x-axis. The count of the correct and incorrect predictions for different classes in the dataset can be identified from the figure.

3.4.2 Amazon Rainforest

The Amazon Rainforest dataset is a multilabel classification and involves tagging images using tags such as tag_primary, tag_clear, tag_agriculture, and 14 others. The comprehensive list is provided in Figure 3.11.

3.4.2.1 Dataset

The Amazon Rainforest dataset is a 17-class satellite-based image dataset with hazy, primary, agriculture, and cloudy classes, with each image measuring 256×256 pixels. The dataset is highly imbalanced, with the primary tag being the most prominent. The distribution of the different classes in the dataset is shown in Figure 3.11. From Figure 3.11, it is evident that tag_primary has the maximum count and the tags such as cloudy, tag_blooming, slash_burn, and blow_down are rare.

3.4.2.2 Dataset Analysis

Our research aims at providing tags to each image in the dataset, which segments a larger image in the Amazon Rainforest. It is observed that each image can have multiple labels, making it a multilabel classification compared to the standard multiclassification. Some images with their tags are shown in Figure 3.12 giving a better overview of the dataset.

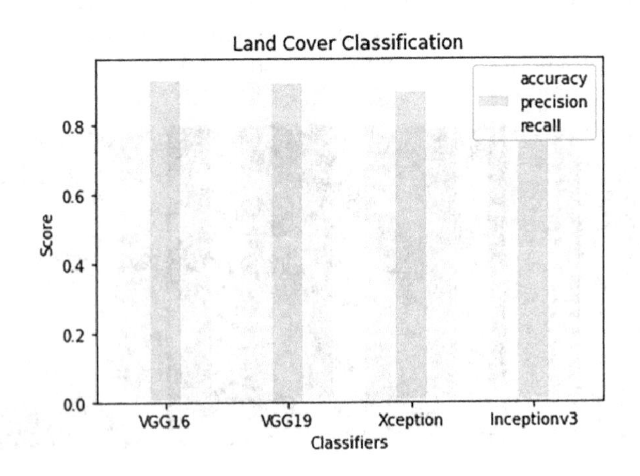

FIGURE 3.9
Classification results for the UC Merced dataset images.

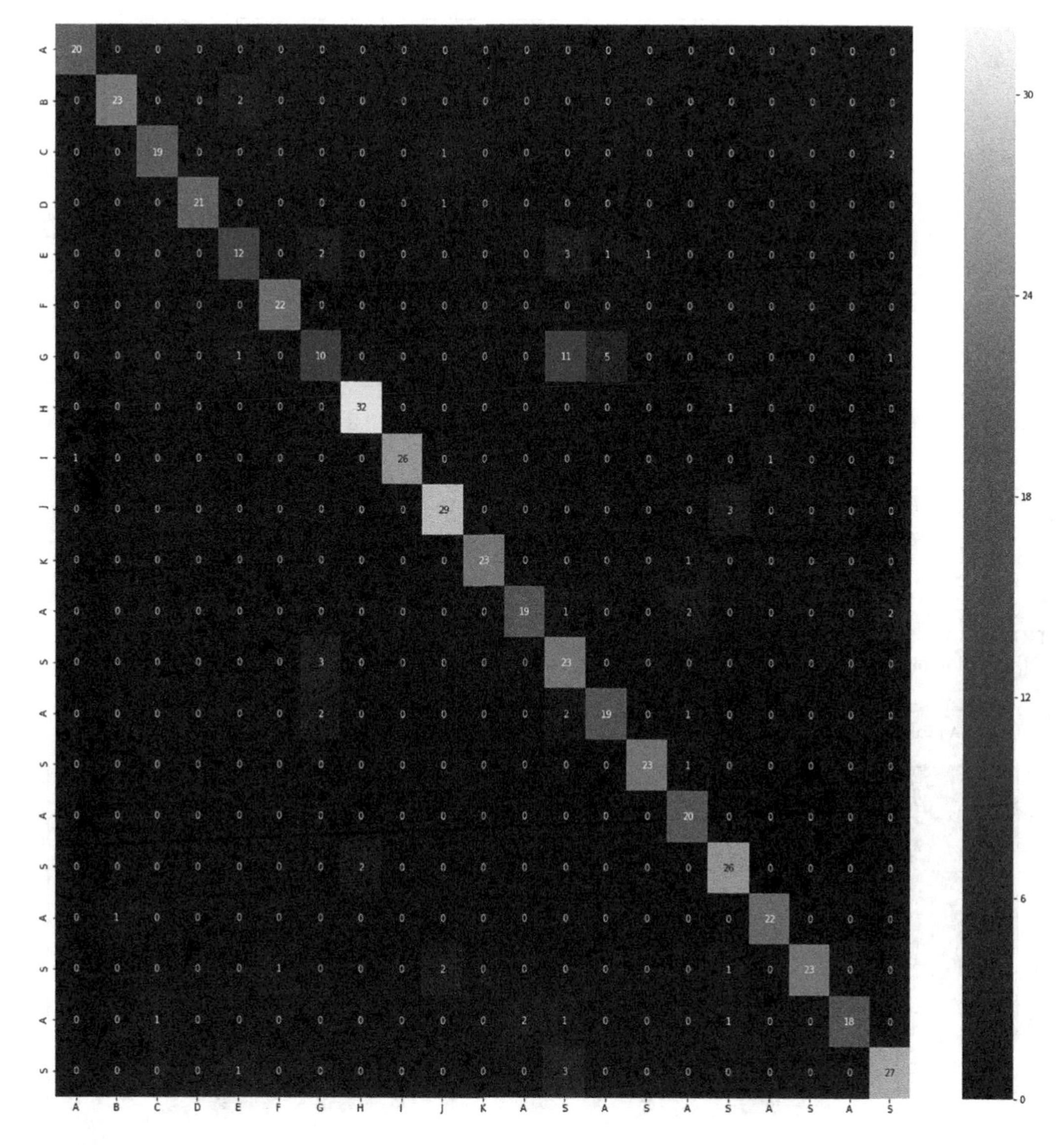

FIGURE 3.10
Confusion matrix for the VGG19 model on the UC Merced dataset.

The t-SNE model (Maaten and Hinton, 2008) has measured the distributions of pairwise similarities of input images. In Figure 3.13, x-axis represents the "t-SNE one" and y-axis represents the "t-SNE two." Many green chunks of boxes can observe most of the images that belong to the rainforest primary tag. We also can observe the rare images of cloudy and hazy weather conditions. The images are generated using the OffsetImage and AnnotationBox from the Matplot library. Since it is a multilabel classification problem, the co-occurrence of different tags in the dataset is worth noting. The images have been clustered by their pixel intensities and computing distances pairwise. Also, the images have been reshaped, normalized, and fitted into a square form.

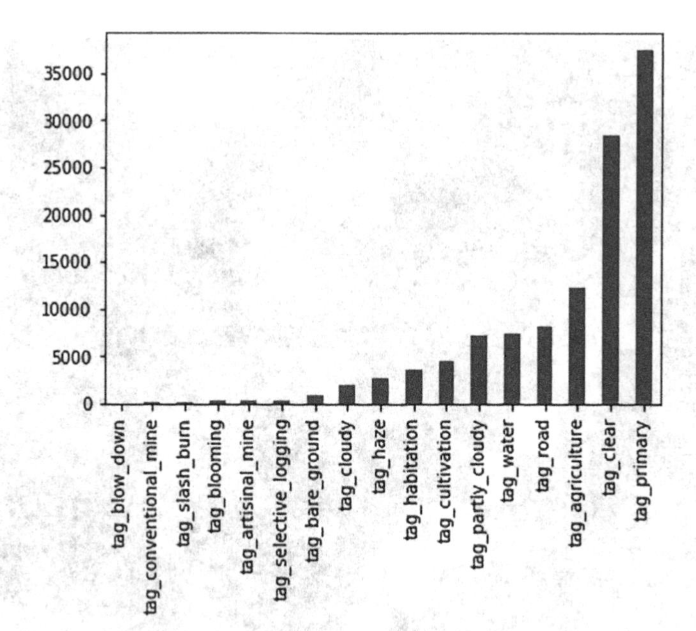

FIGURE 3.11
Distribution of tags in the dataset.

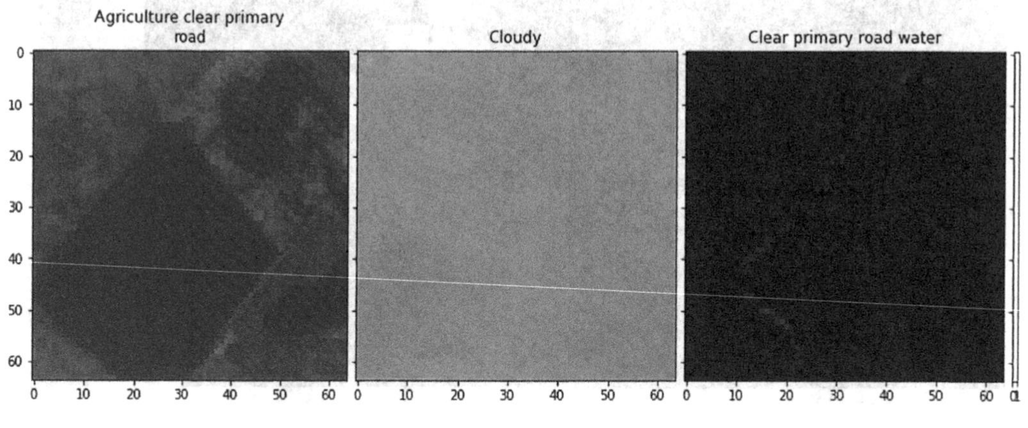

FIGURE 3.12
Images from the Amazon Rainforest dataset.

Figure 3.14 shows the co-occurrence matrix, which depicts what percentage of the *X* label also has the *Y* label. The primary_tag has the highest percentage where it overlaps with the tags at the *Y* label.

A spatial distance library has been used for cluster analysis. Figure 3.15 shows that very few images are dissimilar to all other images by using pixel intensities.

t-SNE is used for data exploration and visualization. Figure 3.16 provides an overview of the outliers and plots the images in 3D format. It provides evidence regarding the dissimilarities and the outliers.

From the distance matrix of the images of pixel intensities, we found the most dissimilar image and the most similar image in the dataset and computed the average distances to

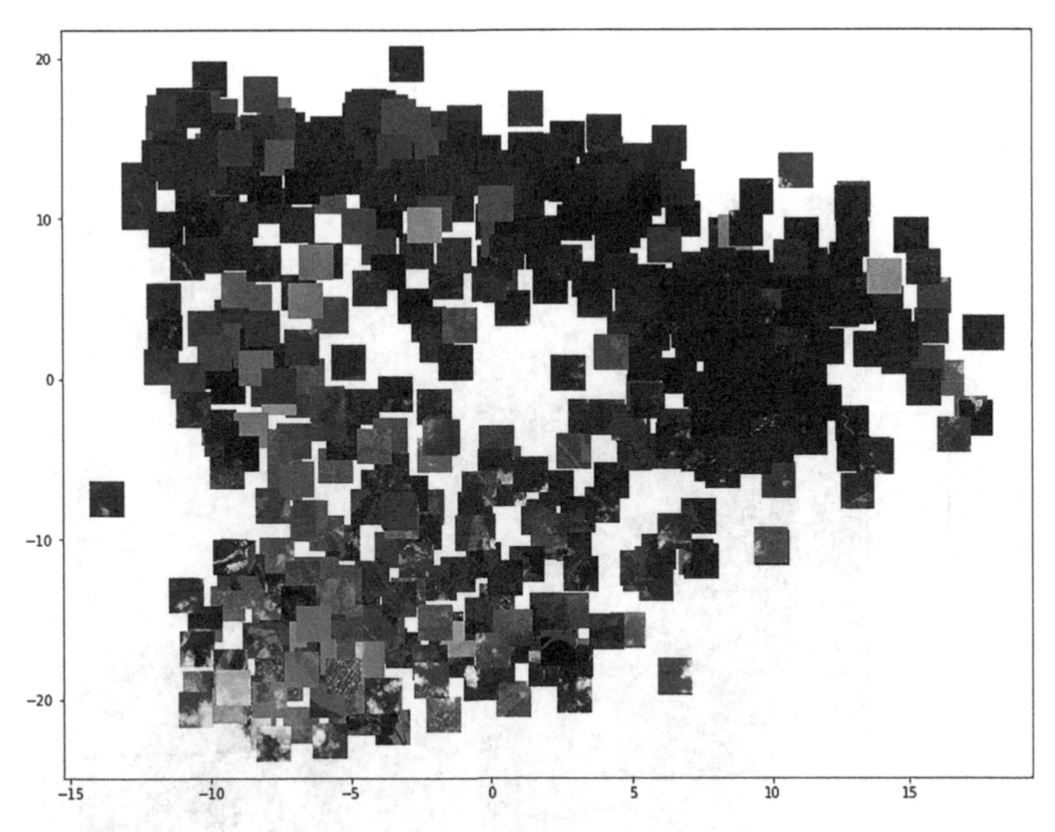

FIGURE 3.13
Overview of the Amazon Rainforest dataset.

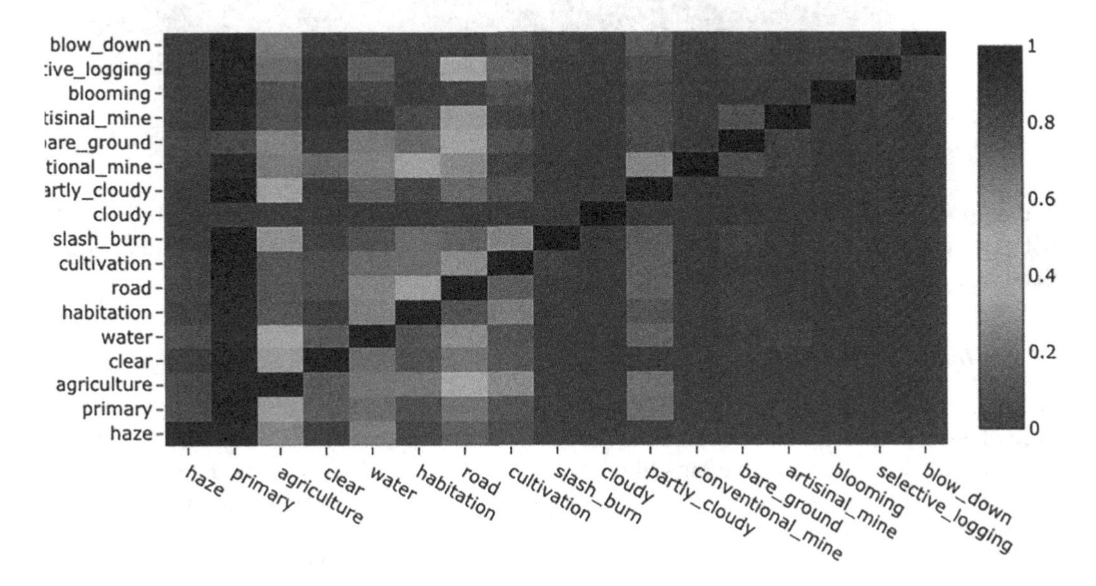

FIGURE 3.14
Co-occurrence matrix for the Amazon Rainforest dataset.

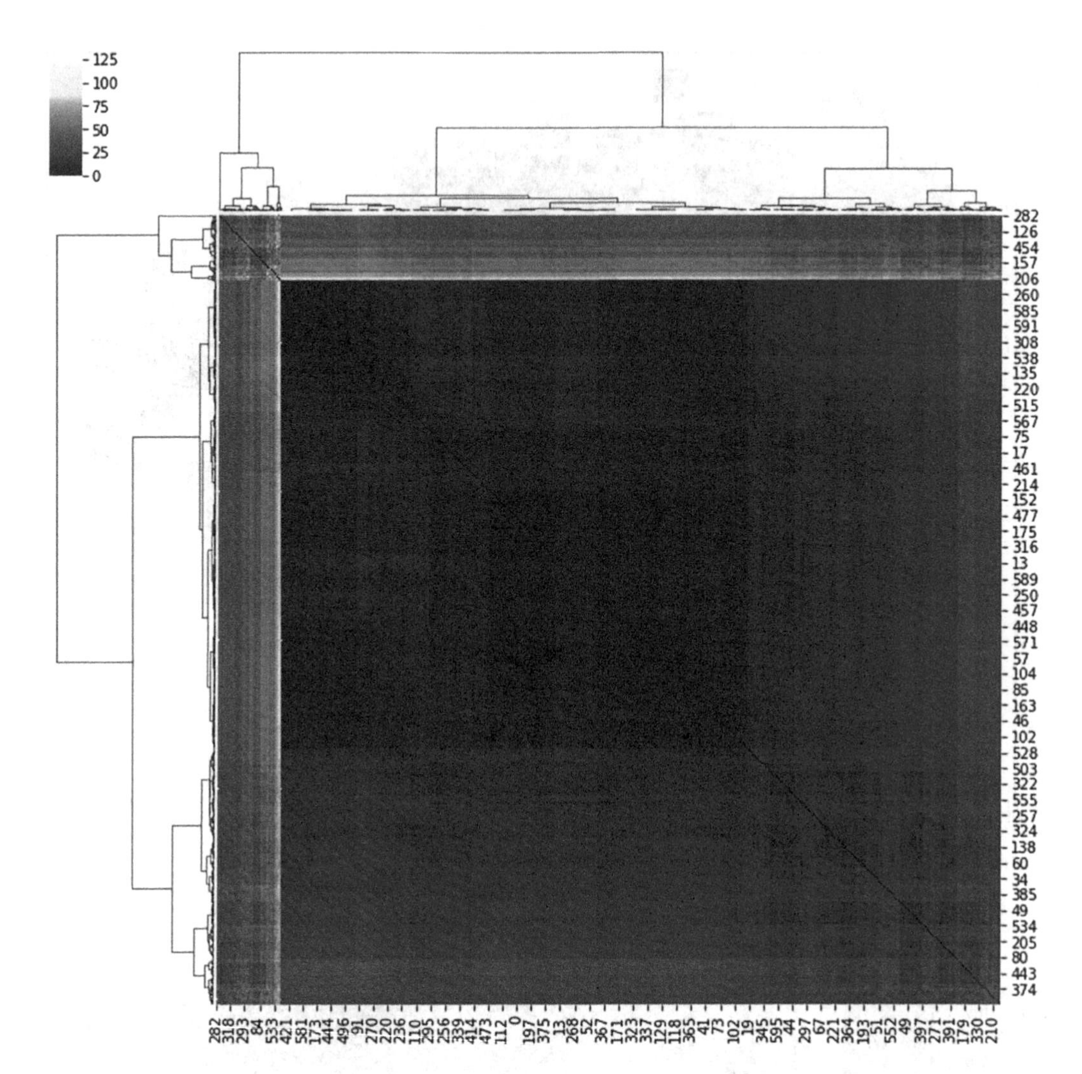

FIGURE 3.15
Cluster Analysis of images in the Amazon Rainforest dataset.

all different images, as shown in Figure 3.17. As we can observe, the image with the least distance cost metric is the primary rainforest, and the image with the maximum average distance depicts clouds and haziness.

3.4.2.3 Implementation

For the Amazon Rainforest dataset, the images in the dataset possibly belong to one or more classes, i.e., an image can have multiple tags. For example, an image train_0.jpg in the dataset belongs to multiple tags like haze, primary, and other images. The tag predictions expected for a test image are also multiple. Every image in the dataset is assigned a list of tags it belongs to, but it needs to be of standard shape for the output labels to be fed into any classifier. Since the count of total possible classes in the dataset is 17, every image is assigned a Boolean array of size 17 where only the values for the image tags are set to 1 and the rest to 0. For the image train_0.jpg, values corresponding to the classes haze and

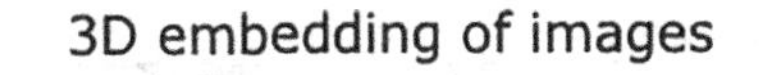

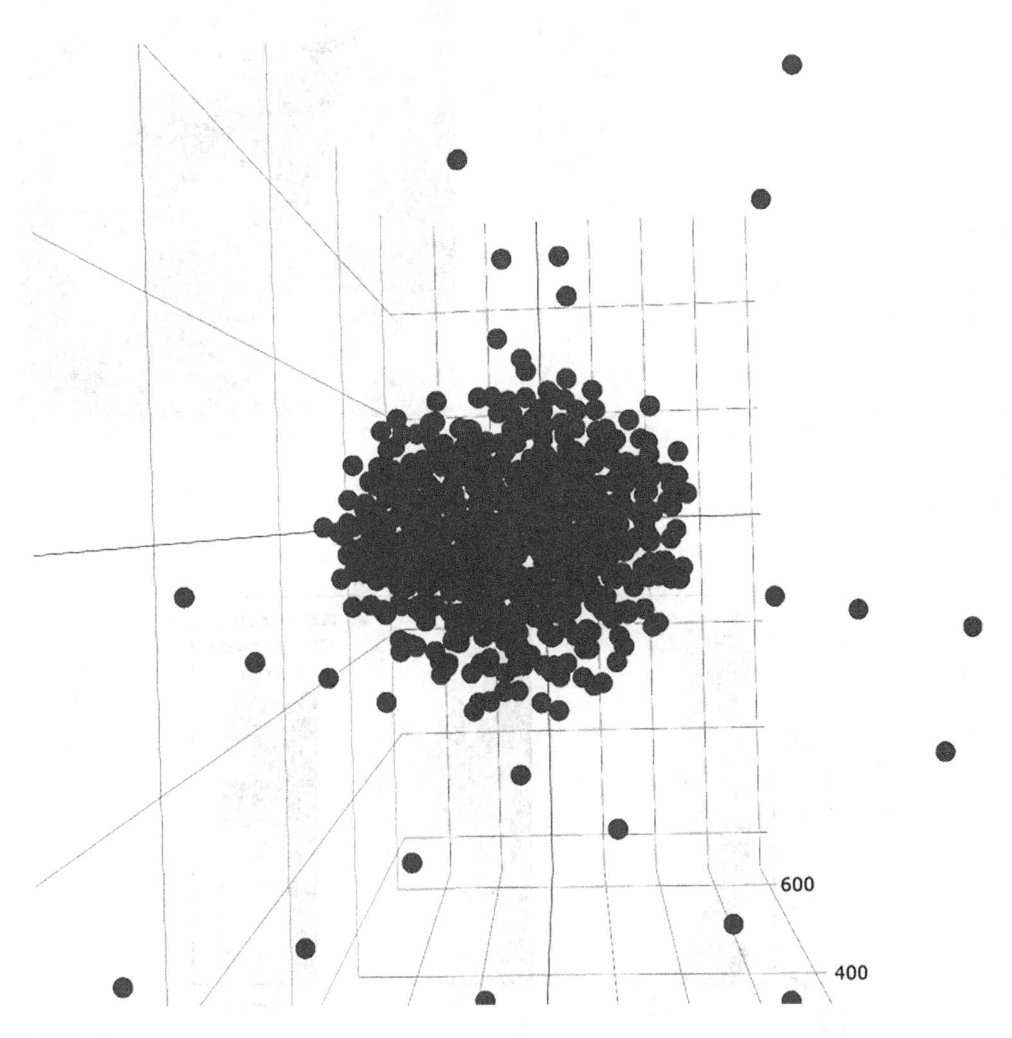

FIGURE 3.16
3D analysis of the Amazon Rainforest images.

primary are 1, and the others are 0 in the array. The assigning of binary values is achieved using the MultiLabelBinarizer class provided by the sklearn library.

Furthermore, after preparing the input and output labels, these images are preprocessed (resized and normalized) for better classification results. Different ensemble learning techniques such as XGBoost and deep learning models like VGG16 and VGG19 have been applied to these final images. The performance of these algorithms is evaluated in terms of metrics like accuracy, precision, and recall.

The Amazon Rainforest dataset fed into the XGBoost classifier returns an accuracy of 88.9%. Figure 3.18 depicts the results obtained by applying the different models on dataset images. It is observed that the VGG16 model gives the best results with an accuracy of 91.4%, whereas the other models VGG19 and Xception show comparable performances with accuracies of 89.9% and 87.11%, respectively.

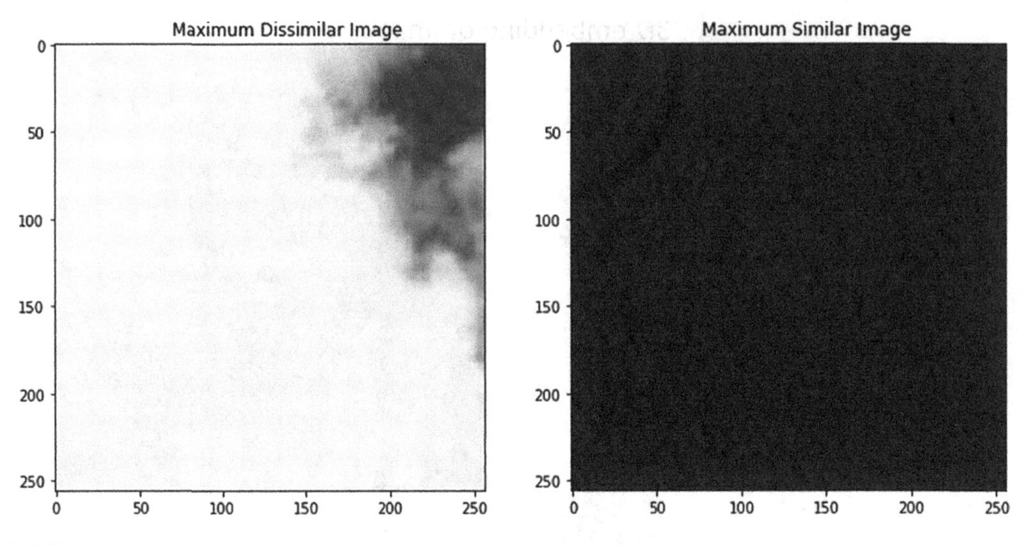

FIGURE 3.17
Maximum similar and dissimilar images.

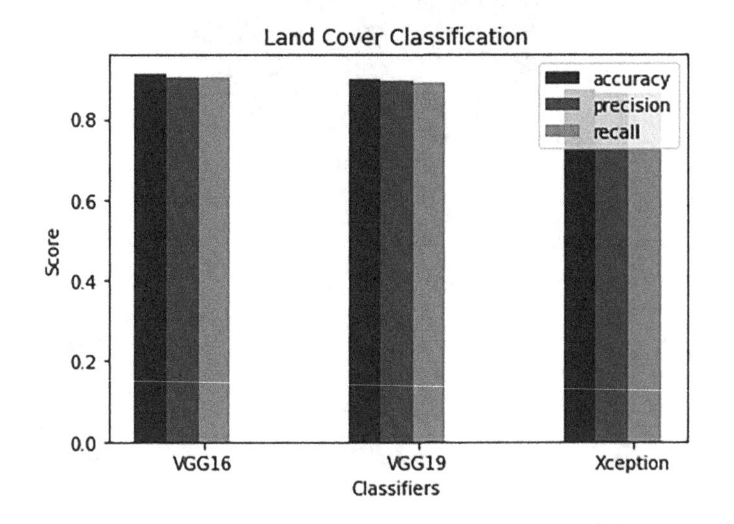

FIGURE 3.18
Classification results for the Amazon Rainforest images.

The NDVI is an indicator of the presence of green vegetation for a target area. It gives a measure and degree of the photosynthetic capacity of vegetation. It works because living green plants absorb solar radiation in the photosynthetically active radiation (PAR) spectral region. The leaf cells of the plants re-emit solar radiations in the near-infrared spectral regions. The NDVI is calculated as follows in Equation (3.1):

$$NDVI = \frac{(NIR - RED)}{NIR + RED} \tag{3.1}$$

Here, NIR stands for spectral reflectance in the red spectral region (the visible spectral region). Red signifies spectral reflectance in near-infrared regions.

Different areas and their usual NDVI index values are as follows:

1. Regions having dense vegetation canopies usually have positive values in the range of 0.3–0.8.
2. A negative NDVI index indicates clouds and snowfields.
3. Water bodies show relatively low positive or slightly negative values, which are attributed to low reflectance in NIR and Red. Examples of such bodies are oceans, seas, and lakes.
4. Barren land or Soil-covered regions generate small positive NDVI values (in the range of 0.1–0.2).

Specific satellite images are fed into the NDVI model to analyze the vegetation of that area better. The NDVI parameter also helps determine the vegetation as and when the climate conditions change throughout the year. The NDVI indexes are depicted from Figures 3.19–3.22.

Figure 3.19 displays farmlands, and they are assigned NDVI values in the range of 0–0.25; in contrast, Figure 3.20 depicts most pixels showing clouds with an NDVI value in the range of −0.02 to 0. As portrayed in Figure 3.21, the model performs well where the farmlands having green vegetation are correctly assigned NDVI in the range of 0.5–1.0, while the urban buildings are assigned NDVI in the range of −0.25 to 0. Hence, the model can somewhat differentiate between the present regions. In Figure 3.22, some areas are misclassified because of the clouds and shadows cast on the land. The model marks the cloud-filled areas with values ranging from −0.15 to −0.05.

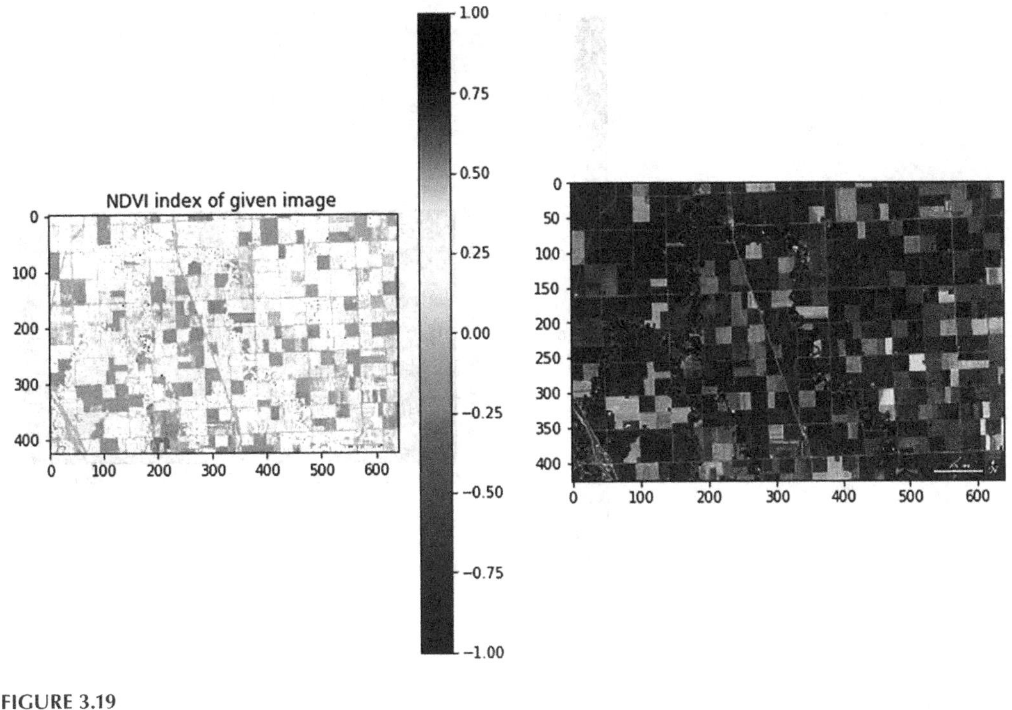

FIGURE 3.19
Farmlands.

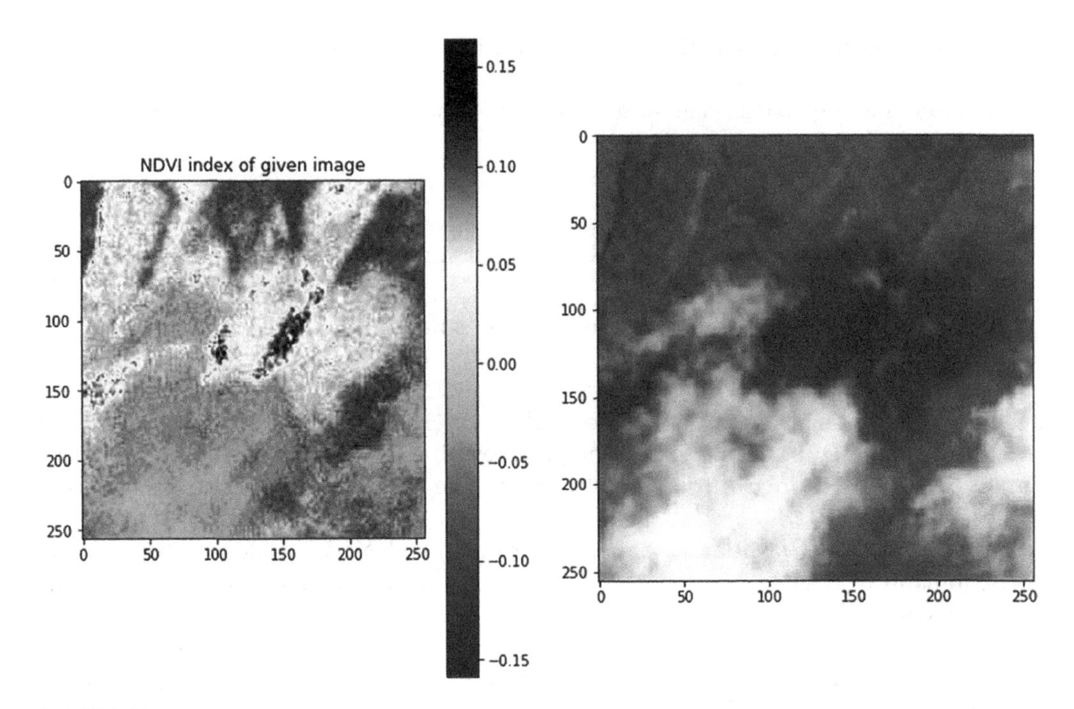

FIGURE 3.20
Clouds.

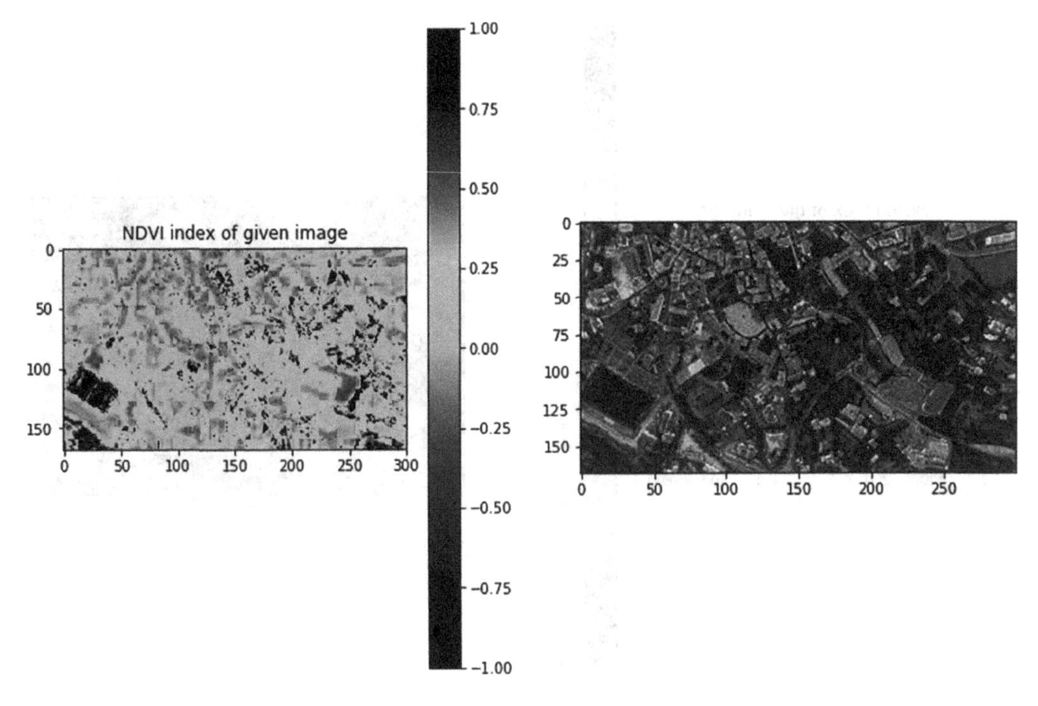

FIGURE 3.21
Farmlands and urban buildings.

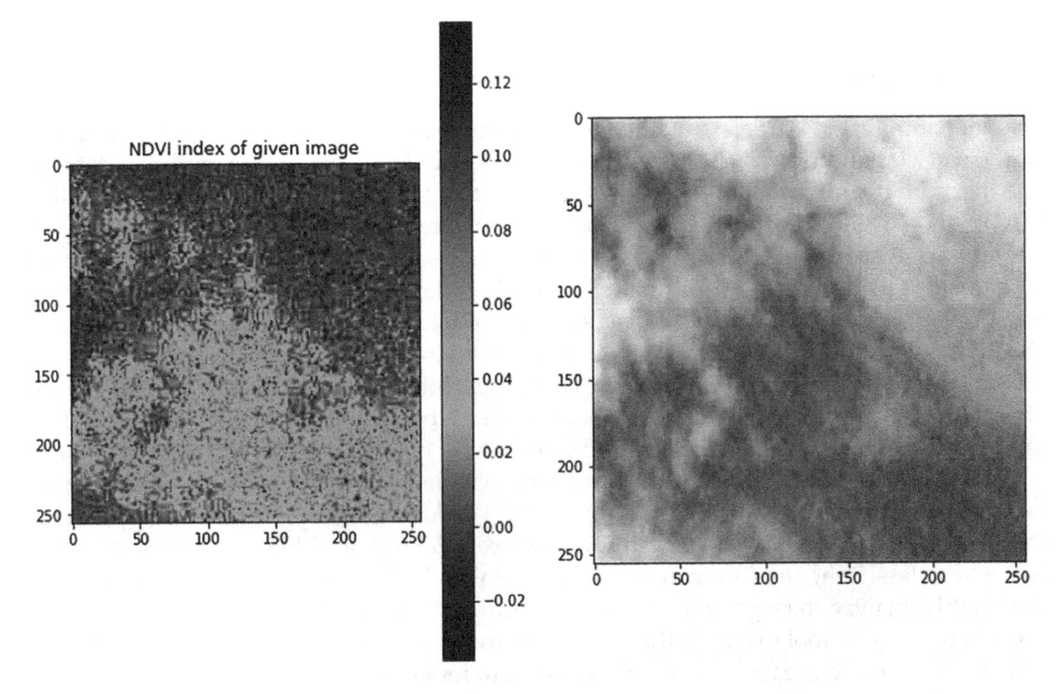

FIGURE 3.22
Dense clouds.

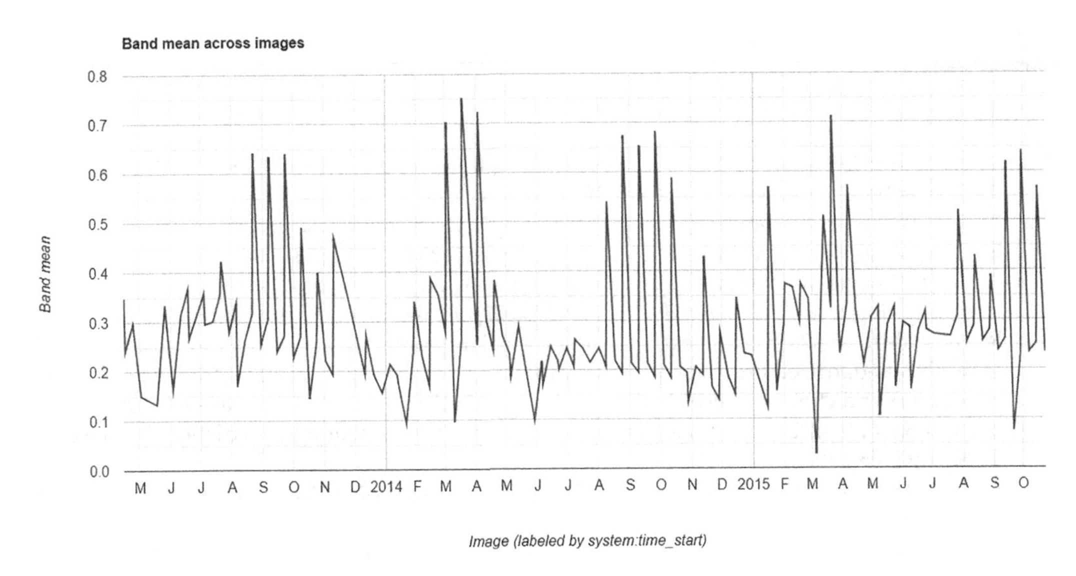

FIGURE 3.23
Year-wise NDVI analysis.

Figure 3.23 was generated using the Google Earth Engine. The NDVI index trend for a selected area is depicted from 2013 to 2016. The trend also provides insights into the months of the year when the vegetation is more abundant compared to the other months.

3.5 Conclusion

The contribution of artificial intelligence is immensely evident in drought prediction, land cover classification of the UC Merced dataset, and the Amazon Rainforest dataset. The machine learning models like KNN, SVM, and Decision Tree and CNN models like VGG19 and VGG16 have helped the research provide excellent results. The study operates in two phases, with the first phase being the drought prediction for the districts of Maharashtra using various machine learning classifiers. It is observed that the decision tree provides the best accuracy among several models applied. Moreover, the technique used for oversampling of data has had a more considerable impact on accuracy. Removing the imbalance is vital as it may lead to overfitting of data of the rare classes. The second phase of the study employs the methods for classifying remotely sensed data or satellite imagery. The later phase is divided into two halves, classifying the UC Merced dataset and analyzing of tagging of the Amazon Rainforest dataset. Deep learning models prove to be the most efficient, and hence several CNN models are explored. VGG19 provides the best accuracy while classifying the UC Merced dataset, and VGG16 proves to be the best when classifying the Amazon Rainforest dataset. Tools like t-SNE and clustering help thoroughly analyze the tags of the Amazon Rainforest dataset. Moreover, the NDVI index acts as an essential tool to assess the vegetation trends in certain regions. Both the phases combined serve as a great recommendation system for farmers helping them increase their productivity and hence a great tool toward sustainable development.

References

Agana, N. A. and Homaifar, A. (2017) A deep learning-based approach for long-term drought prediction. In SoutheastCon 2017, pp. 1–8. IEEE.

Castelluccio, M., Poggi, G., Sansone, C. and Verdoliva, L. (2015) Land Use Classification in Remote Sensing Images by Convolutional Neural Networks. ArXiv, abs/1508.00092.

Chawla, N. V., Bowyer, K. W., Hall, L. O. and Kegelmeyer, W. P. (2002) SMOTE: Synthetic minority over-sampling technique, *Journal of Artificial Intelligence Research*, 16, pp. 321–357.

Deo, R. and Şahin, M. (2015) Application of the artificial neural network model for prediction of monthly standardized precipitation and evapotranspiration index using hydrometeorological parameters and climate indices in eastern Australia, *Atmospheric Research*, 161–162, pp. 65–81.

Kaur, A. and Sood, S. (2020) Deep learning-based drought assessment and prediction framework, *Ecological Informatics*, 57, p. 101067.

Maaten, L. V. D. and Hinton, G. (2008) Visualizing data using t-SNE, *Journal of Machine Learning Research*, 9, pp. 2579–2605.

Marmanis, D., Datcu, M., Esch, T. and Stilla, U. (2015) Deep learning earth observation classification using ImageNet pretrained networks, *IEEE Geoscience and Remote Sensing Letters*, 13(1), pp. 105–109.

Mishra, A. K. and Desai, V. R. (2005) Drought forecasting using stochastic models, *Stochastic Environmental Research and Risk Assessment*, 19(5), pp. 326–339.

Poornima, S. and Pushpalatha, M. (2019) "Drought prediction based on SPI and SPEI with varying timescales using LSTM recurrent neural network", *Soft Computing*, 23(18), pp. 8399–8412.

Rainfall Dataset, Maharashtra Government, Accessed on August 2021 [Online], Available: www.maharain.gov.in, 2001–2021.

Ramos-Giraldo, P., Reberg-Horton, C., Locke, A., Mirsky, S. and Lobaton, E. (2020) Drought stress detection using low-cost computer vision systems and machine learning techniques, *IT Professional*, 22(3), pp. 27–29.

Tejaswini, M., Pranuthi, P., Ravichand, S., and Anuradha, T. (2019) "Land Cover Change Detection Using Convolution Neural Network," 2019 3rd International conference on Electronics, Communication and AerospaceTechnology (ICECA), pp. 791–794. doi: 10.1109/ICECA.2019.8821840.

Time and Date, Pressure and Temperature Data, Accessed on August 2021 [Online], Available: www.timeanddate.com, 1995–2021.

UC Merced Land Use Dataset, University of California Merced, Accessed on August 2021 [Online], Available: www.weegee.vision.ucmerced.edu/datasets/landuse, 2010.

Yang, Y. and Newsam, S. (2010) Bag-of-visual-words and spatial extensions for land-use classification, pp. 270–279, doi: 10.1145/1869790.1869829.

Zhao, X., Gao, L., Chen, Z., Zhang, B., Liao, W. (2018) CNN-based large scale landsat image classification. In 2018 Asia-Pacific Signal and Information Processing Association Annual Summit and Conference (APSIPA ASC), pp. 611–617. Honolulu, HI: IEEE, doi: 10.23919/apsipa.2018.8659654.

4

Electronic Health Record for Sustainable eHealth

Prapti Miglani, Jagriti Bhandari, Ujjwal Alreja, and Adwitiya Sinha
Jaypee Institute of Information Technology

CONTENTS

4.1 Introduction

Our research paper provides the significance of Electronic Health Records (EHRs), highlighting the pros and cons of the EHR system. It also describes the health analysis benefits, which focus mainly on the accessibility of information.

4.1.1 Significance of EHR

Ever since digital technology arrived on the scene with its widely varied applications, it has significantly expanded its domain and set its foothold in almost every industry. It has transformed the way things function on devices such as smartphones or PCs, becoming a necessity in our everyday lives.

DOI: 10.1201/9781003046431-4

The field of medicine is no stranger to this growing trend. It is an industry that essentially functions to maintain and distribute information, including records of diseases, symptoms, treatments, and patients' medical history. Traditionally, all documents are stored in hard copy format, which poses the risk of damage, theft, and the loss of records. Other drawbacks include the need for large storage spaces and maintenance of these records (Singhal et al. 2019). EHR counters these drawbacks and provides a feasible solution by storing all the data and information in digital repositories (Al-Janabi et al. 2016). This enables access to information anywhere and at any time. The time it takes to search for a record can be considerably improved, revolutionizing how we handle information.

4.1.2 Health Analysis Benefits

The benefits of health record analysis are manifold. It saves time for the patient, and the doctor, as all the medical history and patient records can be accessed online from anywhere and makes the entire process much more user-friendly. With all the information accessible at the tap of a button, a patient is not required to make unnecessary visits to the doctor to obtain reports, thus saving time and money. The creation and updating of traditional medical reports are prone to human error. However, bringing technology into the foray drastically reduces the inaccuracies one might find in the information. By relying on digital technology to maintain and access the records, the time it takes to obtain the form is drastically reduced compared to earlier storage techniques (Rathi and Pareek, 2013). It also enables faster communication between the patient and the doctor. The ability to quickly access information sets up a chain reaction as it allows quicker detection of symptoms and disease identification (Rathi and Pareek, 2016). Time is a crucial factor in the medical industry, and this relay of timely information can save many lives. EHR can also help you do away with the hassle of filling the same forms repeatedly every time you pay a visit to the doctor. The providers can use the digitally stored data to compile accurate disease statistics to identify the threat levels and develop policies to combat them (Dewan, 2015). The information is far more reliable and convenient compared with traditional records. It also has the advantage of sending an e-prescription directly to a pharmacy without the need for explicit communication. With the advent of electronic referrals, a patient's information can directly be sent to another doctor specializing in the field in which the patient requires treatment.

4.2 Description

To improve accuracy in applying algorithms on EHRs, we first acquired the heart disease dataset and cleaned and preprocessed it (Mythili, 2013). Then we used primarily five imputation methods, namely mean, multivariate imputation by chained equations (MICE), K-Nearest Neighbor (KNN), Lagrange, and centroid to replace NAN (not-a-number) values and then compared the accuracies obtained upon using each of these methods on application to various machine learning algorithms.

4.2.1 Dataset Descriptions and Parameters

The complete dataset of heart disease was picked from the UCI Repository (David, 1988). This database initially included 76 attributes, of which 14 characteristics were relevant and

TABLE 4.1

Dataset Attributes with Their Descriptions

Number	Attributes	Description
1.	Age	Age in years
2.	Gender	Female: 0 Male: 1
3.	Cp	Chest Pain Types: Value 1: Typical angina Value 2: Atypical angina Value 3: Nonanginal pain Value 4: Asymptomatic
4.	Trestbps	Resting blood pressure in mm Hg
5.	Chol	Serum cholesterol in mg/dl
6.	Fbs	Fasting blood sugar level
7.	Restecg	Resting ECG results: Value 0: Normal Value 1: ST-T wave abnormality Value 2: Left ventricular hypertrophy
8.	Thalach	Max heart rate
9.	Exang	Exercise-induced angina
10.	Oldpeak	ST Depression by exercise
11.	Slope	The slope of the peak exercise section Value 1: Upsloping Value 2: Flat Value 3: Down sloping
12.	Ca	No. of major blood vessels Range: 0–3
13.	Thal	Thallium Heart Scan: Value 3: Normal Value 6: Fixed defect Value 7: Reversible effects
14.	Num	Diagnosis of heart disease (angiographic disease status): Value 0: No heart disease Value 1: Low chance Value 2: Average Value 3: High Value 4: Very high

were retained. The heart column in the dataset has integral values where 0 refers to No Presence, 1 refers to Low, 3 refers to High, and 4 refers to Very High Presence (Table 4.1).

4.2.2 Methodology

The complete dataset of heart disease was picked from the UCI Repository. The data have been preprocessed using different imputation methods to replace missing values and garbage values. For classification, we have implemented various machine learning algorithms (Yadav 2013). Our methodology has been divided into three major stages: data preprocessing, machine learning classifiers, and finally, performance evaluation. The data preprocessing is done by several imputation methods, including mean, MICE, KNN, Lagrange, and centroid. Next, the machine learning algorithms are applied to the datasets as mentioned above and the performance metrics, i.e., the accuracy obtained on different datasets are compared.

4.2.2.1 Imputation Methods

- **Mean:** This method takes the mean of the available values column-wise and replaces the NULL values with the computed standard.
- **MICE:** The MICE algorithm uses multiple predictive instances rather than one (e.g., mean imputation) so that the data are more specific and reliable than the data imputed using other methods. Linear regression is used here for the prediction-based modeling of data that consists of multiple variables.
- **KNN Imputation:** This algorithm computes the distance of the missing values with its KNN. We assign weights to each neighbor, which are inversely proportional to their distance from the missing values. Using the importance of the neighbors, weights are set to them and the majority instance. Using this, we impute the missing value.
- **Lagrange Imputation:** This is a mathematical polynomial interpolation method wherein we determine the dependent variable by generating a polynomial using the independent variable that fits the NA values. Lagrange Interpolation uses Lagrange polynomials to determine the dependent variable values. In this method, we have generated polynomials of different columns by considering the non-null values within those columns. Now, the obtained polynomial is used to fill the NA values of the column by putting the independent column values corresponding to that tuple into the obtained polynomial.
- **Centroid Imputation:** The Centroid method fills the NA values by considering different classes grouped as per the target column. First, we generated a subset of the non-null values of the column taken into consideration and then grouped the remaining tuples of the target column based on their means. NA values of the columns are filled according to the corresponding norm of the classes they belong to.

4.2.2.2 Machine Learning Algorithms

- **Bayesian Logistic Regression:** This supervised learning algorithm registers the contingent likelihood to order new qualities (Hasan et al. 2018). The class with the most significant conditional probability is assigned to the latest data. The conditional probability is determined by utilizing the previously gained training set of data. It is used to group the patient's records and confirm whether they are in danger of coronary illness.
- **Logistic Regression:** This algorithm is used for categorical classification problems (Rajkumar et al. 2010). It is used to predict an instance "x" of heart disease from the binary values [0,1], where 0 is a negative value (not-at-risk) and 1 is a positive value (at-risk). It can also be used to deal with multiple matters like 0, 1, 2, 3, and so on.
- **SVM:** SVM constructs hyperplanes in space to differentiate between different class labels. The better characterization is accomplished when a hyperplane has an enormous separation from the dataset trained (Jabbar, 2016). Both regression and classification algorithms are supported by SVM. An iterative training algorithm is used to minimize the error function for regression.
- **KNN:** In this algorithm, we compute the distance using Euclidean distance. Then the distances are ranked, KNN is selected, and the majority class of those points is depicted as the final class of that point.

- **Extreme Gradient Boosting:** In this algorithm, we build trees sequentially such that the error is reduced in each subsequent tree. Information is taken from all the weak learners for prediction, which enables boosting to make a strong learner, which lowers the bias that was higher earlier (Kumar and Prasad, 2017).
- **Random Forest:** The Random Decision Forest fabricates prescient models and is utilized in classification and regression problems (Zriqat et al. 2017). It is a gathering strategy where numerous learning models are being used, which leads to better expectation results. Uncorrelated decision tree forests are created to arrive at the best answer.
- **Decision Trees:** A tree is built with each leaf node representing a class mark and an attribute (Rathi et al. 2018). We have used the Gini index and information gain to decide the most crucial feature at each level, and that attribute is made the root of the subtree. We see a significant difference in the accuracy of decision trees with different datasets.

4.2.2.3 Clustering

First, we highlight the bottom-up using an agglomerative algorithm in a detailed manner. Here, the distance metric is computed by the Euclidean method on the MICE dataset.

- **Complete Linkage:** It takes pairs of elements, i.e., one from cluster 1 and the other from cluster 2, and calculates their dissimilarity. Finally, the most immense value of discrepancy is taken to be the distance between the clusters. This produces dense clusters.
    - i) MICE Dataset: The distance metric is computed by the Euclidean method on the MICE-generated dataset. The coefficient of complete linkage was 0.979507.
    - ii) KNN Dataset: The distance metric is computed by the Euclidean method on the KNN-generated dataset. The coefficient of complete linkage was 0.9806796.
- **Single Linkage:** Here, we again calculated pairwise dissimilarity; however, instead of considering the most significant disparity value, we regarded the least important to be the distance between the clusters.
    - i) MICE Dataset: The distance metric is computed by the Euclidean method on the MICE-generated dataset. The coefficient of single linkage was recorded to be 0.8687216.
    - ii) KNN Dataset: The distance metric is computed by the Euclidean method on the KNN-generated dataset, and 0.8774571 was the coefficient.
- **Average Linkage:** We also calculated the pairwise dissimilarity and considered the average dissimilarity as the distance between the clusters.
    - i) MICE Dataset: The distance metric is computed by the Euclidean method on the MICE-generated dataset. The coefficient of complete linkage was 0.9517623.
    - ii) KNN Dataset: The distance metric is computed by the Euclidean method on the KNN-generated dataset. The coefficient of complete linkage was 0.9636966.

4.2.2.4 Binning

When using the binning method, the first step was to find the range between the max value and the min value of the age column and accordingly divide that range into five bins

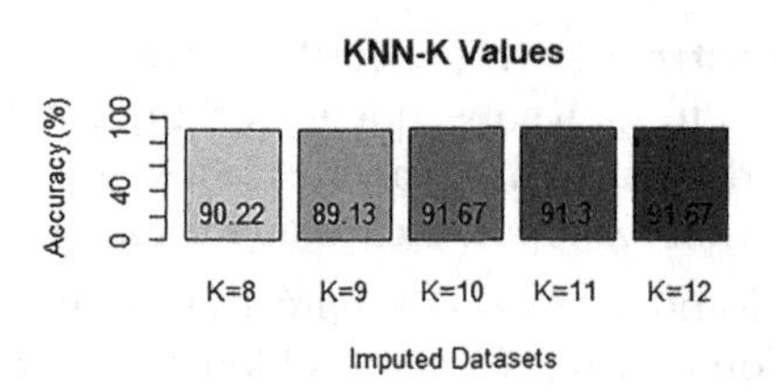

FIGURE 4.1
Accuracy for different K-values by applying the K-Nearest Neighbor method for imputations.

(taking the ceiling value) with a midrange of values that turned out to be 10. The percentage of heart risk was plotted against these five bins ranging from 28 to 38, 38 to 48, 48 to 58, 58 to 68, and 68 to 78, which resulted in bin 3 (48–58) having the highest chance of having heart disease, followed by bin 4 (58–68). Also, we analyzed the extent of heart disease for each age group.

4.2.2.5 *Optimal Value of* K *for KNN*

On comparing different values of K, we observed that 10NN gave us the best accuracy for logistic regression. The trends are highlighted in Figure 4.1, where we see that 91.67% accuracy is associated with $K=10$ (Figure 4.2).

4.3 Experimental Results

The following section summarizes the ROC and AUC results obtained from logistic and Bayesian logistic regression, followed by a comparative study with respect to multiple parameters.

4.3.1 Logistic Regression

In Figure 4.3, we compared the ROC of various datasets and observed that the AUC of the centroid dataset scores the best among all other datasets, followed by the KNN dataset with an AUC score of 94.2%. We also observed that the AUC score of MICE, Mean, and Lagrange datasets lie in the range of 85–90%. The ROC curve consists of a plot of the actual positive rate versus the false-positive rate for the various conceivable cut points of a diagnostic test. If the value of AUC comes out as 0.5 means, it's a lousy classifier and a great classifier. The accuracy of the experiment relies upon how dominantly the test isolates the group being tested into those who have disease and those who do not. The graph is a plot between sensitivity and specificity, two kinds of accuracies, where the first is for actual positive records and the second is for existing negative records. Of all the ROC curves defined above, the one applied on centroid dataset gives the highest AUC as 96.2%, and the one with the least is with the Lagrange dataset as 84.1%.

4.3.2 Bayesian Logistic Regression

In Figure 4.4, we compared the ROC of various datasets and observed that the AUC of the centroid dataset scores the best among all other datasets, followed by the KNN dataset

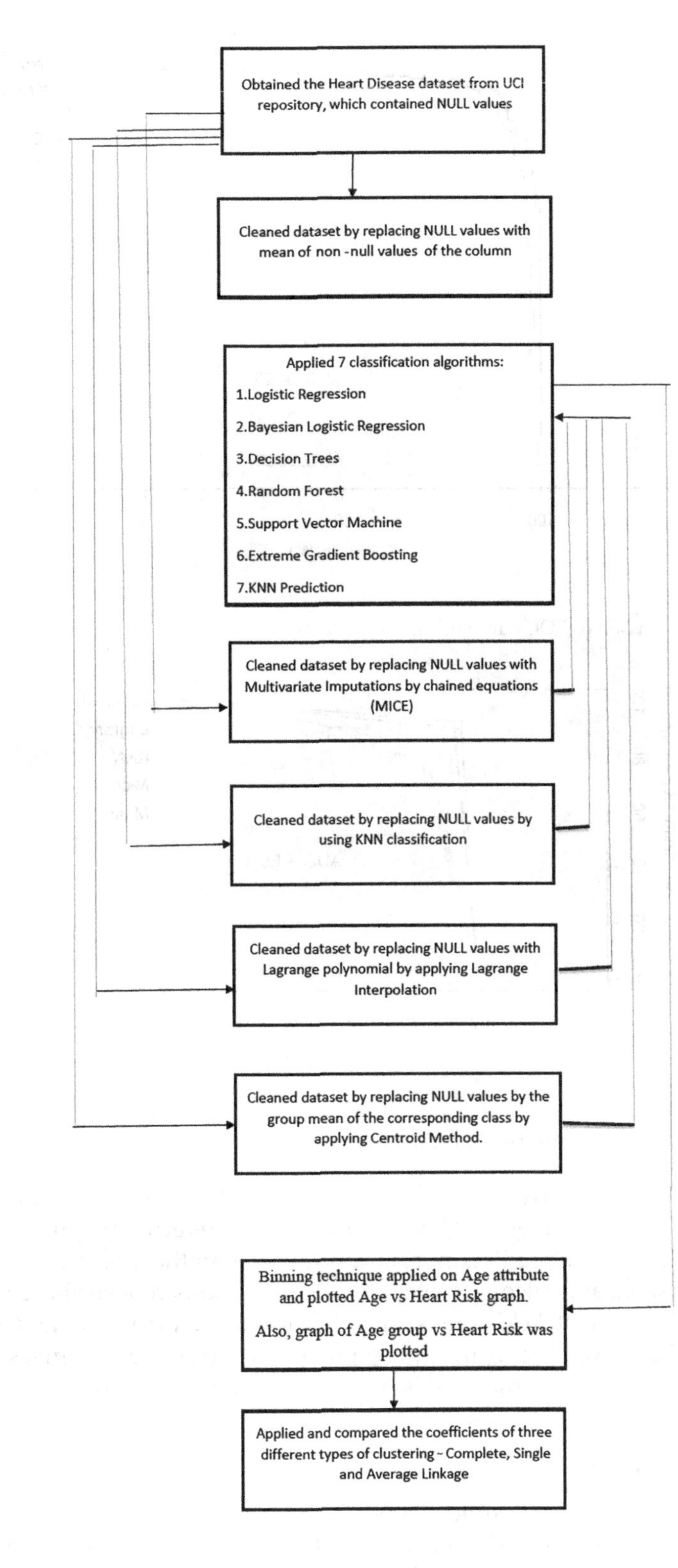

FIGURE 4.2
Flow diagram for automated disease detection.

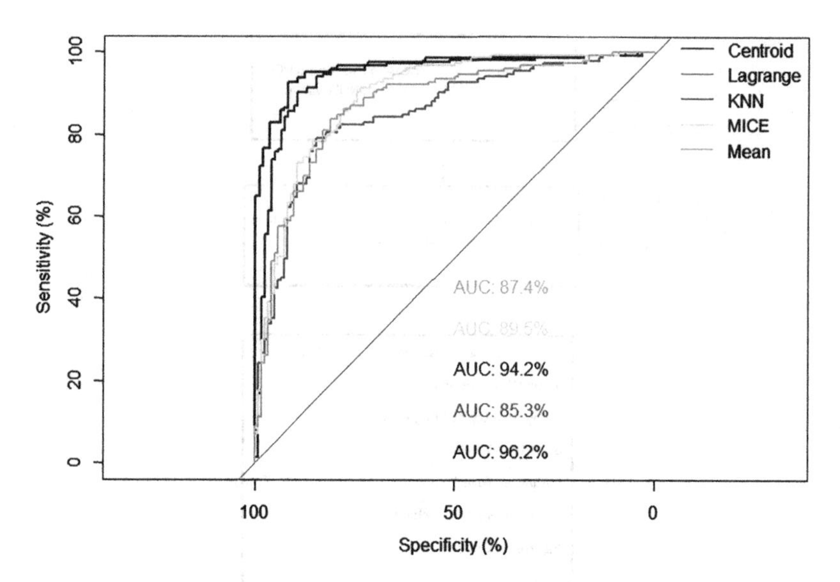

FIGURE 4.3
Logistic regression comparison of ROC curves of different datasets.

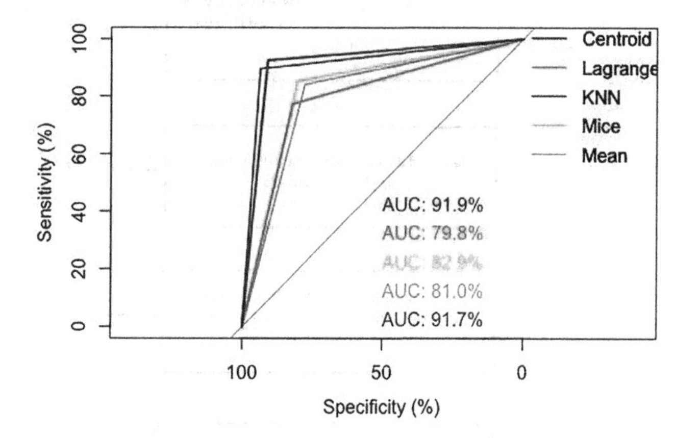

FIGURE 4.4
Bayesian logistic regression comparison of ROC curves for different datasets.

with an AUC score of 91.7%. We also observed that the AUC scores of MICE, Mean, and Lagrange datasets lie in the range of 80–83%. The ROC is preferred because it is independent of scale, i.e., it measures prediction rankings rather than the actual values. It consists of a plot of valid positive and false-positive rates for the various conceivable cut points of a diagnostic test. If the value of AUC comes out as 0.5, it is a lousy classifier and a great classifier. Of all the ROC curves above, the one applied on centroid dataset gives the highest AUC as 91.9%, and the one with the least is on Lagrange dataset as 79.8%.

4.3.3 Comparative Study

The specificity, sensitivity, and accuracy values are highlighted for all five imputed datasets in Tables 4.2–4.4. We observed the accuracy of the several machine learning algorithms on different datasets obtained by imputations based on the mean of the column, Lagrange

TABLE 4.2

Specificity of Several Algorithms on Different Imputed Datasets

	MEAN	LAGRANGE	MICE	KNN	CENTROID
Logistic Regression	0.843	0.862	0.868	0.954	0.879
Bayesian Logistic Regression	0.845	0.715	0.863	0.954	0.937
Decision Trees	0.798	0.869	0.867	0.929	0.943
Random Forest	0.877	0.895	0.914	0.901	0.932
Support Vector Machine	0.881	0.868	0.875	0.907	0.940
KNN Prediction	0.808	0.888	0.891	0.898	0.879
Extreme Gradient Boosting	0.822	0.871	0.883	0.914	0.932

TABLE 4.3

Sensitivity of Several Algorithms on Different Imputed Datasets

	MEAN	LAGRANGE	MICE	KNN	CENTROID
Logistic Regression	0.758	0.709	0.793	0.868	0.881
Bayesian Logistic Regression	0.779	0.715	0.791	0.861	0.895
Decision Trees	0.714	0.756	0.827	0.843	0.881
Random Forest	0.761	0.823	0.805	0.920	0.955
SVM	0.788	0.772	0.837	0.926	0.926
KNN Prediction	0.827	0.862	0.845	0.854	0.881
Extreme Gradient Boosting	0.769	0.911	0.841	0.938	0.955

TABLE 4.4

Accuracy of Several Algorithms on Different Imputed Datasets

	MEAN	LAGRANGE	MICE	KNN	CENTROID
Logistic Regression	81	78.9	84	91.6	92.0
Bayesian Logistic Regression	81.5	79.3	83.3	91.3	92.0
Decision Trees	76.4	81.8	85.1	89.1	91.6
Random Forest	82.9	86.6	86.9	90.9	94.2
SVM	84	82.5	85.8	91.6	93.4
KNN Prediction	80.1	88.7	86.5	92.3	94.2
Extreme Gradient Boosting	81.6	87.8	87.2	88.0	88.0

polynomial, MICE, KNN, and the centroid method. The results from the graph plotted show clear comparisons among the several algorithms and the different models of imputations. We observed that the centroid method proves to be the best method for imputing null values in all seven algorithms. Also, the best results obtained among the algorithms were from Extreme Gradient Boosting.

4.3.4 Gender-Wise Distribution of Heart Diseases

As per the dataset, it can be seen that males have average chances of suffering from heart disease, 36.8%, followed by people who have no chances of developing heart disease (Figure 4.5).

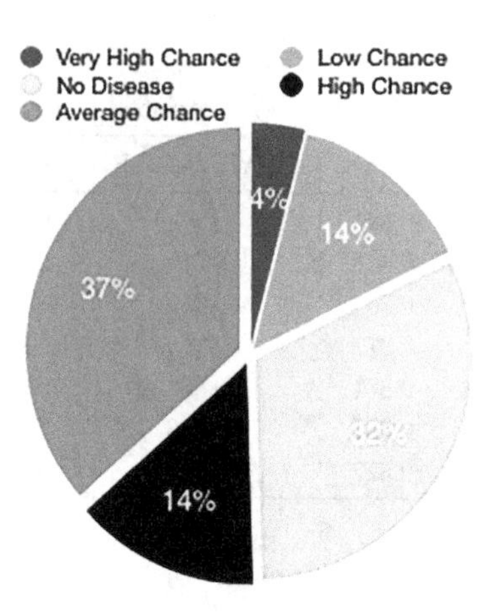

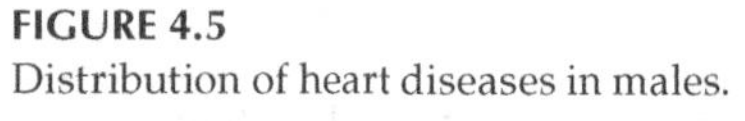
FIGURE 4.5
Distribution of heart diseases in males.

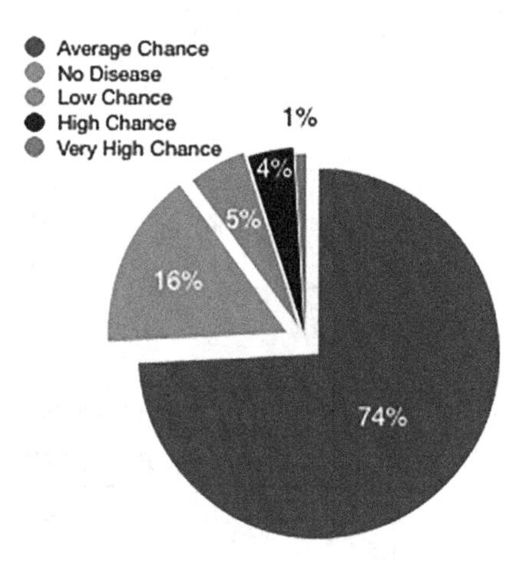

FIGURE 4.6
Distribution of heart diseases in females.

Figure 4.6 confirmed that a more significant part of females have standard odds of experiencing heart diseases. There are under 2% of females who have an exceptionally high possibility of experiencing heart sickness. In contrast with their male partners, females are commonly more averse to experiencing the ill effects of heart illnesses.

4.3.5 Importance of Attributes Using Random Forest

Managing bunches of information and different indicator factors is vital to channel the traits with less commitment level in the exact expectation of heart disease. The graphs in Figures 4.7–4.11 are a dotted outline of variable significance as estimated by the random forest calculation, and factors with the most elevated significance scores are the ones that give the best forecast and contribute most to the model. In all these graphs plotted with the help of various datasets used, the *Ca* attribute is with the highest importance, whereas *F.B.S.* is with the minor matter.

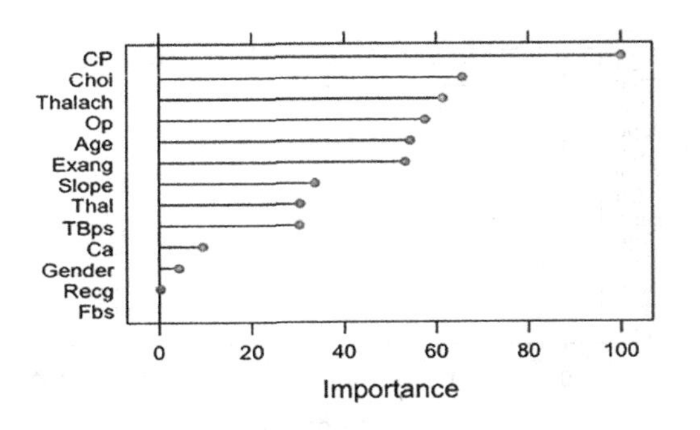

FIGURE 4.7
Variable importance estimated using random forest when applied on mean dataset.

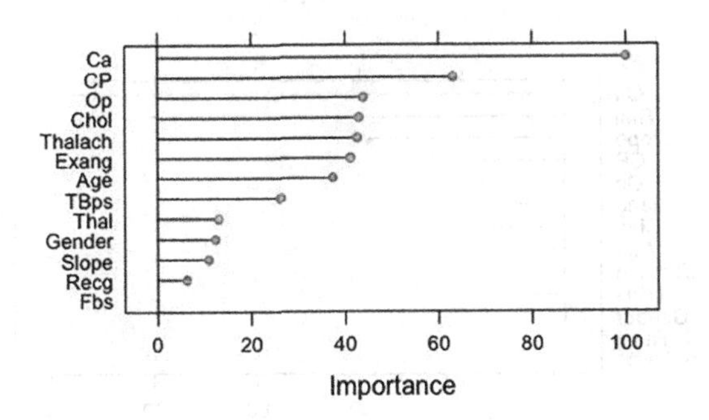

FIGURE 4.8
Variable importance estimated using random forest when applied on MICE dataset.

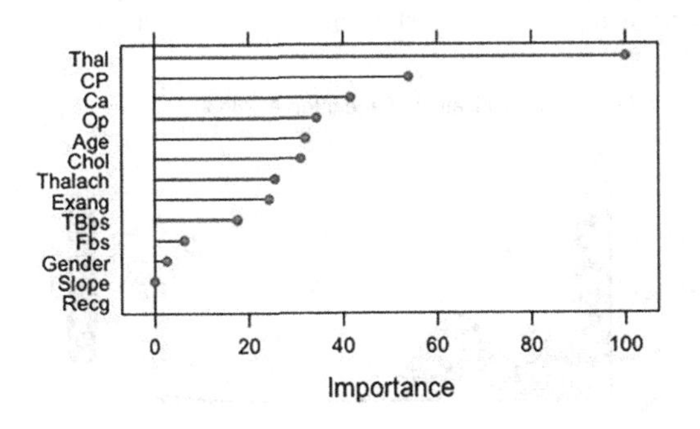

FIGURE 4.9
Variable importance estimated using random forest when applied on Lagrange dataset.

4.3.6 Effect of Binning

MICE has been applied to the dataset to replace the missing values.

In Figure 4.12, we see that the age group between 28 and 38 years has relatively no chance of having any heart disease, denoted by one on the *x*-axis. Very few people in this age group tend to fall in the high heart risk region.

In Figure 4.13, we can apparently determine that some people in every class of heart disease range from no presence to very high heart risk. We can easily say that there are

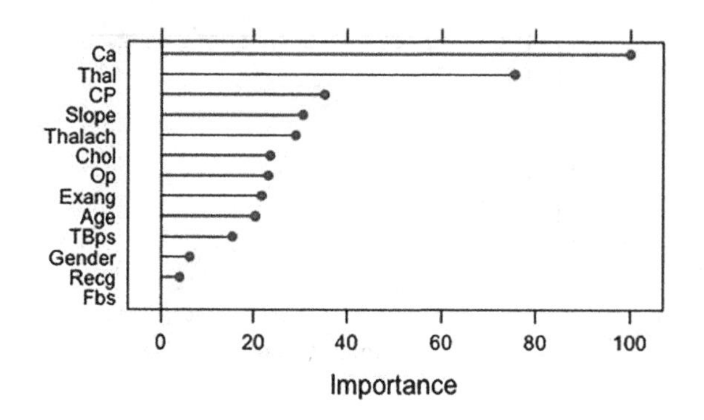

FIGURE 4.10
Variable importance estimated using random forest when applied on KNN dataset.

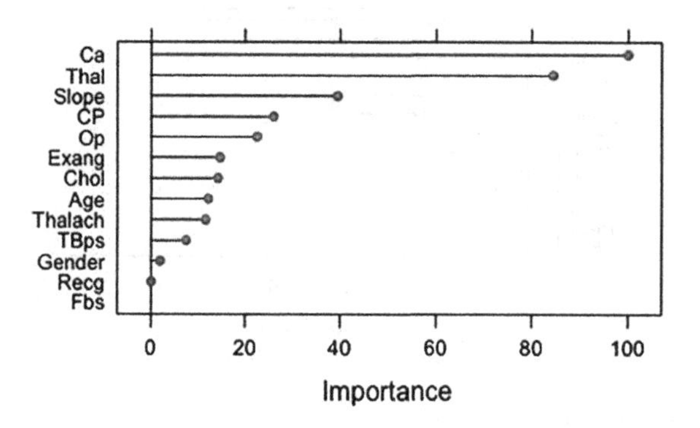

FIGURE 4.11
Variable importance estimated using random forest when applied on centroid dataset.

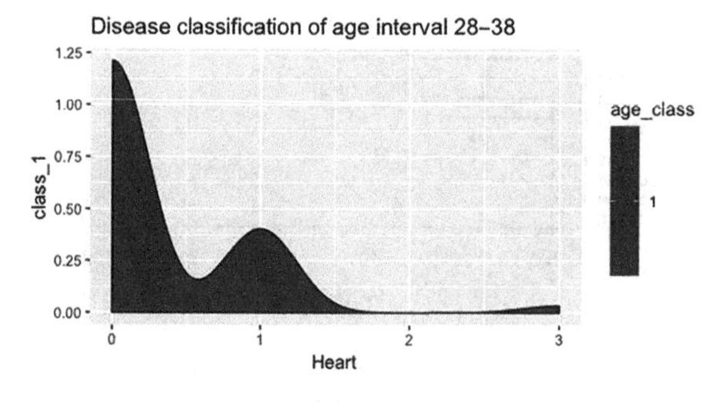

FIGURE 4.12
Disease classification of age interval 28–38.

many people with no heart disease. Next on the list are those with low risk of heart disease followed by those with average risk, high risk, and very high risk of heart disease.

In the plot in Figure 4.14, there are still many people having no heart disease. But many people have a low heart risk, with almost an equal number of people lying under the region denoting average and high risk of heart disease.

Unlike the other plots of age group versus heart risk, in Figure 4.15, there is a clear majority of people having heart disease with low risk. There are still many people who do not

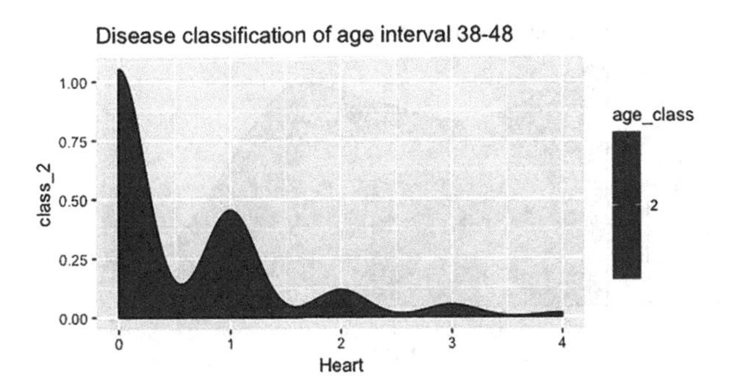

FIGURE 4.13
Disease classification of age interval 38–48.

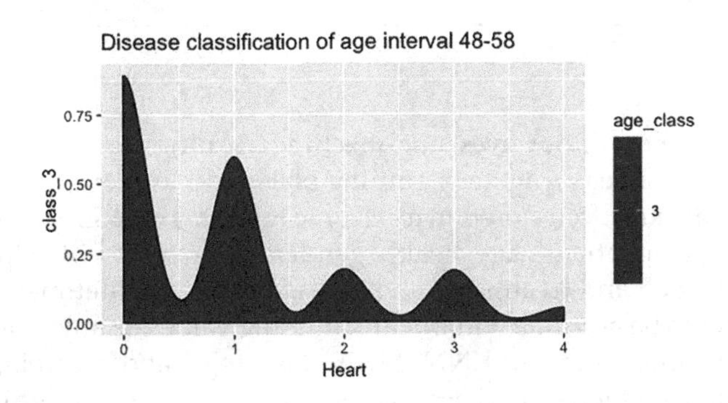

FIGURE 4.14
Disease classification of age interval 48–58.

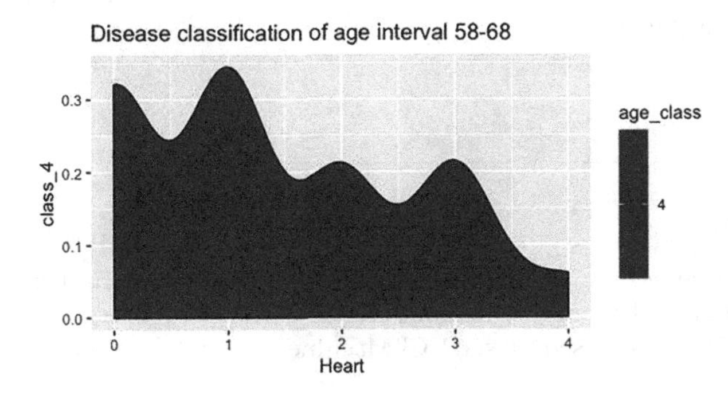

FIGURE 4.15
Disease classification of age interval 58–68.

have any disease. However, there are still people who fall in the region at a high risk of the disease.

By taking a gander at the area under the curve, we can anticipate a decent number of individuals experiencing heart ailments (Figure 4.16). Again, most people with a low risk of heart disease are followed by people with no infection. People are more likely to have a higher heart risk than people with average heart risk.

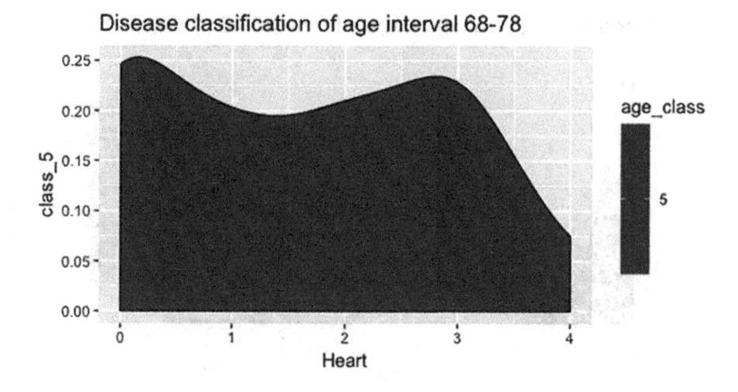

FIGURE 4.16
Disease classification of age interval 68–78.

4.4 Conclusion

Health is an essential part of our lives, and machine learning algorithms can help people diagnose heart-related illnesses at the beginning phase and live longer and healthier lives. Presently, patient data are often stored in electronic format. These EHRs refer to numerical data corresponding to a patient's age, health, gender, etc., which will help predict various diseases. Early detection and treatment can prolong the patient lifetime disease-free. This requires the dataset to be accurate without any missing values or any other discrepancies. Therefore, we have applied mean, KNN, MICE, Lagrange, and centroid, which suitably handles missing data and gave good accuracies. The results show a significant difference between accuracies obtained on applying each of the methods as mentioned earlier. The best imputation methods, in terms of accuracy, include centroid, KNN, MICE, Lagrange, and mean. Of all the algorithms in machine learning, random forest and Extreme Gradient Boosting performed the best.

References

Al-Janabi, M. I., Mahmoud, H. Q. and Mohammad, H. (2016) Machine Learning Classification Techniques for Heart Disease Prediction: A Review.

David, W. A. (1988) Heart Disease Dataset, UCI Machine Learning Repository, archive.ics.uci.edu/ml/datasets/heart+disease.

Dewan, A. and Sharma, M. (2015) "Prediction of heart disease using a hybrid technique in data mining classification." 2015 2nd International Conference on Computing for Sustainable Global Development (INDIACom). IEEE.

Hasan, S. M. M., Mamun, M. A., Uddin, M. P. and Hossain, M. A. (2018) "Comparative Analysis of Classification Approaches for Heart Disease Prediction." In 2018 International Conference on Computer, Communication, Chemical, Material and Electronic Engineering (IC4ME2), pp. 1–4. IEEE.

Jabbar, M. A., Deekshatulu, B. L. and Chandra, P. (2016) "Intelligent heart disease prediction system using random forest and evolutionary approach", *Journal of Network and Innovative Computing*, 2016(4), pp. 175–184.

Mythili, T. (2013) "A heart disease prediction model using SVM-Decision Trees-Logistic Regression (S.D.L.)", *International Journal of Computer Applications*, 68, p. 16.

Parish Venkata Kumar, K. and Prasad, B.D.C.N (2017) "A heart disease prediction model using quality management and SVM and logistic regression", *IJMTST*, 3.08, pp. 2455–3778.

Rajkumar, A. and Reena, G. S. (2010) "Diagnosis of heart disease using the data mining algorithm", *Global Journal of Computer Science and Technology*, 10.10, pp. 38–43.

Rathi, M. and Pareek, V. (2013). Spam mail detection through data mining-A comparative performance analysis, *International Journal of Modern Education and Computer Science*, 5(12), p. 31.

Rathi, M. and Pareek, V. (2016) Hybrid approach to predict breast cancer using machine learning techniques, *International Journal of Computer Science Engineering*, 5(3), pp. 125–136.

Rathi, M., Malik, A., Varshney, D., Sharma, R. and Mendiratta, S. (2018) "Sentiment Analysis of Tweets Using Machine Learning Approach," 2018 Eleventh International Conference on Contemporary Computing (IC3), pp. 1–3, doi: 10.1109/IC3.2018.8530517.

Singhal, S., Jain, S., Rathi, M. and Sinha, A. (2019) Smart technologies to build healthcare models for vision impairment. In Advanced classification techniques for healthcare analysis, pp. 259–285. IGI Global.

Yadav, P. K. (2013) "Intelligent heart disease prediction model using classification algorithms", *IJCSMC*, 3.08, pp. 102–107.

Zriqat, I. A., Altamimi, A. M. and Azzeh, M. (2017) "A comparative study for predicting heart diseases using data mining classification methods." arXiv preprint arXiv: 1704.02799.

5

Team Member Selection in Global Software Development—A Blockchain-Oriented Approach

Chetna Gupta and Megha Rathi
Jaypee Institute of Information Technology

CONTENTS

5.1 Introduction

The current time-to-deliver market is both critical and challenging as it puts pressure on software engineers to deliver software projects within an allocated time and budget. Software faces a number of challenges during its life cycle, such as requirement elicitation, complexity of software, positive schedules, and choosing and managing an appropriate team to cater to unique project demands to name a few. With the advancement of globalization and Internet technology, global companies have expanded their presence around the world. Global software engineering (GSE) enables globalization and endorses increase in distributed software development. This is because many software industries have shifted their focus from on-site development to a distributed development culture, which is called as Global Software Development (GSD). Traditional software engineering does not completely support sustainability. In the last decade, many frameworks and guidelines for sustainable software engineering have been proposed. Distributed software development also contributes in providing some sustainability in SE.

Literature suggest numerous benefits of adopting GSD in overcoming challenges of geographically scattered teams, dissimilar time zones, reduction in development cost and time [Binder, (2009); Kern and Willcocks, (2000)], accessibility of multiskilled human resource, rapid development [Lee-Kelley, (2006); Kommeren et al. (2007); Milewsk et al. (2008); Sooraj et al. (2008), Conchuir et al. (2009); Smite et al. (2010)]. Developing successful projects where teams are geographically distributed puts pressure on project managers

DOI: 10.1201/9781003046431-5

and teams to be collaborative and coordinate effectively. This makes GSD adoption both challenging and complex. Despite the numerous advantages of GSD, some of the barriers in GSD are in the management of task allocation, team coordination and communication, roles and responsibilities, handling of strategic issues, knowledge management, team relationships, and technical issues [Avritzer et al. (2010); Casey et al. (2009); García-Crespo et al. (2010); Colomo-Palacios et al. (2012); Hernandez-Lopez et al. (2010); Islam et al. (2009)]. In reality, every organization has hundreds of developers with various skills and knowledge to support the numerous dimensions of various global projects. The biggest challenge here is to identify and select a set of potential developers (separated geographically) for GSD software development. Stevens (1998) in his research discussed the positive impact of choosing the right skilled team members on success, performance, and productivity in the overall development process. Literature lists various approaches that highlight the significance of team composition, and to the best of our knowledge, there is no benchmark technique that can guarantee sure success by yielding positive results [Gilal et al. (2016); Braun, (2007); Cataldo et al. (2009); Ebert et al. (2008); Lacity et al. (2008)].

The selection of team members is generally done by project leaders or project managers, and their decision is generally based on the project manager's appraisal, their expert judgment, and availability of past experiences of team members. To some extent, this makes the whole process of team selection nontransparent and unjustified. Closely related to the issue of task allocation and task dependency [Sooraj et al. (2008)] is the impact on team members' inter-relationships, communication, and coordination. This causes a lack of team spirit, hampers productivity, software quality, duration, and stakeholder dissatisfaction, leading to rework.

This research presents a novel approach for the selection of team members in a transparent manner so that the issues of coordination, communication, and lack of team spirit can be minimized. Choosing among alternatives impacts both social and physical wellbeing within the team or company. Blockchain-centric software development has gained a lot of attention in the last few years and defines new directions to allow effective software development. The organizations (represented by project managers) can broadcast details of team members with their credentials in the system. The selection committee can vote for a specific employee in an anonymous manner. It also provides authentication of the broadcaster and selector along with support for accuracy, completeness, and robustness. The main contributions of this chapter are listed below:

- Provision for an effective novel solution in selecting team members quantitatively and without ambiguity.
- Use blockchain effectively to achieve the aim of this study, ensuring privacy, verifiability, completeness, uniqueness, and robustness.
- Minimize the challenges and issues faced in GSD particularly communication, coordination, and lack of team spirit.
- Results of initial experimentation, surveys, and interviews conclude that the proposed method can select team members in a transparent manner using blockchain technology in GSD environment.

The rest of the chapter is structured as follows: Section 2 discusses the related work followed by the proposed approach, system architecture, and results of experimentation. Finally, the conclusion is presented.

5.2 Related Work

Among the success factors in software development is how well the objective, scope, and needs of the project are understood and recognized [Imtiaz and Ikram, (2016); Sutanto et al. (2015); Marques et al. (2013); Mahmood et al. (2017)]. Team members should communicate and coordinate effectively, own responsibility as a team, and must have a positive outlook toward the completion of the project. Interaction and coordination among team members are key issues, as coordination allows an effective management of task dependencies among various activities (Strode, 2016). Task dependencies refer to the condition when the progress of an action depends on another action or a piece of information (Strode, 2016). These task dependencies also include application and platform experience. In the study conducted by Li and Ren (2015) and Nguyen (2015) discuss how human characteristics, for example, personality and different geographical personalities, impact software development process and productivity. Parker (2000) states that a group of people cannot be called a team and not every team is always successful. Therefore, the teams should be organized in such a manner that they can influence a positive outcome on overall software development while minimizing issues and challenges. Bohner and Mohan (2009), in their work, clearly state that for developing a successful project, human element is the key component.

The study by Liu and Jin (2012) discusses that software productivity is subjective to the development of environmental features, system product aspects, and project team members' characteristics. Geographically located teams require almost 2.5 times more of the above features than collocated teams for project completion mainly because of the lack of task coordination [Sutanto et al. (2015)]. They conducted a survey with 95 projects and analyzed them toward their performance enhancement for task allocation using various circumstantial factors. Chakraverty et al. (2014) used a genetic algorithm to form pairs of tasks toward optimal task member assignment. Imtiaz and Ikram (2016), Jiang (2016), and Mahmood et al. (2017) conducted a review and classified various factors as influential and ignored task allocation factors depending on their impact on the accomplishment of project completion. Mahmood et al. (2017) conducted a review with 62 practitioners addressing their experience of task allocation. Another important factor in task allocation is process ownership (Edwards et al. 2008). It had a major effect on development, as project owners are responsible for maintaining inter-relationships within the team, ensuring correct execution, and getting work done on time [Larsen and Klischewski (2004)].

5.3 Proposed Approach

The process starts with broadcasting the information in blockchain with a brief description of each prospective team member. It will help the selection committee in deciding whether to select the particular person as a team member. This chapter uses Bitcoin, a cryptography-based digital currency system, and a blockchain using the ring signature algorithm. The ring signature scheme is used with RSA algorithm to encrypt and decrypt messages with a hashing algorithm sha256() to provide security. Each member of the selection committee will have their own public–private key pair. Ring signatures are widely used privacy-preserving blockchain applications. The benefit of using ring signature is

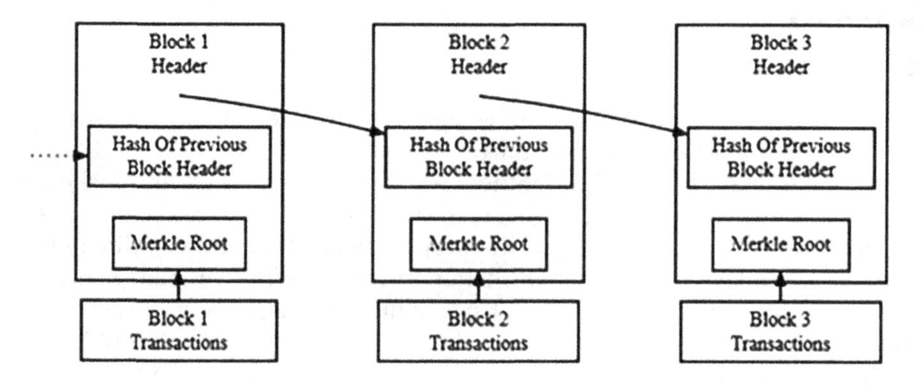

FIGURE 5.1
Basic structure of Bitcoin Block Chain.

that each ring member is equal and anonymous (Maxwell, 2015). It allows users to dynamically choose a set of public keys without disclosing the user's identity. In anonymous e-cash or crypto-currency system, linkable anonymity is more suitable than perfect anonymity since a double-spent payment can be detected. In a linkable ring signature (Liu et al. 2004), it can be verified whether the two signatures are generated by the same signer. The ring signature scheme is split into three portions: key pair generation, ring signature generation, and signature verification. This chapter uses a traceable linkable signature as it provides perfect anonymity and traceability.

A blockchain is an open and distributed ledger that holds transactional records, ensuring security, transparency, and decentralization (Jun et al. 2018). All records of the blockchain are publicly available, and any information once stored in blockchain cannot be altered. Each block has a previous block cryptographic hash, a timestamp, and transaction data. One of the renowned uses and the backbone of blockchain is Bitcoin. The Bitcoin (Böhme, 2015) is a crypto-currency that has cryptographic proof to exchange digital assets online while maintaining anonymity in the transactions. Each transaction is secured by a digital signature. Transactions in blockchain are chained together. The block header stores a Merkle root (Figure 5.1).

In recent years, both Bitcoin and Ethereum (Wood, 2014) have become increasingly popular. Although both are permission-less crypto-currencies, offer anonymous transactions, and are public blockchains, they still differ remarkably in nature as well as functions. Ethereum uses its own currency called gas to assist peer-to-peer agreements and applications. The cost of each transaction depends on storage requirements, bandwidth usage, and complexity. Whereas in Bitcoin, it depends on the size of the block and its address has no relation to its personal identity. Ethereum is more complex than Bitcoin and its complexity makes it more vulnerable to cyber-attacks. This makes Bitcoin a more secure digital currency, although Ethereum is more about smart contract applications. This work uses a Bitcoin-based blockchain.

5.4 System Architecture

The proposed system exhibits the following basic properties:

a. **Project manager/broadcaster organization Authenticity:** This is provided by reputation systems that give a metric for selectors to judge various developers and determine the risk of involvement in an exchange with another particular user. Everyone in the network can view what the broadcaster is posting. The broadcaster organization begins with the registration of potential selectors in their network across the globe. Only registered selection committee members will be eligible to lock their choices or select developers through voting. At the same time, the broadcaster private key Bitcoin address is stored in the system. The two keys (public and private) will then be generated after clicking the random register code link sent in the email for authentication purpose. The newly generated public key is stored, whereas the private key is held privately. The selectors use their private keys to cast a vote using a secure ring signature, broadcast to the blockchain. Not every registered selector is required to choose and cast a vote for selection.

b. **Locking/choosing a selector privacy:** No one can know what the selector has locked. The request is completely hidden.

c. **Individual verification:** The selector can verify their choice correctly even after selection (by voting). If the selector is not satisfied, he/she can turn down his/her choice.

d. **Accuracy/Completeness:** Every locked choice should be counted correctly.

e. **Robustness:** No one can influence/edit the final locked selection list.

5.5 Selection and Verification Process

The following steps help selectors select the most suitable developer as part of the team for the project in hand.

1. Only a registered user (member of the selection committee) can participate in the selection process of voting. He/she can also choose to abstain from voting altogether. In case he/she participates in the voting process, a notification will be sent for his/her choice and some bitcoins will be deducted from his/her account.
2. If the member of the selection committee is not a registered user, he/she will have to first generate his/her private key from his/her organization. After that, he/she can perform Step 1.
3. If the person has already voted, the system will display that the user has already voted and they cannot vote again. For this a smart contract is written, which keeps track of the number of votes cast by each account (and thus, by each person).
4. Verification is performed in two steps. (a) verifying addresses—In this option, the organization can view all the transactions made; from specific address to other specific address and (b) verifying time and date—In this option, an organization can easily spot any transaction made out of the time/date bounds as specified by it. For example, if the organization conducts the selection process on a specific date and time window, then it will be easy for them to spot a transaction that does not adhere to the time boundaries, and the vote outside the specified boundary can be ignored/neglected from the final count list.

5.6 Experimentation

For a comprehensive empirical assessment in software engineering, it is often difficult to select an appropriate empirical method. This is because there are multiple dimensions having uncertainties and complexities, human factors, growing technologies, etc. [Easterbrook et al. (2007)]. The experimentation conducted helps in understanding the tradeoffs between various factors and the dynamics of current team member selection strategies, and whether the proposed approach is feasible and applicable in IT industry.

5.6.1 Experiment

The proposed system is emulated using Windows 10, 3.1 GHz Intel Core i5 systems connected in a way to form a decentralized system on a local network. The selector will use the link sent in their emails and input their private key to choose and lock the module (after link validation). After registering their choice, it will be broadcasted on the blockchain automatically. In the next step a ring signature will be created. The performance analysis of this emulated system is done on sufficient tests conducted using the Bitcoin testnet environment. Bitcoin network uses testnet as its testing environment. These coins do not have any value in this network. These bitcoins can be obtained for free-to-use and for testing purposes from Faucet Kiwi. If the number of the selectors is small, the ring signature will be electively signed and verified. On the other hand, if the number of selectors increases, then the sizes of the public keys, ring signature, and efficiency of signing and verifying will also increase gradually. Table 5.1 summarizes the performance of ring signature.

5.6.2 Result and Discussion of the Experiment Conducted

Referring to Table 5.1 and Figure 5.2, it can be seen that the proposed system is not suitable for more number of users considering the fact that they will have to wait for longer than 1 minute to acquire their ring signatures. Generally there is relatively a very less number of selectors in GSD, hence we have tested it for a maximum of 100 selectors, but a system like this can be supported for around 3000 selectors with a waiting time of less than 1 minute with minimum signature size less than 100 KB. From Figure 5.2, it can be perceived that when the system is emulated on different configurations the performance of the system

TABLE 5.1

Performance Analysis of Ring Signature

Number of Public Keys	Time to Sign-In (ms)	Time to Verify (ms)	Size of Signature (bytes)
1	52.32	1.91	323
5	87.58	4.96	1572
10	100.21	9.50	3098
15	109.09	11.23	5298
25	116.29	19.11	9091
40	121.23	22.38	11093
55	128.76	24.01	15232
75	169.97	34.33	22091
100	212.11	49.99	29873

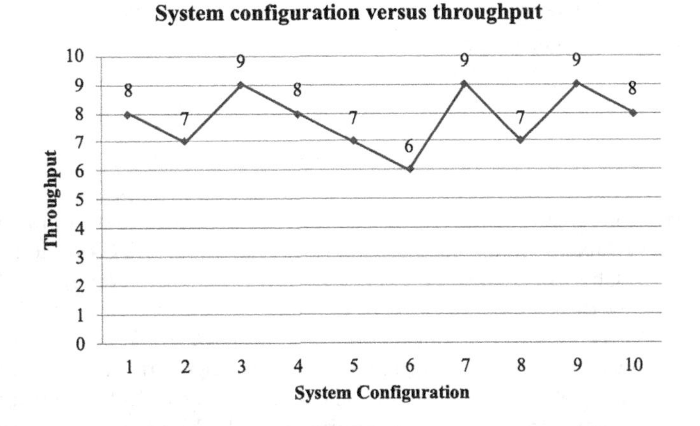

FIGURE 5.2
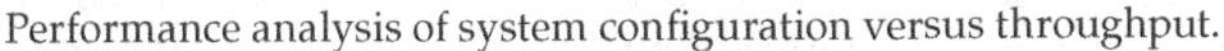
Performance analysis of system configuration versus throughput.

(when measured in throughput) varies showing no specific trend. It is actually influenced by the number of selectors. The throughput of the system is computed as the number of clocks generated in a predefined time limit in blocks per second. Thus, it can be concluded that with a lesser number of selectors, the performance of the system can be improved.

5.7 Conclusion

Blockchain-centric software development has gained a lot of attention in the last few years and defines new directions to allow effective software development. This chapter presents a novel approach for selecting team members in a transparent manner using the concept of blockchain technology in GSD. The proposed concept addresses the issue of fair team member selection, coordination, and communication among team members and the issue of a lack of team spirit. Using the concept of blockchain, the chapter has presented a blockchain-oriented GSD that ensures privacy, verifiability, completeness, uniqueness, and robustness for broadcasting and selection process. The proposed model provides authentication to broadcasters and selectors. The results of initial experimentation conclude that the proposed method is transparent and effective in selecting team members quantitatively and without ambiguity.

References

Avritzer, A., Paulish, D., Cai, Y. and Sethi, K. (2010) Coordination implications of software architecture in a global software development project, *Journal of Systems and Software*, 83(10), pp. 1881–1895. doi: 10.1016/j. jss.2010.05.070.

Binder, J. (2009) Global project management: Communication, collaboration and management across borders, *Strategic Direction*, 25(9).

Böhme, R., Christin, N., Edelman, B. and Moore, T. (2015) Bitcoin: Economics, technology, and governance, *Journal of Economic Perspectives*, 29(2), pp. 213–238.

Bohner, S. and Mohan, S. (2009) Model-Based Engineering of Software: Three Productivity Perspectives. In *Proceedings of the 33rd Annual IEEE Software Engineering Workshop*. IEEE. doi: 10.1109/SEW.2009.19.

Braun, A. (2007) A framework to enable offshore outsourcing. In *Proceedings of the Second IEEE International Conference on Global Software Engineering ICGSE 2007*, pp. 125–129. IEEE.

Casey, V. and Richardson, I. (2009) Implementation of global software development: A structured approach, *Software Process Improvement and Practice*, 14(5), pp. 247–267. doi: 10.1002/spip.422.

Cataldo, M., Shelton, C., Choi, Y., Huang, Y. Y., Ramesh, V., Saini, D. and Wang, L. Y. (2009) Camel: A tool for collaborative distributed software design. In *Proceedings of the Fourth IEEE International Conference on Global Software Engineering ICGSE 2009*, pp. 83–92. IEEE. doi: 10.1109/ICGSE.2009.16

Chakraverty, S., Sachdeva, A. and Singh, A. (2014) A genetic algorithm for task allocation in collaborative software development using formal concept analysis, In Proceedings International Conference Recent Advantage Innovation Engineering (ICRAIE), pp. 1–6.

Colomo-Palacios, R., Casado-Lumbreras, C., Soto-Acosta, P., Misra, S. and García-Penalvo, F. J. (2012) Analyzing human resource management practices within the GSD context, *Journal of Global Information Technology Management*, 15(3), pp. 30–54. doi: 10.1080/1097198X.2012.10845617.

Conchuir, E. O., Holmstrom-Olson, H., Agerfalk, P. J. and Fitzgerald, B. (2009) Benefits of global software development: Exploring the unexplored, *Software Process Improvement and Practice*, 14(4), pp. 201–212. doi: 10.1002/spip.417.

Ebert, C., Murthy, B. K. and Jha, N. N. (2008) Managing risks in global software engineering: principles and practices. In *Proceedings of the IEEE International Conference on Global Software Engineering ICGSE 2008*, pp. 131–140. IEEE.

Edwards, H. K., Kim, J. H., Park, S. and Al-Ani, B. (2008) Global software development: Project decomposition and task allocation. In International Conference on Business and Information.

García-Crespo, A., Colomo-Palacios, R., Soto-Acosta, P. and Ruano-Mayoral, M. (2010) A qualitative study of hard decision making in managing global software development teams. *Information Systems Management*, 27(3), pp. 247–252. doi: 10.1080/10580530.2010.493839.

Gilal, A.R., Jaafar, J., Omar, M., Basri, S., Waqas, A. (2016) A rule-based model for software development team composition: team leader role with personality types and gender classification, *Information Software Technology*, 74, pp. 105–113.

Herna'ndez-Lo'pez, A., Colomo Palacios, R., Garcıa Crespo, A. and Soto-Acosta, P. (2010a) Trust building process for global software development teams, *A review from the Literature. International Journal of Knowledge Society Research*, 1(1), pp. 65–82.

Imtiaz, S. and Ikram, N. (2016) 'Dynamics of task allocation in global software development, *Journal Software Evolution Process*, 29(1).

Islam, S., Joarder, M. M. A. and Houmb, S. H. (2009) Goal and risk factors in offshore outsourced software development from vendor's viewpoint. In *Proceedings of the Fourth IEEE International Conference on Global Software Engineering ICGSE 2009*, pp. 347–352. IEEE

Jiang, Y. (2016) A survey of task allocation and load balancing in distributed systems, *IEEE Transformation Parallel Distribution System*, 27(2), pp. 585–599.

Jun, S., Xiaodong, H. and Jianhua, C. (2018) Study on the Structure of Agricultural Products Traceability System Based on Blockchain vol. 47 (Zhengzhou: Henan Agricultural Sciences), pp. 149–153.

Kern, T. and Willcocks, L. (2000) Exploring information technology outsourcing relationships: Theory and practice, *The Journal of Strategic Information Systems*, 9(4), pp. 321–350. doi: 10.1016/S0963-8687(00)00048-2.

kiwi. Bitcoin testnet sandbox. https://testnet.manu.backend.hamburg/faucet.

Kommeren, R. and Parviainen, P. (2007) Philips experiences in global distributed software development, *Empirical Software Engineering*, 12(6), pp. 647–660. doi: 10.1007/s10664-007-9047-3.

Lacity, M. C. and Rottman, J. W. (2008) The impact of outsourcing on client project managers. *Computer*, 41(1), pp. 100–102. doi: 10.1109/MC.2008.31.

Larsen, M. H. and Klischewski, R. (2004) Process ownership challenges in it- enabled transformation of interorganizational business processes. In System Sciences, 2004. Proceedings of the 37th Annual Hawaii International Conference, p. 11.

Lee-Kelley, L. (2006) Locus of control and attitudes to working in virtual teams, *International Journal of Project Management*, 24(3), pp. 234–243. doi: 10.1016/j.ijproman.2006.01.003.

Li, D. F. and Ren, H. P. (2015) Multi-attribute decision making method considering the amount and reliability of intuitionistic fuzzy information, *Journal of Intelligent & Fuzzy Systems*, 28(4), pp. 1877–1883.

Liu, J. K., Wei, V. K., Wong, D. S. (2004) Linkable spontaneous anonymous group signature for ad hoc groups (extended abstract). In: Wang, H., Pieprzyk, J., Varadharajan, V. (eds.) ACISP 2004. LNCS, vol. 3108, pp. 325–335. Springer.

Liu, P. and Jin, F. (2012) Methods for aggregating intuitionistic uncertain linguistic variables and their application to group decision making, *Information Sciences*, 205, pp. 58–71. doi: 10.1016/j.ins.2012.04.014.

Mahmood, S., Anwer, S., Niazi, M., Alshayeb, M. and Richardson, I. (2017) Key factors that influence task allocation in global software development, *Information Software Technology*, 91, pp. 102–122.

Marques, A. B., Carvalho, J. R., Rodrigues, R., Conte, T., Prikladnicki, R. and Marczak, S. (2013) "An ontology for task allocation to teams in distributed software development," In Proceedings IEEE 8th International Conference Global Software Engineering, pp. 21–30.

Maxwell, G. (2015) Confidential Transactions, https://people.xiph.org/~greg/confidential_ values.txt

Milewski, A. E., Tremaine, M., Kobler, F., Egan, R., Zhang, S. and O'Sullivan, P. (2008) Guidelines for effective bridging in global software engineering, *Software Process Improvement and Practice*, 13(6), pp. 477–492. doi: 10.1002/spip.403.

Nguyen, H. (2015) A new knowledge-based measure for intuitionistic fuzzy sets and its application in multiple attribute group decision making, *Expert Systems with Applications*, 42(22), pp. 8766–8774.

Parker, G., Zielinski, D. and McAdams, J. (2000) *Rewarding Teams: Lessons From the Trenches San Francisco*, California: Jossey-Bass Inc.

Smite, D., Wohlin, C., Gorschek, T. and Feldt, R. (2010) Empirical evidence in global software engineering: A systematic review, *Empirical Software Engineering*, 15(1), pp. 91–118. doi: 10.1007/s10664-009-9123-y.

Sooraj, P. and Mohapatra, P. K. J. (2008) Modeling the 24-h software development process. *Strategic Outsourcing*, 1(2), pp. 122–141. doi: 10.1108/17538290810897147.

Stevens, K. (1998) *The effects of roles and personality characteristics on software development team effectiveness* [Doctor of Philosophy]. Virginia Tech.

Strode, D. E. (2016) A dependency taxonomy for agile software development projects, *Information System Frontiers*, 18(1), pp. 23–46.

Sutanto, J., Kankanhalli, A. and Tan, B. C. Y. (2015) Investigating task coordination in globally dispersed teams: A structural contingency perspective. *ACM Transformation Management Information System*, 6(2), pp. 1–31.

Wood, G. (2014) Ethereum: A secure decentralised generalised transaction ledger, *Ethereum Project Yellow Paper*, 151(2014), pp. 1–32.

6

Machine Learning in Sustainable Healthcare

Dipanwita Thakur
Banasthali Vidyapith

Suparna Biswas
MAKAUT

CONTENTS

6.1 Introduction

Sustainable healthcare is defined as "Care delivered in a way that does not adversely affect the health of the population and does not use resources in a way that may compromise the ability of those in the future to provide high quality care to their population or increase their burden of illness." Although healthcare providers strive to shift toward an integrated and sustainable model of healthcare delivery, an IT-enabled or e-Healthcare approach is increasingly being implemented. Figure 6.1 shows the four principles of sustainable healthcare (Mortimer, 2010).

- Prevention—Encouraging health and restraining disease by handling the causes of illness and inequalities.
- Patient Self-Care—Enabling patients to play more of a part in maintaining their own health and healthcare.
- Lean Service Delivery—Rationalization of treatment services to reduce inefficient tasks.
- Low-Carbon Alternatives—Organizing medications and advancements with a lower ecological effect.

DOI: 10.1201/9781003046431-6

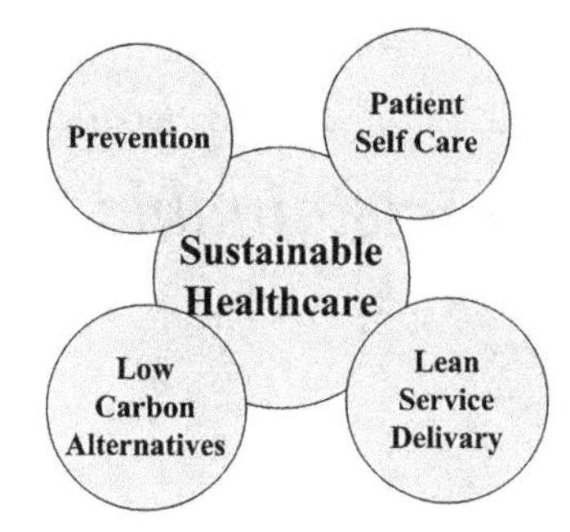

FIGURE 6.1
Principles of sustainable healthcare.

Of these four principles, in this chapter, we emphasize on the first two where machine learning (ML) plays a huge role.

With the increasing problem of a growing population and continuous improvement in living standards, there is an intense demand for e-healthcare, which can improve diagnostic technology and prolong human life. According to the World Health Organization (WHO), in India, the doctor–patient ratio is 1:1000. So, there is a huge crisis of medical assistance in India. In addition, as indicated by United Nations Population Fund and Help Age India, the number of elderly people is estimated to increase by up to 173 million by 2026. Moreover, the study has shown that about 89% of the elderly will probably be living alone. In addition, a medical research survey report says that 80% of people aged more than 65 years suffer from some form of chronic disease. So, most of the elderly people are incapable of taking care of themselves. The WHO (Organization, 2015) has assessed that the general pervasiveness pace of burdensome issues among the older for the most part differs somewhere in the range of 10% and 20% depending on the social situations; this condition breaks down when they lose their accomplices and are completely disregarded in the family. As indicated by the last registration, around 15 million of the elderly live in solitude and three-fourths are women. The elderly require more medical assistance. Monitoring system for elderly people has increased impressive impact in the field of remote health monitoring. Along these lines, in the home and assisted-living environments, e-healthcare is relied upon to live longer.

Various healthcare systems have been proposed by researchers. The motivation behind this research is to maintain a trade-off between the need for healthcare services and the need to control costs and guarantee the sustainability of the system. According to research by the BBC: "Prevention must become a cornerstone of the healthcare system rather than an afterthought." So, remote monitoring of patients for blood glucose, blood pressure, heart rate can provide information to a medical practitioner (Lake et al. 2013). Moreover, these measurements are taken under normal conditions, which helps the medical practitioner to take appropriate measures for the patients. In this regard, we need sensors to monitor the different activities of people. The sensors can be mobile sensors such as accelerometers, gyroscopes, or wearable sensors or image sensors. In real time, for prediction analysis, the results are gathered for the analysis of physical activities. Each parameter has a threshold value. If the obtained result exceeds any of the parameter limits of the given threshold value, then an alarm is sent to the medical practitioner (Gupta et al. 2017). The predictive analysis is performed using various ML approaches because of which we are able to obtain sustainable healthcare. For many years, several researchers have been proposing various research studies for disease prevention and early disease prediction using various ML approaches for sustainable healthcare. For instance, Manogaran et al. (2018) proposed a cancer detection mechanism through big data using ML. In sustainable development, big data has had a major impact (Dash et al. 2019; Manogaran and Lopez, 2017; Thota et al. 2017). The authors Thakur and Biswas (2020) presented an exhaustive survey to show the impact of ML in human activity monitoring and recognition. Nowadays, people are very health conscious. They want to monitor their vital health activities constantly and get early disease prediction (Ahmadi et al. 2015; Alshurafa et al. 2014). Thus, human activity monitoring and recognition is very important for early disease prediction. Odhiambo and Muganda Ochara (2018) emphasized on the importance of an intelligent healthcare ecosystem for prevention, early disease prediction, and personalized treatment through health data integration.

6.2 Health Monitoring

Prevention and patient self-care can be achieved using health monitoring. There are different categories of health monitoring. They are as follows.

6.2.1 Cyber-Physical System in Healthcare

Cyber-physical system in healthcare is a combination of digitally acquired data and various sensors and smart devices that efficiently collect data so the medical practitioner can take proper decisions (Haque et al. 2014). Mobile healthcare cyber-physical systems using ML perform a crucial role in current systems of medical surveillance, such as diagnosis, medical care, attention, and emergency relief, etc. (Costanzo et al. 2016; Hossain, 2017; Jaimes et al. 2015; Zhang et al. 2017).

Figure 6.2 depicts an example of a health cyber-physical system, where end-to-end delay is the main problem in case of some emergency medication.

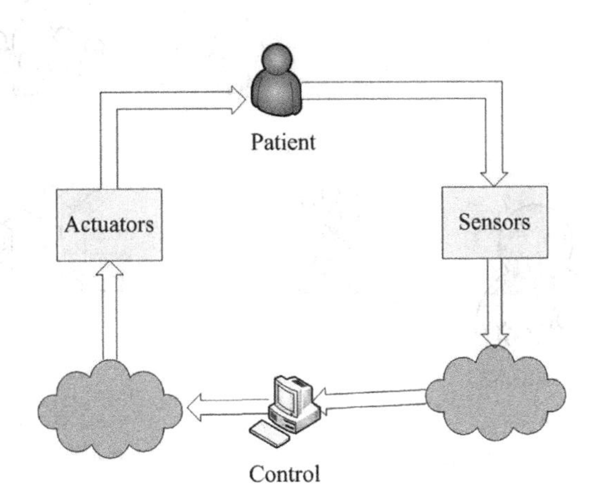

FIGURE 6.2
Cyber-physical system in healthcare.

6.2.2 Mobile Health Monitoring

Recently, Smartphone-based *mobile health-monitoring* system using ML is gaining popularity due to its in-built sensors (Chao et al. 2016; Wang et al. 2015). Using smartphones, the signals of physical human activities are gathered to identify early disease (Li et al. 2016; Lin et al. 2015; Moser and Melliar-Smith 2015; Zhang et al. 2015). Recently, smartphones are popularly used to recognize human activities as smart phones consist of some in-built sensors such as accelerometers, gyroscopes, cameras, etc. Figure 6.3 shows the framework of the mobile health-monitoring system. Using mobile health monitoring an elderly person's health can be monitored by any family member or any medical authority when the person is staying alone or outside the home. The in-built sensors of the smartphone are capable enough to capture the data of different human activities. Using various ML approaches, researchers have provided different ML models to monitor and predict various human physical activities such as walking, cycling, sitting, falling, lying, standing, jogging, stairs-up, downstairs, etc. After the prediction of various activities, we can predict the abnormal behavior of the person.

6.2.3 Internet of Things in Healthcare

Internet of Things (IoT) is an emerging field in modern healthcare. Real-time health monitoring, healthcare management, and medical data management are the major applications of the IoT that help both the patient and the medical practitioner (Gope and Hwang, 2016). The IoT facilitates fast and efficient communication between various types of devices, such as sensors, cameras, home appliances, etc. (Gope and Hwang, 2015). Therefore, the IoT is very effective in the healthcare sector. In the healthcare system, the IoT comprises multiple varieties of sensors (wearable, embedded, and environment) by which people are able to take advantage of modern healthcare services anytime, anywhere. It also significantly improves the quality of life of elderly people. Figure 6.4 shows the IoT-based healthcare system using ML, where various sensors are used to collect data. Using IoT, we are able to achieve a smart healthcare system, which is part of a smart environment, i.e., sustainable healthcare. A combination of ML and IoT plays a vital role in IoT healthcare.

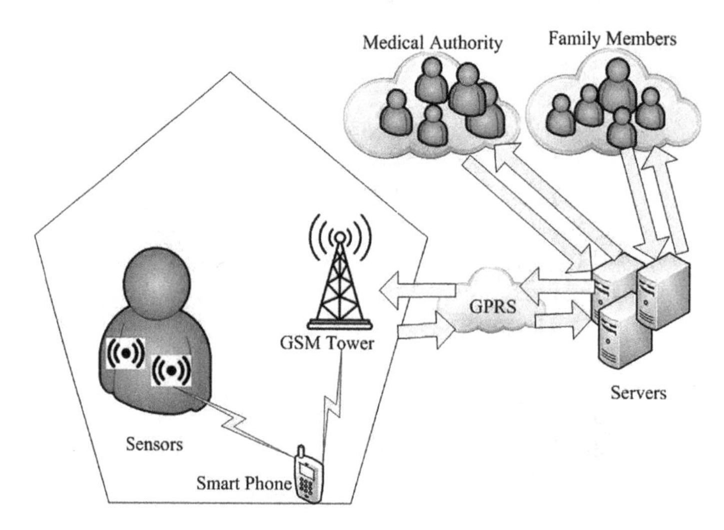

FIGURE 6.3
Mobile health-monitoring system.

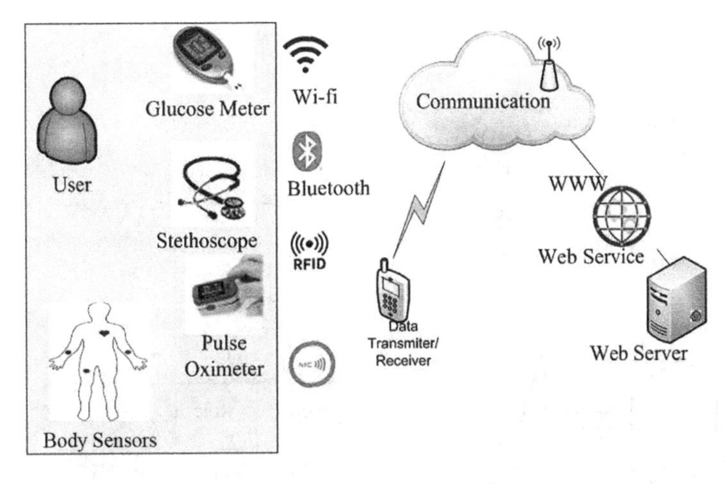

FIGURE 6.4
IoT- and ML-based healthcare system.

6.2.4 Wearable Computing for Health Monitoring

Wearable sensor technology is moving forward and creating major opportunities to enhance personalized healthcare (Meng and Kim, 2011; Rodgers et al. 2015; Sultan, 2015). Various body wear sensors, smart watches, and smart phones are used to measure pulse rate, body temperature, etc. for health monitoring. Various wearable computing devices are available that can be used to monitor our vital health signs. Implementation of wearable computing for health monitoring involves some of the following issues:

- Interaction of the system with the environment using various sensors assigned in various segments of the body or any wearable clothing. It creates some awareness with regard to the health activities of a person who has worn the sensors.
- User interface, which must be very simple in such a way that it requires very little human intervention.
- The system must be able to automatically perform various activities without human intervention. For example, the system should send alerts when there is some type of health issue.
- The system should not interfere with the normal daily activities of the person who has worn the sensors.

Figure 6.5 shows the framework of wearable computing in healthcare, where the user wears various sensors by which health data are monitored.

6.2.5 Ambient Assisted Living

From a human services viewpoint, ambient assisted living incorporates any item, procedure, or administration that can help legitimately or in a roundabout way to keep people healthy in their home conditions, while giving them the fundamental help required considering their conditions or pathologies. Ambient assisted-living arrangements intend to balance out or even increase the self-governance of the people, encourage clinical medications, also, observe from home, as well as facilitate the completion of

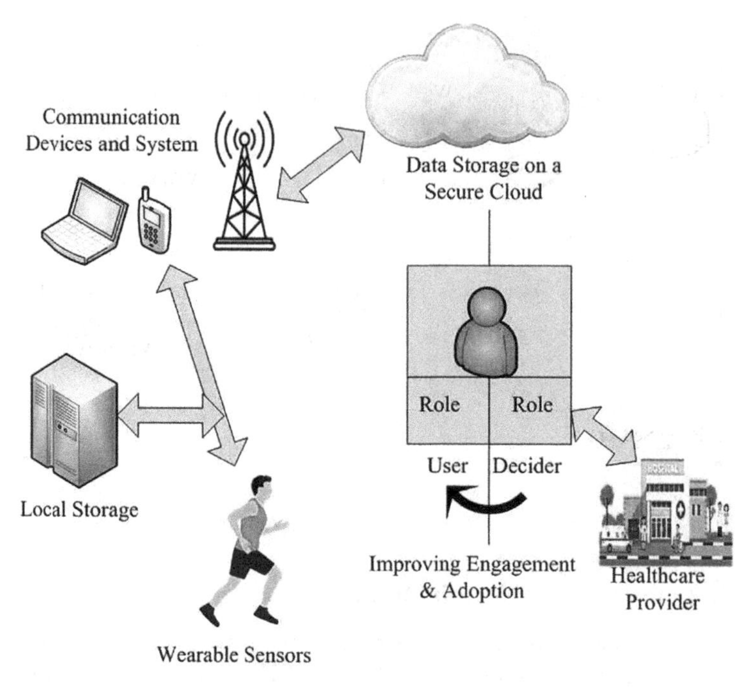

FIGURE 6.5
Wearable computing for health-monitoring system.

everyday assignments. Ambient assisted-living arrangements are in light of the craving of people to stay inside their home surroundings to the extent that this would be possible and to defer their passageway into organized situations. These methodologies provide answers for the elderly; weakened people; people recuperating from a condition, surgery, or sickness; or individuals with cognitive, motor, or neurological needs (Avila and Sampogna, 2011). Figure 6.6 shows the framework for ambient assisted living in healthcare.

6.2.6 Body Area Network

One of the most significant developments in contemporary healthcare systems is the body area network (BAN). BANs are created using low-power gadgets and biosensors that are worn on or embedded in the human body. The improvement of BANs is developing as one of the fundamental research patterns, especially to gather and mutually process natural information for consistent, long-haul checking of health conditions (Fourati, 2014; Jovanov et al. 2005; Seyedi et al. 2013). From a clinical perspective, the BAN rises as a key innovation in providing constant health monitoring and judgments of numerous hazardous infections. This is exceptionally delicate in the elderly or patients with interminable conditions, yet additionally following the presentation of competitors, just to make reference to certain applications. BAN comprises sensors and actuators around the human body so as to screen human organs or to convey either motivations or medication on the body or inside the body, otherwise called embedded clinical gadgets, such as pacemakers or case endoscopes with a remote correspondence connected to a passageway or center point. Figure 6.7 shows the framework for BANs in healthcare.

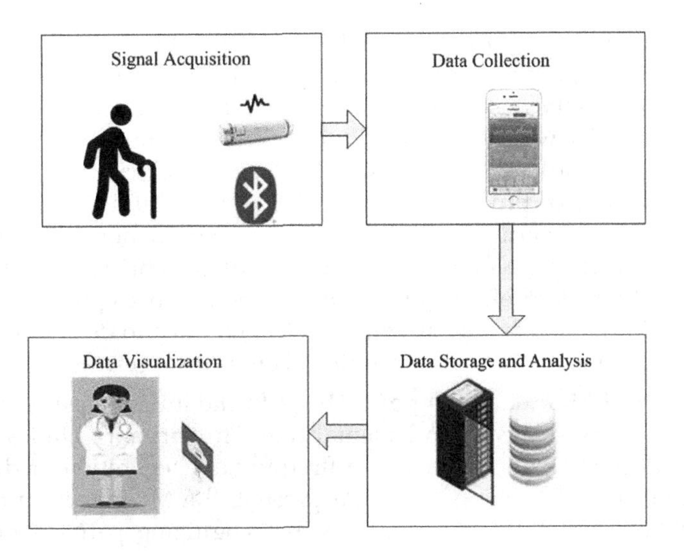

FIGURE 6.6
Ambient assisted-living system for remote patient monitoring.

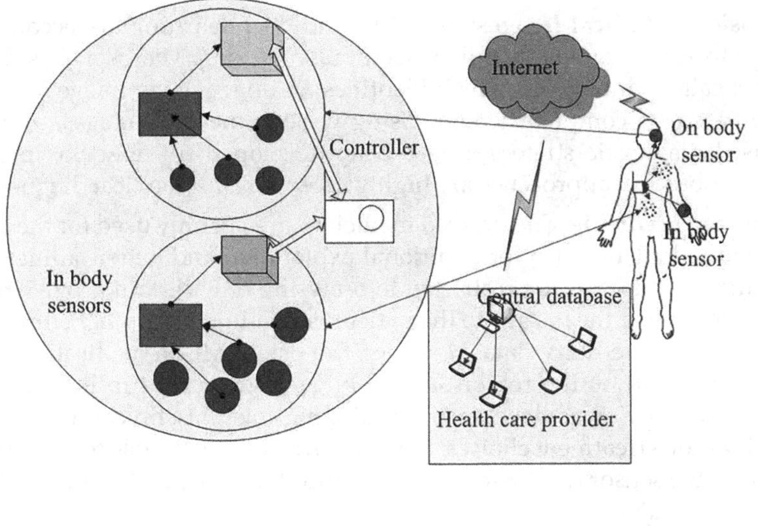

FIGURE 6.7
Body area network.

6.3 Significance of Machine Learning in Sustainable Healthcare

One of the most popular applications of artificial intelligence (AI) technology is ML, which offers numerous routes in achieving key health policy objectives. It incorporates "creating conditions that guarantee healthy body" (Navarro, 2007) and collective responsibility for the whole population by incorporating preventive measures, disease prevention, encouraging healthy lifestyles, and community screening. In healthcare, there is a significant application of ML.

- **Identify and Diagnosis of Disease**: It is very difficult to identify and diagnose a disease only on the basis of symptoms without analyzing the vital health parameters. This is one of the major applications of ML. This can integrate anything from malignant growths, which are difficult to diagnose, to underlying phases and to variant genetic disorders. "IBM Watson Genomics" is a primary instance of coordination of subjective processing with genome-based tumor sequencing, which can assist in rapid determination. "Berg, the biopharma monster" has been using AI in manufacturing life-saving drugs in various categories, for instance, cancer treatment. "P1vital's PReDicT (Predicting Response to Depression Treatment)" plans to establish a mechanically conceivable technique to deal with break down and provide a remedy under routine clinical circumstances.
- **Realization and Manufacturing of a Drug**: In the initial stage of the drug discovery process, ML plays a fundamental role. This primary clinical application of ML includes R&D advancements in finding next-generation medicine for the treatment of multifactorial ailments. At present, the ML-based frameworks use unsupervised learning. Its advantage is in recognizing patterns of data without giving any predictions. Microsoft developed a project called "Hanover," in which ML-based technologies have been used for various strategies, including the treatment of carcinoma and modifying medicine blend for AML (acute myeloid leukemia).
- **Diagnosis of Medical Images**: Both ML and deep learning are accountable for the leap forward revolution called "Computer Vision." Microsoft has developed a project called "InnerEye," which identifies an image using image analysis procedure based on computer vision. Using various medical images, a number of ML-based diagnostic strategies have been developed till now. So, in computer vision, various ML approaches are highly acceptable for medical diagnosis.
- **Customized Medicine**: Customized medicines are not only used for the treatment of the individual, but also for additional exploration and better ailment evaluation. Currently, doctors are restricted to browsing only certain provisions of conclusions regarding the hazard to the patient depending on his/her clinical history and accessible hereditary data. However, the use of ML in medication has been extraordinary. ML is the reason for the huge development in healthcare. "IBM Watson Oncology" was developed by utilizing tolerant clinical data, which has created various treatment choices. In the future, we will be able to acquire prompt and accessible sensor data using more advanced devices, which will be beneficial for healthcare advancement.
- **ML-Based Behavioral Modification**: Social alteration is a significant piece of preventive medication, and since the time of the multiplication of AI in human services, incalculable new businesses are springing up in the fields of malignant growth anticipation and ID, quiet treatment, and so forth. Somatix is a B2B2C-based information investigation organization that has released a ML-based application to perceive signals, which we make in our everyday lives, permitting us to comprehend our oblivious conduct and roll out vital improvements.
- **Smart Health Records**: Continuing with cutting-edge health records is a thorough procedure, and keeping in mind that innovation has had its impact in facilitating the information section process, in all actuality even now, a larger part of the procedures set aside a great deal of effort to finish. The primary job of AI in medicinal services is to ease procedures to spare time, exertion, and cash. Archive

characterization strategies utilizing vector machines and ML-based OCR acknowledgment procedures are gradually gaining steam, for example, Google's Cloud Vision API and MATLAB's AI-based penmanship acknowledgment innovation. MIT is currently at the front line of building up and coming of age of canny, brilliant well-being records, which will consolidate ML-based tools starting from the earliest stage to help with analysis, clinical treatment proposals, and so on.

- **Smart Data Collection**: Public support has been extremely popular in clinical field nowadays, permitting scientists and specialists to gather a huge amount of data transferred by individuals out of their own initiative. This live well-being information has had extraordinary repercussions in the manner that medication will be seen going forward. Apple's Research Kit permits clients access to intelligent applications that apply ML-based facial acknowledgment in an attempt to treat Asperger's and Parkinson's sickness. IBM as of late has collaborated with Medtronic to translate, amass, and make accessible diabetes and insulin information progressively dependent on publicly supported data. With the progress being made in IoT, the human services industry is still discovering new methods by which to utilize this information and handle intense-to-analyze cases and help in the general improvement of analysis and medicine.
- **Outbreak Prediction**: Computer-based intelligence and ML are currently being put to use in observing and anticipating plagues the world over. Today, researchers access a lot of information gathered from satellites, continuous Web-based life refreshes, site data, and so on. Counterfeit neural systems assist in gathering these data and foreseeing everything from jungle fever episodes to serious, ceaseless irresistible sicknesses. Foreseeing these flare-ups is particularly useful in underdeveloped nations as they need significant clinical foundation and instructive frameworks. An essential case of this is the ProMED-mail, an Internet-based revealing stage that screens advancing sicknesses and developing ones and gives episode reports continuously.

6.4 Case Studies

Some case studies are discussed in this section to show the application of ML in healthcare.

A case study in identifying rheumatoid arthritis based on primary health data is proposed by Zhou et al. (2016). A ML-based approach is used to recognize rheumatoid arthritis using primary data using the following steps: (i) Primary data with a disease case is compared to a control to select the parameters. (ii) Using a random forest method reduces the predictor parameters. (iii) Using a decision tree, a decision rule is designed. (iv) The proposed method is validated on an independent dataset. (v) The performance is compared with two existing deterministic algorithms used for rheumatoid arthritis, which are implemented using expert clinical research.

Essential consideration EHRs were accessible for 2,238,360 patients beyond 16 years of age. Out of these, 20,667 were additionally connected in the optional consideration rheumatology clinical framework. In the related dataset, 900 predictors (out of a total of 43,100 variables) were discovered more often in those without RA than those with primary care records. These predictors were decreased to 37 various classes of similar clinical code by which a decision tree model was developed. Ultimately the algorithm recognized eight predictors correlated to RA diagnosis codes, medication codes, such as anti-rheumatic drug-modifying disease codes, and lack of alternative diagnoses such as psoriatic arthritis. The proposed ensemble ML approaches are able to define the most helpful predictors in a cost-effective and efficient way to classify rheumatoid arthritis or other medical complications, accurately, and efficiently using primary electronic health records.

Zou et al. (2018) used a ML approach for the prediction of diabetes mellitus. Decision trees, random forests, and neural network models are used for prediction. The dataset of 14 attributes was collected from a hospital situated in Luzhou, China. To validate the model five-fold cross-validation method was used. The researchers randomly collected data from 68,994 healthy people and diabetic patients, respectively. The data were used as a training set. Principal component analysis and minimum redundancy maximum relevance methods were used for dimensionality reduction. As a result, the random forest method has given the highest accuracy to predict diabetes mellitus.

Early detection of lung cancer was proposed by Petousis et al. (2019). The NLST dataset was accessed from the National Cancer Institute. The dataset contained clinical, demographic, and imaging information of more than 25,000 candidates. Here, the data preprocessing was done with the Tammemgi model to impute missing values using a variety of "multiple clustering imputation" approach. In this work, a five-fold stratified cross validation is performed with a dataset, where the training and testing data ratio is 80%:20%. Each fold is randomly generated while preserving the NLST study of relative ratio of cancer to noncancer instances noticed at each investigation. Combinations of multiple ML approaches are used to learn a partially measurable Markov decision mechanism that simultaneously optimizes the detection of lung cancer while improving the test specificity. In this, the researchers trained a dynamic Bayesian network as an empirical model and used reverse reinforcement learning to discover a mechanism of rewards based on decisions made by experts. The resulting predictive model minimized the false-positive rate while keeping a high, true-positive rate at a stage comparable to that of human experts. The proposed model observed a variety of earlier lung cancers.

Liu et al. (2020) proposed a new ML model for predicting Alzheimer's. Alzheimer's is a disease mainly found in elderly people. Various works have been conducted to identify this disease using medical images with the use of different ML approaches (Ezzati et al. 2019; Zhang et al. 2015). As collecting medical images is a tedious job, in this work, the author used speech data to predict Alzheimer's with a logistic regression approach of ML. The framework comprises a wearable IoT gadget that records the individual's voice constantly (Chen and Hu, 2013). The voice information is transmitted to the cloud server, where the first discourse information is put away. The obtained new information will be perceived utilizing the prepared model for distinguishing the Alzheimer's disease manifestations. Right off the bat, we gather the first sound information of a considerable number of subjects. The sound information can be separated from a diverse number of sound sections because the individual can't talk constantly, which must contain the quietness of sound fragments. Also, separate every sound fragment into additional multifragments of 1 second. Then, extricate the spectrogram highlights from the sound portions. The acquired spectrogram information can be accustomed to preparing models. Finally, utilize this prepared model to recognize a set of new information with regard to Alzheimer's disease.

Recently, in this year "COVID-19" is a widely spread virus, which is popularly known as "Corona Virus". WHO characterized it as pandemic. Researchers are continuously searching for the remedy from it in their own ways. Till date literature shows that different researchers are using ML approaches massively to predict the virus or to predict the medicines or to predict the chances of spreading the virus all over the world. In March

2020 publication of "IEEE Spectrum," an article demonstrates that there are five different companies, such as Deargen, a South Korea-based company; Insilico Medicine, a Hong Kong-based company; SRI Biosciences, a Menlo Park-based research center; Benevolent AI, a British AI-startup; and Iktos, an AI company, all using deep learning model to discover the drug to fight with this virus.

Recently, Xu et al. (2020) suggested a model based on deep learning to screen this virus disease. In this study, the authors have taken CT images of the chest of the pulmonary region. For the feature extraction, they have taken the classical ResNet-18 network structure. For the report generation, they have used the "Noisy-or Bayesian function," and the overall accuracy measure for early virus prediction was 87.6%.

Another very interesting study was done by Magar et al. (2020), in which the authors discovered the antibodies for novel coronavirus using the ML approach. The authors tried to predict whether the antibodies are able to fight with the novel coronavirus. In this work, they used diverse classical ML strategies such as decision tree, support vector machine, etc. This ML model was trained with more than 14 different types of viruses and achieved over 90% five-fold test accuracy.

Therefore, from the above-mentioned case studies, we were able to understand that the various ML approaches have a significant impact in sustainable healthcare.

6.5 Conclusion

This chapter discusses the application of ML in sustainable healthcare. Addressing the sustainability of healthcare and how it can be achieved is discussed. Different case studies show the efficiency of ML in achieving sustainable healthcare. In conclusion, in the near future, we will be able to achieve more sustainable health care using various ML approaches.

References

Ahmadi, A., Mitchell, E., Richter, C., Destelle, F., Gowing, M., OConnor, N. E. and Moran, K. (2015) Toward automatic activity classification and movement assessment during a sports training session, *IEEE Internet of Things Journal*, 2(1), pp. 23–32.

Alshurafa, N., Xu, W., Liu, J. J., Huang, M., Mortazavi, B., Roberts, C. K. and Sarrafzadeh, M. (2014) Designing a robust activity recognition framework for health and exergaming using wearable sensors, *IEEE Journal of Biomedical and Health Informatics*, 18(5), pp. 1636–1646.

Avila, N. and Sampogna,C. (2011) e-health ambient assisted living and personal health systems, Kramme, R., Hoffmann, K. P., Pozos, R. S. (eds) *Springer Handbook of Medical Technology*.

Chao, H.-C., Zeadally, S. and Hu, B. (2016) Wearable computing for health care, *Journal of Medical Systems*, 40(4), p. 87. doi: 10.1007/s10916-016-0448-y.

Chen, Y. and Hu, H. (2013) Internet of intelligent things and robot as a service, *Similation Modelling Practice and Theory*, 34, pp. 159–171.

Costanzo, A., Faro, A., Giordano, D. and Pino, C. (2016) Mobile cyber physical systems for health care: Functions, ambient ontology and e-diagnostics. In *2016 13th IEEE Annual Consumer Communications Networking Conference (CCNC)*, pp. 972–975. doi: 10.1109/CCNC.2016.7444920.

Dash, S., Shakyawar, S. K., Sharma, M. and Kaushik, S. (2019) Big data in healthcare: Management, analysis and future prospects, *Journal of Big Data*, 6(1), p. 54. doi: 10.1186/s40537-019-0217-0.

Ezzati, A., Zammit, A. R., Harvey, D. J., Habeck, C., Hall, C. B., Richard, B. (2019) Lipton, and for the Alzheimer's Disease Neuroimaging Initiative, Optimizing machine learning methods to improve predictive models of alzheimer's disease, 71, pp. 1027–1036. doi: 10.3233/JAD-190262. 3.

Fourati, L. C. (2014) Wireless body area network and healthcare monitoring system. In *2014 IEEE International Conference on Healthcare Informatics*, pp. 362–362. doi: 10.1109/ICHI.2014.57.

Gope, P. and Hwang, T. (2015) Untraceable sensor movement in distributed IoT infrastructure. *IEEE Sensors Journal*, 15(9), pp. 5340–5348. doi: 10.1109/JSEN.2015.2441113.

Gope, P. and Hwang, T. (2016) Bsn-care: A secure IoT-based modern healthcare system using body sensor network, *IEEE Sensors Journal* 16(5), pp. 1368–1376. doi: 10.1109/JSEN.2015.2502401.

Gupta, P. K., Maharaj, B. T. and Malekian, R. (2017) A novel and secure IoT based cloud centric architecture to perform predictive analysis of users activities in sustainable health centres, *Multimedia Tools and Applications*, 76(18), pp. 18489–18512. doi: 10.1007/s11042-016-4050-6.

Haque, S., Aziz, S. and Rahman, M. (2014) Review of cyber-physical system in healthcare. *International Journal of Distributed Sensor Networks* 2014, p. 20. doi: 10.1155/2014/217415.

Hossain, M. S. (2017) Cloud-supported cyberphysical localization framework for patients monitoring, *IEEE Systems Journal*, 11(1), pp. 118–127. doi:10.1109/JSYST.2015.2470644.

Jaimes, L. G., Calderon, J. Lopez, J. and Raij, A. (2015) Trends in mobile cyberphysical systems for health just-in time interventions. In *Southeastcon 2015*, pp. 1–6. doi: 10.1109/SECON.2015.7132887.

Jovanov, E., Milenkovic, A., Otto, C. and de Groen, P. C. (2005) A wireless body area network of intelligent motion sensors for computer assisted physical rehabilitation, *Journal of NeuroEngineering and Rehabilitation*, 2(1), p. 6. doi: 10.1186/1743-0003-2–6.

Lake, D., Milito, R., Morrow, M. and Vargheese, R. (2013) Internet of Things: Architectural framework for ehealth security, *Journal of ICT*, 1(3), pp. 301–328.

Li, Y., Dai, W., Ming, Z. and Qiu, M. (2016) Privacy protection for preventing data over-collection in smart city. *IEEE Transactions on Computers*, 65(5), pp. 1339–1350. doi: 10.1109/TC.2015.2470247.

Lin, K., Wang, W., Wang, X., Ji, W. and Wan, J. (2015) Qoe-driven spectrum assignment for 5g wireless networks using sdr, *IEEE Wireless Communications*, 22(6), pp. 48–55. doi: 10.1109/MWC.2015.7368824.

Liu, L., Zhao, S., Chen, H. and Wang, A. (2020) A new machine learning method for identifying alzheimer's disease, *Simulation Modelling Practice and Theory*, p. 99.

Magar, R., Yadav, P. and Farimani, A. M. (2020) Potential Neutralizing Antibodies Discovered for Novel Corona Virus Using Machine Learning.

Manogaran, G. and Lopez, D. (2017) Disease surveillance system for big climate data processing and dengue transmission, *International Journal of Ambient Computing and Intelligence*, 8, pp. 88–105. doi: 10.4018/IJACI.2017040106.

Manogaran, G. V., Varatharajan, V. R., Kumar, P. M., Sundarasekar, R. and Hsu, C. H. (2018) Machine learning based big data processing framework for cancer diagnosis using hidden markov model and gm clustering, *Wireless Personal Communications*, 102(3), pp. 2099–2116. doi: 10.1007/s11277-017-5044-z.

Meng, Y. and Kim, H. (2011) Wearable systems and applications for healthcare. In *2011 First ACIS/JNU International Conference on Computers, Networks, Systems and Industrial Engineering*, pp. 325–330.

Mortimer, F. (2010) The sustainable physician. *Clinical Medicine Journal* 10(2), pp. 110–111.

Moser, L. E. and Melliar-Smith, P. M. (2015) Personal health monitoring using a smartphone. In *2015 IEEE International Conference on Mobile Services*, pp. 344–351. doi: 10.1109/MobServ.2015.54.

Navarro, V. (2007) What is a national health policy?, *International Journal of Health Services* 37(1), pp. 1–14. doi: 10.2190/H454–7326-6034-1T25. PMID: 17436983.

Odhiambo, J. N. and Muganda Ochara, N. (2018) Precision health care for sustainable patient centric solutions. In *2018 Open Innovations Conference (OI)*, pp. 194–197. doi: 10.1109/OI.2018.8535905.

Petousis, P., Winter, A., Speier, W., Aberle, D. R., Hsu, W. and Bui, A. A. T. (2019) Using sequential decision making to improve lung cancer screening performance, *IEEE Access*, 7, pp. 119403–119419. doi: 10.1109/ACCESS.2019.2935763.

Rodgers, M. M., Pai, V. M. and Conroy, R. S. (2015) Recent advances in wearable sensors for health monitoring, *IEEE Sensors Journal*, 15(6), pp. 3119–3126. doi: 10.1109/JSEN.2014.2357257.

Seyedi, M., Kibret, B., Lai, D. T. H. and Faulkner, M. (2013) A survey on intrabody communications for body area network applications, *IEEE Transactions On Biomedical Engineering* 60(8).

Sultan, N. (2015) Reflective thoughts on the potential and challenges of wearable technology for healthcare provision and medical education, *International Journal of Information Management*, 35(5), pp. 521–526.

Thakur, D. and Biswas, S. (2020) Smartphone based human activity monitoring and recognition using ml and dl: a comprehensive survey, *Journal of Ambient Intelligence and Humanized Computing*. doi: https://doi.org/10.1007/s12652-020-01899–y.

Thota, C., Manogaran, G., Lopez, D. and Vijayakumar, V. (2017) Big data security framework for distributed cloud data centers, *Cybersecurity Breaches And Issues Surrounding Online Threat Protection, IGI Global.*

Wang, J., Qiu, M. and Guo, B. (2015) High reliable real-time bandwidth scheduling for virtual machines with hidden markov predicting in telehealth platform, *Future Generation computer Systems*, 49, pp. 68–76

World Health Organization. (2015) *World report on ageing and health*, Luxembourg: World Health Organization.

Xu, X., Jiang, X., Ma, C., Du, P., Li, X., Lv, S., Yu, L., Chen, Y., Su, J., Lang, G., Li, Y., Zhao, H., Xu, K., Ruan, L. and Wu, W. (2020) Deep Learning System to Screen Coronavirus Disease 2019 Pneumonia.

Zhang, Y., Dong, Z., Phillips, S., Wang, S., Ji, G., Yang, J. and Yuan, T. F. (2015) Detection of subjects and brain regions related to alzheimer's disease using 3d mri scans based on eigenbrain and machine learning, *Frontiers in Computational Neuroscience*, 9, p. 66. doi: 10.3389/fncom.2015.00066. https://www.frontiersin.org/article/10.3389/fncom.2015.00066.

Zhang, Y., Qiu, M., Tsai, C., Hassan, M. M. and Alamri, A. (2017) Health-cps: Healthcare cyber-physical system assisted by cloud and big data, *IEEE Systems Journal*, 11(1), pp. 88–95. doi: 10.1109/JSYST.2015.2460747.

Zhang, Z., Wang, H., Wang, C. and Fang, H. (2015) Cluster-based epidemic control through smartphone-based body area networks, *IEEE Transactions on Parallel and Distributed Systems*, 26(3), pp. 681–690. doi: 10.1109/TPDS.2014.2313331.

Zhou, S.-M., Fernandez-Gutierrez, F., Kennedy, J., Cooksey, R., Atkinson, M., Denaxas, S., Siebert, S., Dixon, W. G., O'Neill, T. W., Choy, E., Sudlow, C. (2016) UK biobank follow-up group, outcomes, and sinead brophy. Defining disease phenotypes in primary care electronic health records by a machine learning approach: A case study in identifying rheumatoid arthritis. *PloS One*, 11(5), pp. 0154515–0154515. doi:10.1371/journal.pone.0154515. 27135409[pmid]. https://pubmed.ncbi.nlm.nih.gov/27135409.

Zou, Q., Qu, K., Luo, Y., Yin, D., Ju, Y. and Tang, H. (2018) Predicting diabetes mellitus with machine learning techniques, *Frontiers in Genetics*, 9, p. 515. doi: 10.3389/fgene.2018.00515. https://www.frontiersin.org/article/10.3389/fgene.2018.00515.

7

Multimedia Audio Signal Analysis for Sustainable Education

Megha Rathi, Aditya Lahiri, Ayushi Aggarwal, Parul Jindal, and Adwitiya Sinha
Jaypee Institute of Information Technology

CONTENTS

7.1 Introduction

Today, data in the audio form are available and present everywhere, especially for video conferencing, online video lectures, etc., to promote sustainable education. The main problem associated with this field is an unstructured and unorganized format of the audio content. Hence, it is mandatory to regularize, arrange, and segregate audio features to extract useful information easily. The deliberate change of the audio signals usually through the audio signal unit is known as audio signal processing. Audio signals are of two types: digital and analog. Processing of audio signal occurs on the electric signal directly, whereas digital processors function on the signal mathematically.

Speech comprises numerous voiced as well as unvoiced regions. Its classification into these components provides a preliminary acoustic segmentation useful in several speech-related processing applications such as speech enhancement, speech synthesis, and speech recognition (Shete et al. 2014). Audio feature extraction and classification methods have been developed by various people in the last decade and have hence gained a lot of importance in the field of research and development. Using the extraction method, we can compute a numerical representation that can further be useful in characterizing the audio using the existing toolbox (Mitra and Saha, 2014). Over the years, a lot of approaches and numerous audio features have been put forth having different degrees of success. However, a generic methodology includes extricating different features from the input audio file and forwarding the same to a pattern classifier. The extraction of data and meaning from audio files for examination, classification, etc., are known as audio analyses. These features can be extricated in two ways: one is through a time-domain signal, and the other is through a transformation domain, which depends on the chosen signal analysis methodology. Zero-crossing rate (ZCR), Mel-frequency cepstral coefficients (MFCCs), spectral similarity,

DOI: 10.1201/9781003046431-7

entropy, etc. are some of the many audio features that have led to successful audio classification. In this chapter, we have introduced a three-step approach to the process of feature extraction. In the first phase, the input audio data is classified into an .mp3 or .wav format. Then, in the second phase, it is checked whether the input is stereographic or not, i.e., whether several channels are being used to transmit the audio signal. If yes, then it is converted to monaural input, meaning there is only one source for the audio. And in the third step, we normalize the output obtained from the previous step between 0 and 1 and then use it further for feature extraction (Theodoros Giannakopoulos).

Audio feature extraction plays a vital part in examining, distinguishing, and characterizing the content of audio input. Efficient audio feature extraction is required in, for example, audio scene analysis, indexing, audio fingerprinting, etc. Also, audio feature extraction plays a pivotal role in applications in areas such as processing speech, multimedia file distribution, and management, biometrics, security, etc. It can be further extended to do sentiment analysis on audio speeches, depression detection by analyzing the voice of a person. An audio classification system goes through the input signals and then defines them under a label that best describes the input audio signal. During the analysis of the input audio file, the classifier extracts important information regarding the content of the given audio, which is also referred to as the audio content analysis, and can be further extended to extract content information from the given signals. (Eyben et al. 2015).

A category of audio classification is musical genre classification, which is booming at the moment. Classification into musical genres is done using the members' common characteristics, typically allied to its rhythmic structure, instrumentation, and harmonic content. Using this feature set, there is a 61% classification for ten musical genres. Using a similar approach, a set of useful features can be jotted down for other categories of signals (G. Tzanetakis et al. 2015). Another such category is the classification and analysis of the content of scenes of a sequence of videos, which are essential for indexing based on content and multimedia databases' retrieval. Zhu Liu reported in their research that they used the corresponding audio information for the classification of video scenes. They described several audio features that have been deemed effective in differentiating the audio characteristics of various scene classes. A neural net classifier obtained using these features was sufficiently effective in separating audio clips from various television programs (Liu et al. 1997).

7.2 Related Work

Tremendous work has been done in this field using Python and highly advanced libraries have already been developed; some of them are as follows:

- Audio: extracts high-level features, such as beat tracking, tempo, onset detection, melody, etc. (David Moffat et al. 2015);
- Librosa: well-known API put to use for extricating features and data processing using Python (P. M. Brossier, 2006);
- LibXtract: the basic aim is to efficiently extract low-level features in real-time. Although originally written in C but now shifted to vamp formats, SuperCollider, PureData, and Maz-MSP (B. McAfee et al., 2014);

- YAAFE: low-level feature extraction designed with the basic aim of efficiency while computation and batch processing (J. Bullock et al., 2007);
- Pyaudioanalysis: covers a vast range of tasks related to audio analysis such as extraction, classification, segmentation, and visualization.

For data analysts and scientists, the two most popular and user-friendly programming languages are Python and R. Both of them are open source and free. They were developed around the early 1990s. R was mainly developed for statistical purposes. For working in the field of complex data visualizations, or working with large datasets, or pursuing projects in the field of machine learning, these two languages are godsends (B. Mathieu et al. 2010).

However, no such work on feature extraction has been done in R programming. Therefore, the aim is to make a library for audio analysis and segmentation in R programming, which is user-friendly and offers a variety of general features that can be extracted most efficiently and accurately. Another yet to be explored stream in audio analysis is forgery detection. In today's world, most of our opinions are formed from what we read, hear, or see on the Internet. As a consequence, the authentication of multimedia data has become a major concern now. Although some amount of research has been done on image and video, audio remains untapped. To illustrate this, let us consider a transaction conducted over the phone. There is a possibility to change certain bits of the phone call and replay it, but currently, we don't have any solid method to prove it. Methods to detect such kind of forgery are clearly in demand. And this can be achieved with enough knowledge and research on audio signal analysis (Namdev et al. 2015).

In the late 1990s, E. Scheirer developed a computer system that could bifurcate music signals from speech signals over a very wide range of signals. A set of about 13 features were used to make this classification. These conceptual features were then arranged into various multidimensional classification frameworks. A lot of information was gained from this (Scheirer, 1998).

J. Foote put forth a simple method for recognizing the features of the audio input provided by making use of a tree-based vector-supervised quantizer, which is trained to maximize mutual information. This method has been very useful in talker identification. Not only that, but it has also helped in extending speech to generic audio. To sum up, it is a classifier that is used to differentiate speech from nonvocal sounds. Results from various experiments indicate that very accurate and perfect classification can be obtained on a very tiny corpus of the test audio file. These methodologies can also be put to use for other applications and various domains, such as segmentation of continuous audio and similarity-based audio retrieval (Foote, 1997). Anssi Klapuri in his research work explains that once the audio-based model is computed, a computationally economical strategy is formulated for its implementation. Simulations were conducted using mixtures of speech samples and musical sounds. The resulting model provided better results as compared to two reference methods and was found to be highly robust in processing signals wherein essential components of the audible spectrum were omitted to simulate band-limited interference (Klapuri et al. 2008). The outcome of several types of nonlinear distortion on the apparent quality of music and speech signals was studied. Results demonstrated moderately strong negative correlations between the objective measure DS and the subjective ratings. It was concluded that a multitone signal-based objective measure of nonlinear distortion can predict the perceptual effects of nonlinear distortion satisfactorily (Tan & Jiang, 2018). The following features were extracted:

- Energy
- Entropy of Energy
- Zero-Crossing Rate
- Spectral Entropy
- Spectral Spread
- Spectral Centroid
- Spectral Roll-Off
- Spectral Flux
- Chroma Vector
- MFCCs

We selected the features based on information obtained from existing Python packages such as Pyaudioanalysis and from academic literature that was available with regard to audio classification. When successive samples of the audio input have different algebraic signs, then a ZCR occurs. The measure of the frequency content of a signal is known as the art of zero crossings (Zhang and Kuo, 1999). It is the number of times the amplitude of the audio input crosses zero in a given time frame (Sharma et al. 2019), in other words, how frequently it changes its sign from positive to negative or vice versa. High frequencies indicate high ZCRs, while lower ones point to low ZCRs, indicating that energy distribution and frequency and ZCRs are strongly correlated (Shete et al. 2014). Voiced parts demonstrate low ZCRs and unvoiced parts demonstrate high rates, whereas the energy demonstrates a reverse trend with regard to voiced and unvoiced parts (Bachu et al. 2009). Also, the analysis suggests that a high crossing rate implies the speech signal is unvoiced and a low crossing rate implies the signal is voiced. The entropy of energy is the entropy of normalized energies of the subframes. It can be used as an indicator of sudden changes. Speech signals appear more unordered compared with music considering the observations of spectrograms and signals (Umapathy et al. 2007). A spectral centroid is used to determine where the center of mass of the given spectrum lies. It is closely related to the brightness of a sound. Mathematically, it is computed by taking the weighted average of all the frequencies in the given signal, which in turn is computed with the help of Fourier transform by using the magnitudes as weights. Sometimes spectral centroid is used about the median of the input because both are measures of central tendency. So at times both display similar behavior in certain situations. Higher centroid values indicate much brighter textures having high rising frequencies. The centroid also gives shape to the sharpness of sound, which is again related to the high frequency of sound (Zahid et al. 2015). A technique used for transmitting telecommunication or radio signals is known as spectral spread. The higher the value of spectral spread, the more distributed the spectrum is on both sides of the centroid, whereas lower values imply that the spectrum is highly confined near the centroid. Spectral entropy is a quantitative analysis of spectral disorder of given input audio. Mainly it is used to determine voiced and silent regions of speech. It finds use in speech recognition because of this discriminatory property. Entropy can also be used to capture the peaks and formants of a distribution. The spectral flux is the squared difference of the normalized magnitudes of successive spectral distributions of corresponding successive signal frames. Flux is a significant feature for differentiating music from speech. This feature is specifically used to regulate the timbre of the audio signal (Sharma et al. 2020). Spectral roll-off is the critical frequency below which 85% of the magnitude distribution of the input is concentrated. Similar to that of the spectral centroid, it is a measure of spectral shape that yields values for frequencies in high ranges. Therefore, it can be concluded that there is a strong correlation between the spectral centroid and spectral roll-off (Sharma et al. 2020). MFCCs are a compact representation of an audio signal's spectrum that accounts for the nonlinear human perception of pitch, according to the Mel scale. The Short-Time Fourier Transform (STFT) coefficients of each frame are grouped into a set of 40 coefficients. MFCC is calculated using a set of 40 weighting curves that in turn stimulate the frequency perception of the human hearing system. The next step in the process is to take the logarithm of the coefficients. After that, de-correlation is done by doing a discrete cosine transform (DCT). Usually, the five first coefficients are taken as features. Small musical units are detected as a combination of chords and MFCCs.

Results of experiments demonstrate that the proposed application of using MFCC for gesture recognition has satisfactory accuracy and therefore is applicable for sign language recognition or other household applications when coupled with other techniques such as DWT and Gabor filter to increase the accuracy rate and efficiency (Shikha Gupta et al. 2013). Various experiments and studies conducted show that it is viable to extract features using MFCCs (Lindasalwa Muda et al. 2010). Chroma vector refers to a set of twelve pitch classes represented as a vector. The distribution of energy is calculated for every Chroma vector, and as a result, we get an updated audio signal of the twelve-dimensional Chroma distribution vector (Sharma et al. 2020). Also, currently, for audio signal processing, audio feature extraction toolboxes have been the main field of focus for research and development purposes. There are an innumerable number of toolboxes available for the same but in different formats, as mentioned below (David Moffat et al. 2015), including stand-alone applications, host application plugins, and software function libraries.

The main purpose of feature extraction is to overcome the overfitting of data, increase model efficiency and performance, and cater to providing cost-effective and faster models. The advantage of feature selection is that it decreases the noise, which leads to improvement of classification accuracy and dimensionality reduction as well. Feature extraction is nothing but a particular form of a dimensional reduction in which the features selected are such that they will only provide us with the relevant information from the humongous dataset available to us and hence help us in reaching our goal (Mitra and Saha, 2014).

7.3 Proposed Methodology and Model Assumptions

Our package first obtains relevant features from the specified directories in the function call and then used them to train a model of the specified type. The audio test train file calls the audio feature extraction file. In each directory, it first segregates valid files based on their extensions, selects the files that have a valid extension, and then parses them if they are large enough to be considered. Then the midterm features are evaluated, which are aggregations of short-term features. To calculate the short-term features, feature extraction is applied, which calls each function that uses the frame to calculate the features for each window length, which are then combined into a matrix for further use. This matrix is obtained for each directory and is then returned to the audio-feature-and-train file.

In the audio-feature-and-train file, the function calculates the number of columns, naming them by appending an index to "feature" and then writing them along with the names to an ARFF file. Then the rows with NaN values are removed from the matrix as they cannot be interpreted by R. Then the evaluate classifier function is called, which then computes the best model to use for the model. The evaluate classifier function first normalizes the features. Then it calculates the number of experiments to be performed based on the number of samples. Then, the features are split into training and testing portions randomly. Then the model is trained using the training data and the parameters being tested. For each class, precision, recall, and F1 statistic are calculated, and the parameter that gives the most optimum value of the above is selected as the parameter to be used (Algorithm 7.1).

```
Algorithm 7.1: Proposed Classification Model
1.  Function featureAndTrain()
2.  {
3.  Function dirWavFeatureExtraction()
4.  {
5.  Function audioFeatureExtraction()
6.  {
7.  while directories:
8.  {
9.  result<-Paths of mp3/wav obtained
10. while the result:
11. {
12. Mtfeatures calculated
13. }
14. }
15. //Training data saved to aiff file
16. Features<-Evaluate(classifier,Parameters)
17. Eliminate_Nan_values(Features)
18. Train_Classifier()
19. Save_Classifier()
20. End
21. Function fileClassification()
22. {
23. bool x<-Check_if_model_name_exists()
24. If x==True:
25. bool y<-Check_if_filename_exists()
26. if y==True:
27. file<-Obtain_file_features()
28. return Classify_file(file)
29. }
30. end
```

7.4 Results and Observations

The dataset that was used for our testing was the TESS (Toronto emotional speech set) in which the stimuli were modeled on the NU-6 test (North-Western Auditory Test). A set composed of 200 words was spoken in the form of "Say the word___" by two actresses and seven emotions were portrayed by the actresses for each of the words. A total of 2800 stimuli were used. Ten audio clips were chosen from each category for the testing process from each emotion: Angry, Calm, Disgust, Fearful, Neutral, Happy, Sad, and Surprised.

Figures 7.1–7.9 provide the various statistics obtained for the individual classes of the dataset, while the final figure, Figure 7.10, provides the overall statistics for the dataset.

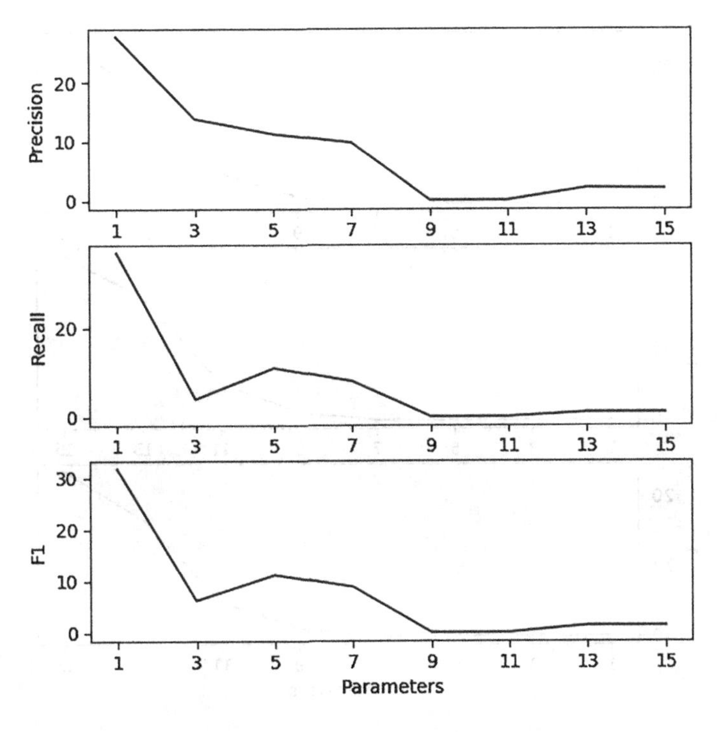

FIGURE 7.1
Emotion Dataset: Sad.

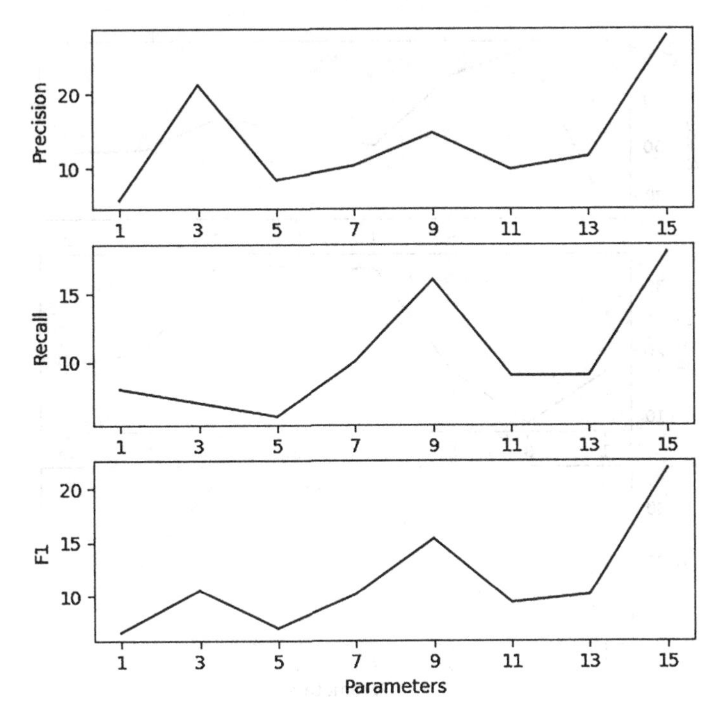

FIGURE 7.2
Emotion Dataset: Neutral.

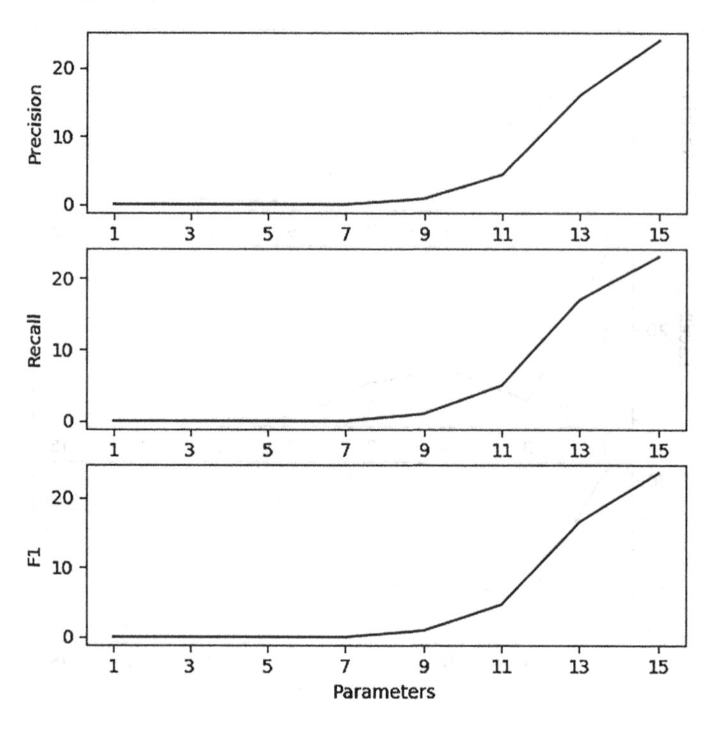

FIGURE 7.3
Emotion Dataset: Happy.

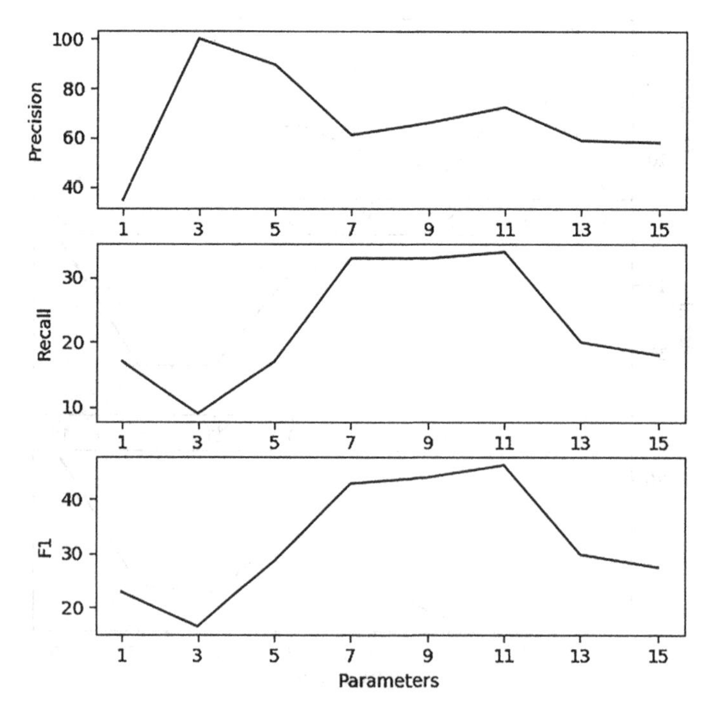

FIGURE 7.4
Emotion Dataset: Fearful.

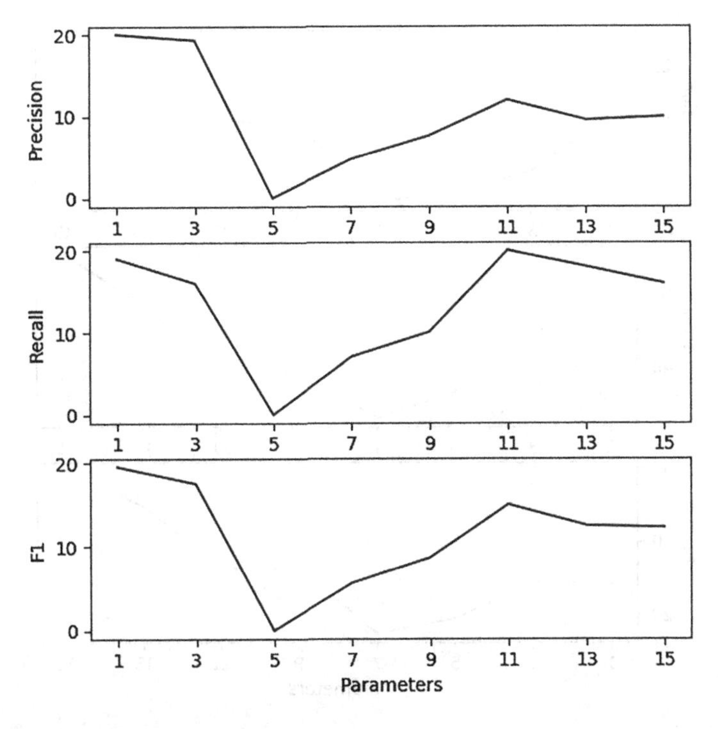

FIGURE 7.5
Emotion Dataset: Disgust.

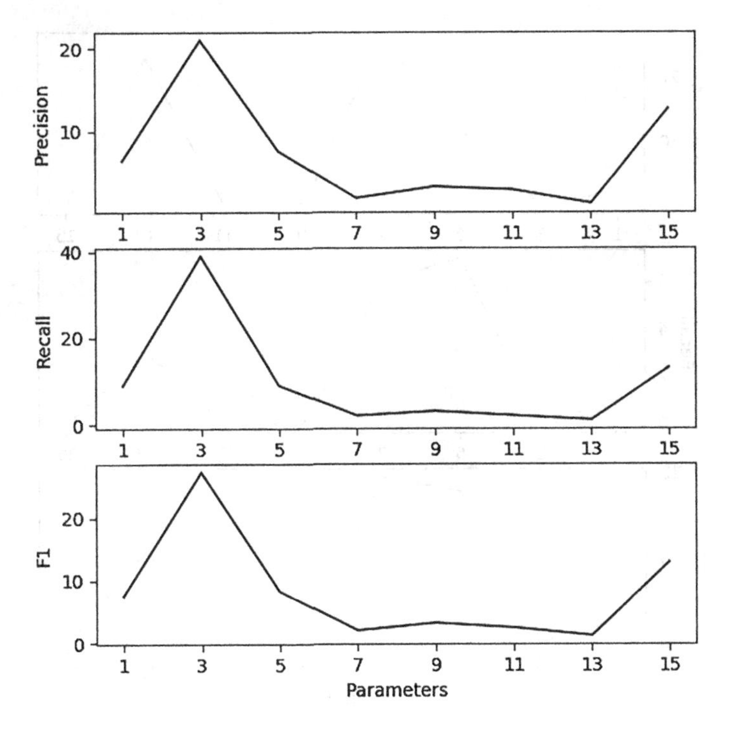

FIGURE 7.6
Emotion Dataset: Calm.

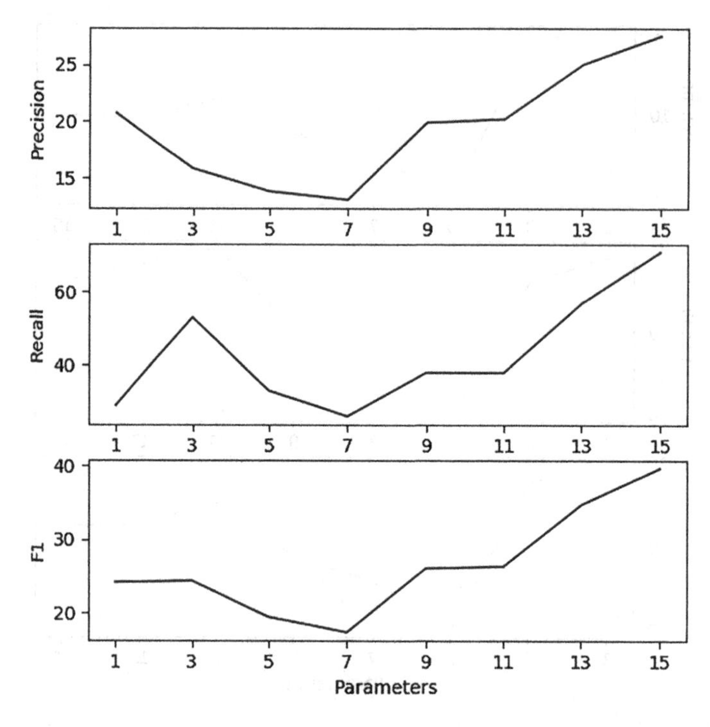

FIGURE 7.7
Emotion Dataset: Angry.

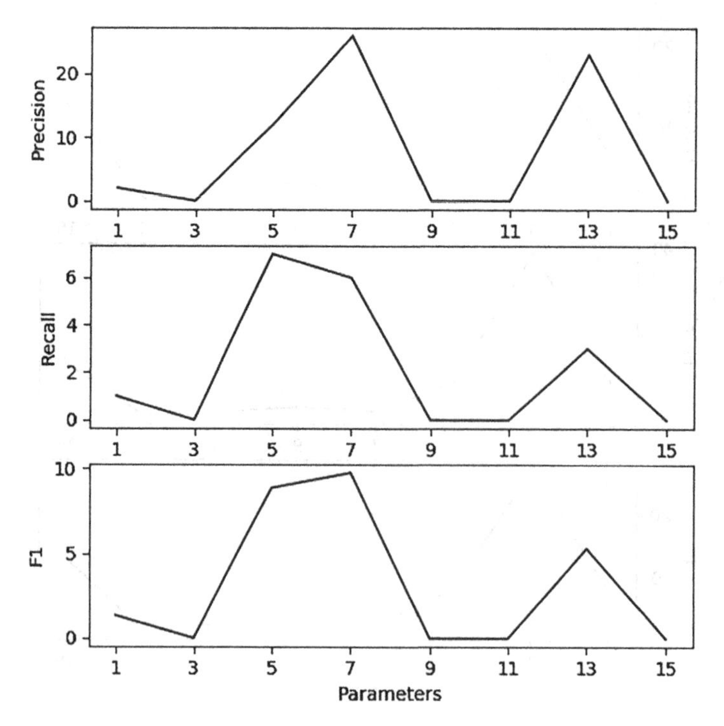

FIGURE 7.8
Emotion Dataset: Surprise.

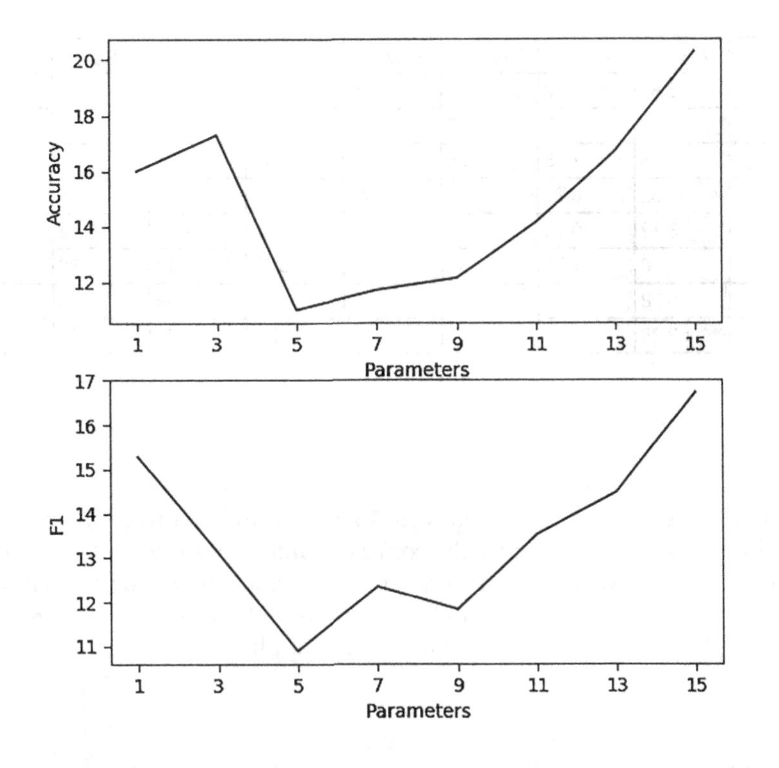

FIGURE 7.9
All Emotions.

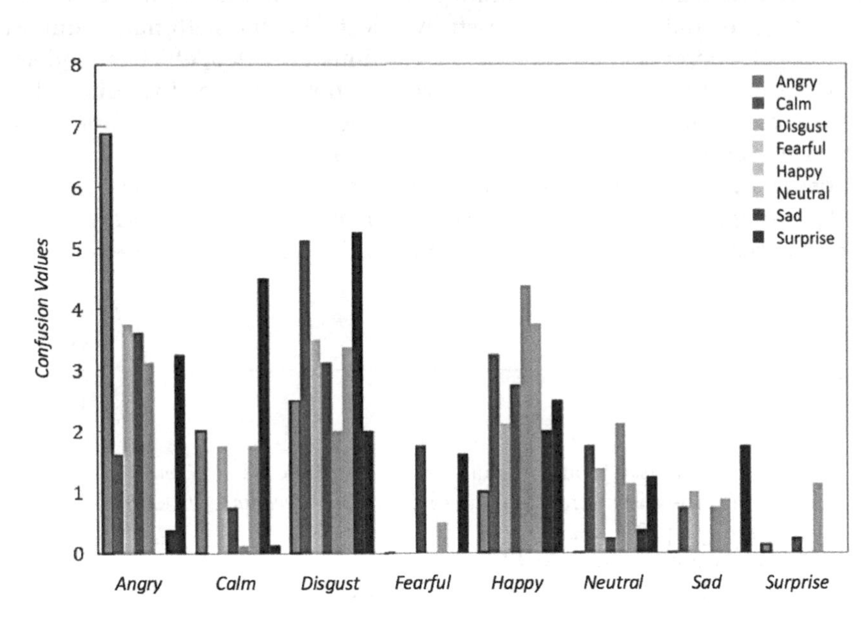

FIGURE 7.10
Confusion Matrix Plotted.

	Angry	Calm	Disgust	Fearful	Happy	Neutral	Sad	Surprise
Angry	6.87	2	2.5	0	1	0	0	0.13
Calm	1.62	0	5.12	0	3.25	1.75	0.75	0
Disgust	3.75	1.75	3.5	0	2.12	1.38	1	0
Fearful	3.62	0.75	3.12	1.75	2.75	0.25	0	0.25
Happy	3.12	0.13	2	0	4.37	2.12	0.75	0
Neutral	0	1.75	3.37	0.5	3.75	1.13	0.88	1.13
Sad	0.38	4.5	5.25	0	2	0.38	0	0
Surprise	3.25	0.13	2	1.62	2.5	1.25	1.75	0

FIGURE 7.11
Model Confusion Matrix

The parameter selected for these data were 15.00 as is evident from the performance in the figures above. Figure 7.11 provides the confusion matrix for the model, and it is evident that the model performs with low efficiency for a specific application like emotion detection. However, the foundation for a model has been laid. All that is required is a better algorithm to implement and features suitable to the application.

7.5 Conclusion

The package aims at calculating the validity of a set of parameters, as described in the chapter, for the given dataset, and successfully selects the one with maximum accuracy and F1 score. With appropriate features and algorithms provided, efficient models can be created in the R programming language that rival libraries such as Pyaudioanalysis. The main objective of this undertaking was to extract the features from an input audio clip. That has been achieved using the R language. This provides hope for a package in R dedicated to audio analysis and classification in the future. This package will be of importance to those attempting to perform machine learning in audio-related applications. The classification part of the package can be implemented with relative ease in the future.

References

Bachu, R., Kopparthi, S., Adapa, B. and Barkana, B. (2009) Voiced/Unvoiced decision for speech signals based on zero-crossing rate and energy, *Advanced Techniques in Computing Sciences and Software Engineering*, pp. 279–282.

Brossier, P. M. (2016) The audio library, *MIREX*.

Bullock, J. and Conservatoire, U. (2007) Libxtract: A lightweight library for audio feature extraction, *International Computer Music Conference*.

Chin-Tuan, T. and Moore, B. (2003) The effect of nonlinear distortion on the perceived quality of music and speech signals, *Audio Engineering Society*, 51(11), pp. 1012–1031.

Dupuis, K. and Pichora-Fuller, M. (2010) University *of Toronto, Psychology Department*.

Eyben, F. (2015) Introduction. Springer Theses Real-time Speech and Music Classification by Large Audio Feature Space Extraction, pp. 1–7.
Foote, J. (1997) A similarity measure for automatic audio classification, *AAAI Technical Report.*
Giannakopoulos, T. (2015) pyAudioAnalysis: An open-source python library for audio signal analysis, *Plos One*, 10(12), p. e0144610.
Gupta, S., Jaafar, J., Ahmad, W. F. W. and Bansal, A. (2013) Feature extraction using Mfcc. *Signal & Image Processing: An International Journal.*
Klapuri, A. (2008) Multipitch analysis of polyphonic music and speech signals using an auditory model. *IEEE Transactions on Audio, Speech, and Language Processing.*
Liu, Z., Huang, J., Wang, Y., Chen, T. (1997) Audio feature extraction and analysis for scene classification, *Proceedings of First Signal Processing Society Workshop on Multimedia Signal Processing.*
Mathieu, B., Essid, S., Fillon, T., Prado, J. and Richard, G. (2010) YAAFE, easy to use and efficient audio feature extraction software.
McFee, B., McVicar, M., Raffel, C., Liang, D. and Repetto, D. (2014) librosa. v0.3.1.
Mitra, J. and Saha, D. (2014) An Efficient Feature Selection in Classification of Audio Files.
Moffat, D., Ronan, D. and Reiss, J. C. (2015) An evaluation of audio feature extraction toolboxes. *Proceedings of the 18th International Conference on Digital Audio Effects*
Muda, L., Begam, M. and Elamvazuthi, I. (2010) Voice Recognition Algorithms Using Mel Frequency Cepstral Coefficient (MFCC) and Dynamic Time Warping (DTW) Techniques. *A Determination of the Hubble Constant from Cepheid Distances and a Model of the Local Peculiar Velocity Field.*
Namdev, D. and Bansal, A. (2015) Frequency domain analysis for audio data forgery detection. *Fifth International Conference on Communication Systems and Network Technologies.*
Scheirer, E. D. (1998) Tempo and beat analysis of acoustic musical signals, *The Journal of the Acoustical Society of America.*
Sharma, B., Gupta, C., Li, H., and Wang, Y. (2019, May). Automatic lyrics-to-audio alignment on polyphonic music using singing-adapted acoustic models. In *ICASSP 2019-2019 IEEE International Conference on Acoustics, Speech, and Signal Processing (ICASSP)* (pp. 396-400). IEEE.
Sharma, G., Umapathy, K., and Krishnan, S. (2020). Trends in audio signal feature extraction methods. *Applied Acoustics, 158*, 107020.
Shete, D. and Patil, P. S. (2014) Zero crossing rate and energy of the speech signal of devanagari script, *IOSR Journal of VLSI and Signal Processing.*
Tan, L., and Jiang, J. (2018) *Digital Signal Processing: Fundamentals and Applications.* Academic Press.
Tzanetakis, G. and Cook, P. (2002) Musical genre classification of audio signals, *IEEE Transactions on Speech and Audio Processing.*
Umapathy, K., Krishnan, S., and Rao, R. K. (2007). Audio signal feature extraction and classification using local discriminant bases. *IEEE Transactions on Audio, Speech, and Language Processing, 15*(4), 1236–1246.
Zahid, S., Hussain, F., Rashid, M., Yousaf, M. H., and Habib, H. A. (2015). Optimized audio classification and segmentation algorithm by using ensemble methods. *Mathematical Problems in Engineering, 2015*, 1–11.
Zhang, T., and Kuo, C. C. (1999, March). Hierarchical classification of audio data for archiving and retrieving. In *1999 IEEE International Conference on Acoustics, Speech, and Signal Processing. Proceedings. ICASSP99 (Cat. No. 99CH36258)* (Vol. 6, pp. 3001–3004). IEEE.

8

Smart Health Analytics for Sustainable Energy Monitoring Using IoT Data Analytics

Sherry Garg
Jaypee Institute of Information Technology

Rajalakshmi Krishnamurthi
Jaypee Institute of Information Technology

CONTENTS

8.1 Introduction

Internet of Things (IoT) generates time-series data that are collected and arranged sequentially against regular time intervals. The volume, velocity, and variety or heterogeneity of IoT data are the same as big data. An additional parameter associated with IoT data is a timestamp. So, the analysis and predictions are closer to accuracy and relevance when this time factor is handled appropriately.

Moreover, improving data analysis by using deep learning models, which handle voluminous data better, is the best thing that can happen to IoT data analysis. There are various models available for predictions in deep learning, depending on the types of data and analysis requirements. Multiple applications and sectors have opted and

DOI: 10.1201/9781003046431-8

demonstrated these models on their respective datasets. First, this chapter introduces the IoT, its booming need in day-to-day life, and the type of IoT data that are generated. Second, after providing an overview of data analytics and its various levels, the relation between IoT and data analytics is discussed. Third, this chapter elaborates on the deep learning models and optimizers available. Multiple models available and used for IoT time-series data are listed and explained, along with a comprehensive elaboration of deep learning models. Fourth, it discusses how deep learning models and sustainable energy data can predict and enhance raw data value and transform that meaningless data into valuable information. Finally, the challenges and opportunities for IoT data analytics using deep learning are discussed for future growth. In this chapter, (i) the IoT analytics and its scopes are explored; (ii) the applications of deep learning in IoT analytics are explained; (iii) specifically LSTM algorithm in deep learning is discussed and used; (iv) comparison of different optimizers is made; and (v) experiment using Stochastic Gradient Descent (SGD) and Adaptive Moment Estimation (ADAM) is done on wind turbine dataset. This chapter aims to find out how the best predictions can be made, for the time-series data, which IoT devices generate, in terms of accuracy and time. For this big time-series data, using deep learning study of optimizers and their behavior is shown to minimize error factors.

8.2 Internet of Things

IoT comes under the umbrella of the Internet of everything, as shown in Figure 8.1. IoT is communication between IoT devices linked by wired or wireless networks to transfer data without human intervention. A thing in IoT can be anything having sensors or actuators connected. It can be a living being having RFID attached, an automobile having sensors to detect seat belts, windows open/close, tire pressure, etc., electrical appliances having sensors to build a smart home, and so on. When data analytics is applied to IoT, it becomes IoT analytics.

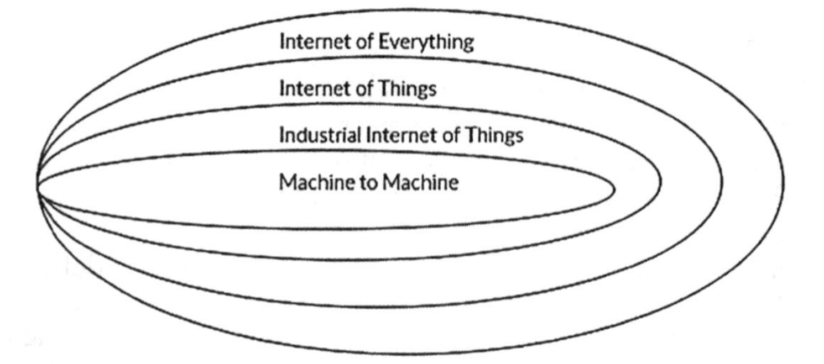

FIGURE 8.1
IoT scenario.

8.3 Data Science vs. Data Analytics

Data Science is to formulate questions and find solutions to those questions that will benefit the business. Data Analytics is finding a solution to existing business queries using existing data. The new thing data analysts look at is just the perspective on existing data and questions. Still, data scientists have to think about new insights with newly formulated data with the answers that will help the business grow. Data Science is a cap over Data Analytics.

a. Data Analytics
   In Data Analytics, from planning analytics, different analytics one can adapt to, as shown in Figure 8.2. There is descriptive analytics to describe from past data about what has happened. After descriptive analytics, a diagnosis can be made on the over-described data to see the reason behind what has happened. In predictive analytics, forecasting is done based on previous trends about what will happen next. And prescriptive analytics helps to prescribe measures to make the required thing happen in the future.
b. IoT Analytics
   IoT Analytics is the analysis of data retrieved from IoT devices. IoT data are generally distributed over geographical locations and are too dependent on external conditions. Both spatial and temporal factors are included in the data. Hence, IoT data are also named time-series data. Data collection converted into business information involves various stages as shown in Figure 8.3.
c. IoT Data Science
   Traditional data science involves analyzing static data on hand. That static data might be the most recent one, but it will be available offline. IoT Data Science involves streaming data coming in every time unit live from IoT devices.

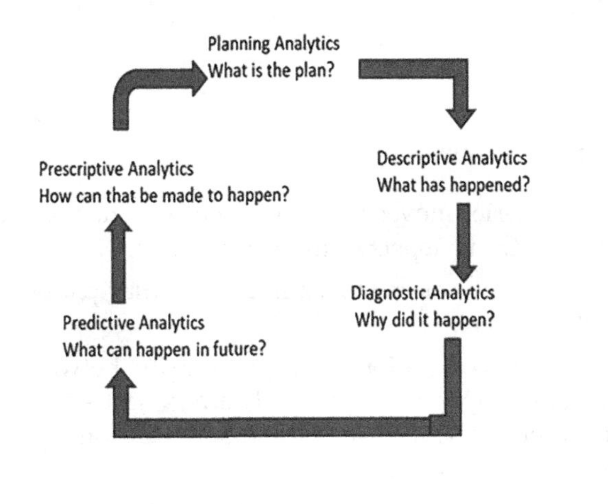

FIGURE 8.2
Types of analytics.

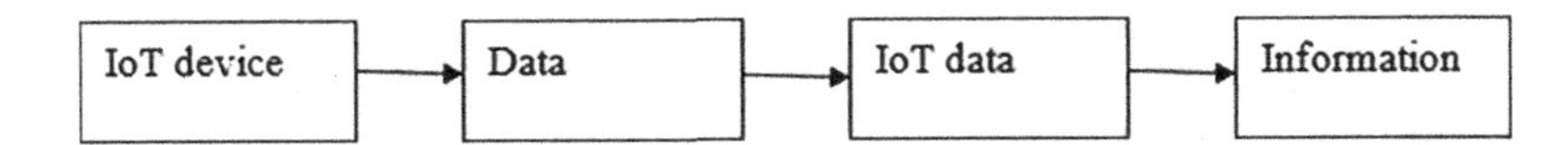

FIGURE 8.3
Stages of IoT analytics.

d. IoT Analytics on Streaming Data and Data in Rest

Streaming data will produce data as and when they are generated and perform analysis over it. Batch processing of data in rest will collect, store, and prepare data for an analysis tool, and perform analysis over it. However, both types of data are sequential. The sequence is because of the time factor in data. Hence, IoT data are considered synonymous with time-series data.

8.4 Time-Series Data

The difference between time-series data and nontime-series data is *time*. Time-series data are ordered over sequential intervals of time. An additional dependency from standard data that needs to be taken care of is time. According to features and variation, time-series data can be univariate or multivariate.

8.4.1 Univariate Time Series

Only one variable is varying over time; for example, data collected from a sensor measuring the pressure of a room every second. Therefore, each second, you will only have a one-dimensional value, which is the pressure.

8.4.2 Multivariate Time Series

Multiple variables vary over time; for example, data collected from a sensor measuring the temperature and pressure of a room every second. Here, both the temperature and the pressure change over time.

8.5 Aspects of Time Series

- **Trend:** When time series moves in a particular direction without repeating itself in any pattern is the trend aspect of time-series data.
- **Seasonality:** When time series moves in a particular pattern, repeated patterns are the seasonality aspect.
- **Autocorrelation:** Autocorrelation is the similarity between observations as a function of the time lag between them. There is no trend or seasonality in the data shape, but some deterministic relation between some previous copy of itself regarding time named lag time.
- **Noise:** Data having no pattern, relation, and completely random to be able to predict anything. That is called complete white noise.

When data are processed and analyzed sequentially in a stream instead of being stored in batches and then processed is called stream analytics. The processing is quite fast to be in line with the upcoming data. Challenges faced in IoT Analytics due to data coming live from the source and applying predictive analytics over it are:

- Considerable size of data
- Noisy data
- Security
- Unpredictable devices
- Faster processing
- Geographical/Seasonal changes

8.6 Predictive Analytics with Time-Series Streaming Data

Predictive analytics can be done by artificial intelligence, machine learning, and deep learning models. Artificial intelligence is an umbrella over machine learning and deep learning and is a rule-based approach. As shown in Figure 8.4, machine learning models are suitable for predictive analytics where the computational power required is limited to CPU. Time-series streaming data are big data in themselves, and applying predictive models over this big data requires enormous computational power. Hence, deep learning models are preferred for this type of big data.

Various deep learning models used for IoT Analytics include:

- Restricted Boltzmann Machine (RBM)
- Recurrent Neural Network (RNN)
- Deep Belief Network
- RNN+RBM
- Stacked Denoising Autoencoder
- Multilayer-Extreme Learning Machine (ML-ELM)

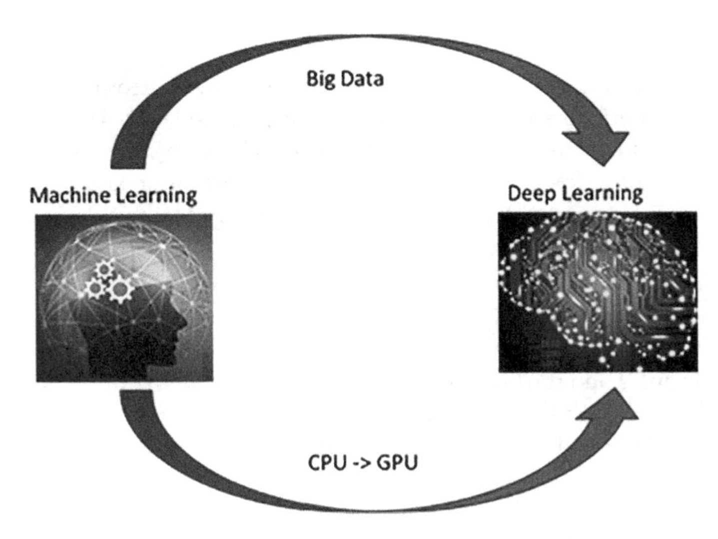

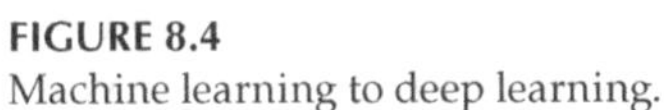
FIGURE 8.4
Machine learning to deep learning.

- Convolutional Neural Network (CNN)
- Long Short-term Memory (LSTM)
- CNN+LSTM
- Autoencoder

8.7 Long Short-Term Memory

This chapter explains and uses LSTM out of various listed deep learning models available for IoT analytics. Long-term memory data saving and retrieving concerning the context requirement is a requirement for predicting time-series data. A RNN can save only the previous state. That leads to a lack of long-term context prediction, one of the critical requirements of time-series data. LSTM is the solution to this shortcoming of RNN, with the same recurrence chain as that of RNN, repeated at every stage. Unlike RNNs, one neural network layer uses four neural network layers for a single output. The initial LSTM block version did not have to forget gate and peephole connections but had cells, input, and output gates. Real-Time Recurrent Learning and Back Propagation Through Time were used for training purposes. The same gradient was missing for training, and only the cell gradient was fed back through time. Also, this version used full gate recurrence, which disappeared from different papers. RNNs combined with LSTMs are an effective and scalable model for several learning problems for time-series data. Prediction of time-series data through earlier proposed methods did not consider long-term dependencies and contexts. LSTMs emerged to be an efficient solution toward extended context saving in its cells and using those remembered states with the help of different gates against the current required context. The inflow and outflow of information are regularized into the cells, which act as a conveyer belt in LSTM. The decision about what information is preserved and what information is rejected is made using gates. Three gates that contribute are forget gate, input gate, and output gate.

a. Forget Gate

   Forget gate decides which information to forget away from the cell state. The sigmoid layer performs the task inside the forget gate layer. The input is the last cell's output hidden state (h) and the current cell's input (x); the output is between 0 and 1. A value of x closer to 1 indicates that the information should be kept, and x closer to 0 points to forget the information. This output from the sigmoid function is multiplied by the cell state.

b. Input Gate

   The input gate is responsible for adding some information to the cell state. The forget sigmoid gate function regulates the previous cell's hidden state and the current cell's input. Then the tanh function is applied to get an output between –1 and 1 and multiplied with the output from the sigmoid function, and sent to be added to the cell state. All this is to preserve important input and discard redundant ones.

c. Output Gate

   Cell state generated after multiplying the previous cell's cell state to the current cell's forget gate's output and adding that to the current cell's input gate output is passed as an input to the output gate's tanh function. Another information will be sigmoid of the previous cell's output hidden state (h) and the current cell's input(*x*). The output is a number between 0 and 1. The tanh function generates values between –1 and 1. The output from the tanh function and sigmoid function are multiplied to create the output of the current state.

8.8 Optimizers

Using these predictive models predicts output as correct and reduces the difference between actual output and expected output. This difference between real and anticipated output is called loss function. The optimizer will update the weight value to minimize this loss function and converge to global optima as quickly as possible. The types of optimizers are explained further with the algorithms and variables used.

a. Gradient Descent

The requirement is to minimize the error or loss function. So, first, this loss function has to be defined for expected and predicted value. The purpose of the loss function is to quantify the error between the actual value and the predicted value. The sum of squared differences between two points, the first point being the actual value and the second point being the predicted value, will tell how far the predicted value is from the actual value. The aim should be to minimize this distance between the two points for the predicted value to be as close to the actual value. As per Equation 8.2, ($L(x,y)$), named loss function, is the sum of the squared difference between the expected value and predicted value. Each data entry out of the N-sized dataset is plotted on a graph with the y-axis as power generated and the x-axis as the timestamp. Every point at ith instance (x_i,y_i) will signify power generated (y_i) at a particular time (x_i). y_i is the actual value expected at x_i time. While predicting y_{pi} against given x_i having b as y-intercept, we will use

$$y_{pi} = b + mx_i \tag{8.1}$$

$$L(x,y) = (\text{Expected power} - \text{Predicted Power})^2 \tag{8.2}$$

So, at the ith instance

$$L(x_i, y_i) = (y_{pi} - (mx_i + b))^2 \tag{8.3}$$

for one update per epoch, we calculate this loss function for all points using Equation 8.4 and average it out over N

$$L(X,Y) = \left(\frac{1}{N}\right)\sum_{i=1}^{n}(y_i - (mx_i - b))^2 \tag{8.4}$$

The error function first differentiates this error function, taking m as a variable first and then b as a variable to minimize this error function. Ultimately, values of

```
# Gradient Descent
For (j = 1 to epoch) :
    #Minimization of error
    for(i = 1 to length(dataset)) :
        d Error(m,b)/db += -2(y_i - (mx_i + b))
        d Error(m,b)/dm += -2 * x_i(y_i - (mx_i + b))
    Step Size (m) = d Error(m,b)/dm * learning rate
    Step Size (b) = d Error(m,b)/db * learning rate
    b_updated = b_old - Step Size (b)
    m_updated = m_old - Step Size (m)
```

FIGURE 8.5
Pseudocode for gradient descent algorithm.

m and b need to be updated every time to converge to the final solution, which is as close to the expected output as possible. This reduction in loss or error will, in turn, improve the performance metrics. Figure 8.5 describes the steps of gradient descent optimizers.

Here, b and m are updated as b_{updated} and m_{updated}, every time for one epoch for all the points in a dataset: i^{th} iteration or i,s loop is to calculate gradient for all the N points in the dataset. This gradient is then used for j^{th} iteration or epoch for calculating b_{updated} and $m_{\text{updated, leading}}$ closer to optima.

b. Stochastic Gradient Descent (SGD)

To converge faster toward an optimal solution, SGD suggests iterating over a randomly selected single sample from the dataset, unlike N iterations in gradient descent. When the nature of dataset entries is too redundant, SGD leads to better and faster convergence.

c. Mini-Batch SGD

Choosing a single value from a large dataset does not look at the bigger picture, which a huge dataset wants to convey. So, to optimally decide on the speed of convergence and quality of the result, mini-batch SGD is useful. Mini-batch SGD uses a predefined number of samples called batch size instead of one in SGD and all in gradient descent.

d. Momentum

The idea behind introducing momentum in optimizations is to speed up convergence in a horizontal direction. The exponential weighted moving average can include a weighted average of previous multiple predictions instead of immediately prior one (Figure 8.6).

e. Exponential Weighted Average instead of taking just the previous value for updating a new one by moving weighted average over the last days or times is taken into consideration. The formula computes exponential weighted moving average in Equation 8.5.

$$F_t = \beta F_{t-1} + (1-\beta)\ A_t \tag{8.5}$$

Here, F_t refers to the prediction for t time, β lies between 0 and 1, A_t is the actual value at t time, and $F_1 = A_1$.

 # Initialization for Momentum Algorithm

$$V_{\Delta_m} = 0$$
$$V_{\Delta_b} = 0$$

For tth time and mini batch size bs (i.e. loop over bs times) Calculate:

$$\Delta_b \rightarrow \frac{d\ Error(m,b)}{db} += -2(y_i - (mx_i + b))$$

$$\Delta_m \rightarrow \frac{d\ Error(m,b)}{dm} += -2 * x_i(y_i - (mx_i + b))$$

Using Exponential Weighted Moving Average, calculate:

$$V_{\Delta_m} = \beta V_{\Delta_{m-1}} + (1-\beta)\Delta_m$$
$$V_{\Delta_b} = \beta V_{\Delta_{b-1}} + (1-\beta)\Delta_b$$

Update next m and b

$$m_{next} = m_{old} - learning\ rate * V_{\Delta_m}$$
$$b_{next} = b_{old} - learning\ rate * V_{\Delta_b}$$

FIGURE 8.6
Pseudocode for momentum algorithm.

V_{Δ_m}—Exponential Weighted Moving Average of m, i.e., slope
V_{Δ_b}—Exponential Weighted Moving Average of b, i.e., y-intercept

8.8.1 Root Mean Square Propagation

Momentum fastens the gradient convergence but does not deal with movement in the vertical direction while converging to optima.

Here, R_{Δ_m} is Exponential Weighted Moving Average of m, and R_{Δ_b} denotes Exponential Weighted Moving Average of b differential. As described in Figure 8.7, Root Mean Square Propagation (RMSProp) will take care of this issue by increasing Δ_b and reduced Δ_m. So, while calculating m_{updated} and b_{updated}, the denominator will increase, leading to faster convergence, and b will decrease, leading to fewer vertical updates.

 # Initialization for RMSProp

$$R_{\Delta_m} = 0$$
$$R_{\Delta_b} = 0$$

For tth time and mini batch size bs(i.e. loop over bs times) Calculate:

$$\Delta_b \rightarrow \frac{d\ Error(m,b)}{db} += -2(y_i - (mx_i + b))$$

$$\Delta_m \rightarrow \frac{d\ Error(m,b)}{dm} += -2 * x_i(y_i - (mx_i + b))$$

Using Exponential Weighted Moving Average, calculate:

$$R_{\Delta_m} = \beta R_{\Delta_{m-1}} + (1-\beta){\Delta_m}^2$$
$$R_{\Delta_b} = \beta R_{\Delta_{b-1}} + (1-\beta){\Delta_b}^2$$

Update next m and b

$$m_{updated} = m_{old} - learning\ rate * {\Delta_m}/{\sqrt{R_{\Delta_m}}}$$
$$b_{updated} = b_{old} - learning\ rate * {\Delta_b}/{\sqrt{R_{\Delta_b}}}$$

FIGURE 8.7
Pseudocode for RMSProp algorithm.

 # Initialization for Adam

$V_{\Delta_m} = 0 \quad V_{\Delta_b} = 0 \quad R_{\Delta_m} = 0 \quad R_{\Delta_b} = 0$

For t^{th} time and mini-batch size b_s (i.e., loop over bs times) calculate:

$$\Delta_b \rightarrow \frac{d\ Error(m,b)}{db} += -2(y_i - (mx_i + b))$$

$$\Delta_m \rightarrow \frac{d\ Error(m,b)}{dm} += -2 * x_i(y_i - (mx_i + b))$$

Using Exponential Weighted Moving Average, calculate Momentum and RMSProp:

$$V_{\Delta_m} = \beta 1 V_{\Delta_{m-1}} + (1 - \beta 1)\Delta_m$$

$$V_{\Delta_b} = \beta 1 V_{\Delta_{b-1}} + (1 - \beta 1)\Delta_b$$

$$R_{\Delta_m} = \beta 2 R_{\Delta_{m-1}} + (1 - \beta 2){\Delta_m}^2$$

$$R_{\Delta_b} = \beta 2 R_{\Delta_{b-1}} + (1 - \beta 2){\Delta_b}^2$$

Add Bias Correction

$$V^{bias}{}_{\Delta_m} = {}^{V_{\Delta_m}}\!/(1 - \beta 1^t)$$

$$V^{bias}{}_{\Delta_b} = {}^{V_{\Delta_b}}\!/(1 - \beta 1^t)$$

$$R^{bias}{}_{\Delta_m} = {}^{R_{\Delta_m}}\!/(1 - \beta 2^t)$$

$$R^{bias}{}_{\Delta_b} = {}^{R_{\Delta_b}}\!/(1 - \beta 2^t)$$

Update next m and b

$$m_{updated} = m_{old} - learning\ rate * \left(\frac{V^{bias}{}_{\Delta_m}}{R^{bias}{}_{\Delta_m}}\right)$$

$$b_{updated} = b_{old} - learning\ rate * \left(\frac{V^{bias}{}_{\Delta_b}}{R^{bias}{}_{\Delta_b}}\right)$$

FIGURE 8.8
Pseudocode for ADAM algorithm.

f. Adaptive Moment Estimation (ADAM)

A combined approach of momentum and RMSProp generates the ADAM optimizer as described in Figure 8.8.

8.9 Comparison of SGD and ADAM

SGD and ADAM are compared based on some distinguishing factors in Table 8.1.

a. Dataset Description

Six years of hourly wind power energy dataset from NREL is taken for analysis. Description of data contains timestamp, air temperature (celcius), pressure (atm), wind direction (deg), wind speed (m/s), and power generated (kilowatt). The features used for prediction are time-series variants, i.e., timestamp and power generated. Figure 8.9 shows the flow starting from data sourcing from sensors in wind turbines, collecting that data at the base station, transferring it to the analytics team who apply predictive and prescriptive analysis, and helping businesses with the solutions provided. Stage 3, which is the analytics stage, is the point where this work will be helpful.

TABLE 8.1
Theoretical Comparison of SGD and ADAM

	GD	SGD (with Mini-Batch)	SGD with Momentum	RMSProp	ADAM
Step size	$learning\ \ rate * \frac{d\ Error(m,b)}{dm}$ For all data points	$learning\ \ rate * \frac{d\ Error(m,b)}{dm}$ For a batch size of $b_s <<< N$	$learning\ rate * V_{\Delta_m}$	$learning\ \ rate * \Delta_m / \sqrt{R_{\Delta_m}}$	$learning\ \ rate * \left(\frac{V^{bias}{}_{\Delta_m}}{R^{bias}{}_{\Delta_m}} \right)$
Convergence	Behaves slowly for flat surfaces, i.e., will take more iterations to converge on flatter surfaces	Time taken for a single iteration is reduced, rest behaves the same as GD	By taking a weighted average of previous predictions, the step size toward local optima is optimized to converge faster	Unnecessary vertical movements are controlled, and horizontal directions are enhanced	A bias correction on a combination of momentum and RMSProp leads to better convergence
Technique	Basic converging steps for data size N	Updation is faster because of $b_s < N$	Exponentially weighted average over previous learning is taken into account as momentum while updating step size	Along with the exponentially weighted average over previous learning, the step size is decreased in the vertical direction and increased in the horizontal direction	Combined momentum and RMSProp generates ADAM's step size with a bias correction

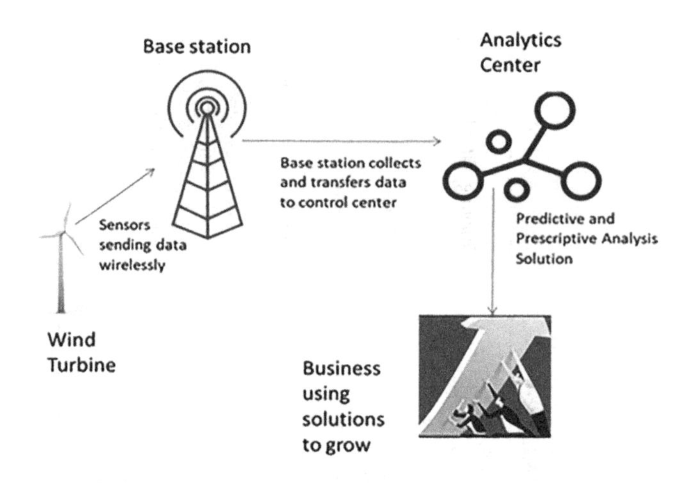

FIGURE 8.9
Basic representation of process flow in wind power energy generation.

TABLE 8.2
Dataset Parameters for Experimentation

Date Time	Air Temperature | (,C)	Pressure | (atm)	Wind Speed | (m/s)	Wind Direction | (deg)	Power Generated by System | (kW)
2007-01-01 00:00:00	10.926	0.979103	9.014	229	33688.1
2007-01-01 01:00:00	9.919	0.979566	9.428	232	37261.9
2007-01-01 02:00:00	8.567	0.979937	8.700	236	30502.9
2007-01-01 03:00:00	7.877	0.980053	8.481	247	28419.2
2007-01-01 04:00:00	7.259	0.979867	8.383	256	27370.3

In the experiment conducted to test LSTM with different optimizers, 6 years of hourly data are divided into 70–30 ratio, which means 70% or 4 years of data are used for training the model and 30% or 2 years of information are used for testing. A sample extracted from the complete dataset is summarized in Table 8.2. The size of training and testing data entries is 36792 and 15768, respectively.

8.10 Performance Metrics

Different performance metrics used to measure SGD and ADAM are as follows:

a. Mean Squared Error (MSE)

MSE, as calculated using Equation 8.6, is the average of the sum of the squared difference between actual and predicted values.

$$\text{MSE} = \frac{1}{N}\sum_{i=1}^{N}\left(v_{i\_\text{actual}} - v_{i\_\text{predicted}}\right)^2 \tag{8.6}$$

b. Root Mean Square Error (RMSE)
   RMSE, as calculated using Equation 8.7, is the square root of the average sum of the squared difference between actual and predicted values.

$$\text{RMSE} = \sqrt{\frac{1}{N}\sum_{i=1}^{N}\left(v_{i\_\text{actual}} - v_{i\_\text{predicted}}\right)^2} \tag{8.7}$$

c. Mean Absolute Error (MAE)
   As calculated using Equation 8.8, MAE is the average absolute difference between actual and predicted values.

$$\text{MAE} = \frac{1}{N}\sum_{i=1}^{N}\left|v_{i_\text{actual}} - v_{i_\text{predicted}}\right| \tag{8.8}$$

d. Variance
   As calculated using Equation 8.9, variance is to figure out how far the predicted values are from the mean.

$$\text{Variance} = \frac{1}{N}\sum_{i=1}^{N}(v_i - \mu)^2 \tag{8.9}$$

8.11 Result Analysis

Two cases (i) for multiple epochs on SGD and ADAM and their respective performance metrics in Table 8.3 and (ii) for single epochs on SGD and ADAM and their respective performance metrics in Table 8.4 are listed in two cases below:

a. Case 1: Multiple Epoch Characteristics of SGD and ADAM
b. Case 2: Single Epoch Behavior of Algorithms

TABLE 8.3
Performance Metrics on Multiple Epoch Size Using SGD and ADAM

Optimizer	EPOCH	MSE	RMSE	Variance	MAE
SGD	300	24.110	4.910	0.694	3.796
ADAM		1.247	1.117	0.984	0.764
SGD	200	12.640	3.555	0.840	2.795
ADAM		1.317	1.148	0.983	0.791
SGD	100	19.042	4.364	0.759	3.445
ADAM		1.181	1.087	0.985	0.752
SGD	50	45.082	6.714	0.429	5.371
ADAM		1.387	1.178	0.982	0.828

TABLE 8.4

Performance Metrics on Single Epoch Size Using SGD and ADAM

Optimizer	EPOCH	MSE	RMSE	Variance	MAE
SGD	300	24.110	4.910	0.694	3.796
ADAM		1.247	1.117	0.984	0.764
SGD		7.970	2.823	0.899	2.226
ADAM		1.205	1.098	0.985	0.750
SGD		23.891	4.888	0.697	3.863
ADAM		1.301	1.141	0.984	0.784
SGD		8.822	2.970	0.888	2.363
ADAM		1.330	1.153	0.983	0.781
SGD		10.966	3.311	0.861	2.534
ADAM		1.296	1.139	0.984	0.770

8.12 Conclusion

In this chapter, IoT analytics and its scopes were explored, along with the applications of deep learning in IoT analytics. Specifically, the LSTM algorithm in deep learning was studied, and comparison of different optimizers was made experimentlly using SGD and ADAM on Wind turbine dataset. From the experimentation, the results obtained on SGD and ADAM are summarized in Tables 8.3 and 8.4. Graphical comparison in Figure 8.10 shows that the behavior of SGD is independent of epoch size. When the epoch size is 300, the MSE is coming 24.110; on 200, it is 12.640; then on 100, it again increases to 19.042, and when the epoch size is 50, MSE is 45.082. Even when it runs multiple times on the identical epoch size, it is also not consistent in the convergence. On the other hand, when ADAM is observed, it is consistent when running on the identical epoch size, i.e., 300. Also, it is converging with a mean square error of approximately 1 every time on varying epoch sizes. It clearly states ADAM to have better convergence power with a more accurate result with a mean square error of approximately 1.

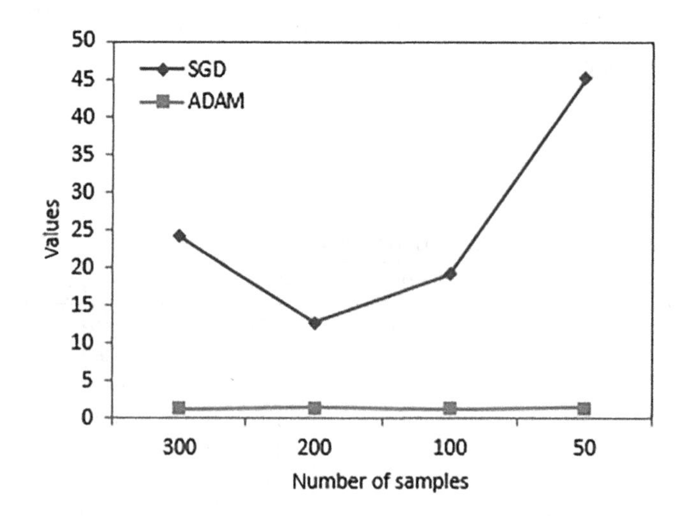

FIGURE 8.10
Epoch and mean square error plotted for SGD and ADAM.

References

Boden, M. (2002) A guide to recurrent neural networks and backpropagation. *The Dallas Project.*

Chen, G. (2016) A gentle tutorial of recurrent neural network with error backpropagation. *arXiv preprint arXiv:1610.02583.*

Gers, F. A., Schmidhuber, J. and Cummins, F. (1999) Learning to forget: Continual prediction with LSTM.

Goodfellow, I., Bengio, Y. and Courville, A. (2016) *Deep Learning*, MIT Press.

Hochreiter, S. and Schmidhuber, J. (1997) Long short-term memory, *Neural Computation*, 9(8), pp. 1735–1780.

Jozefowicz, R., Zaremba, W. and Sutskever, I. (2015) An empirical exploration of recurrent network architectures. *In International conference on machine learning*, pp. 2342–2350.

Karim, F., Majumdar, S., Darabi, H. and Chen, S. (2017) LSTM fully convolutional networks for time series classification, *IEEE Access*, 6, pp. 1662–1669.

Kuremoto, T., Kimura, S., Kobayashi, K. and Obayashi, M. (2014) Time series forecasting using a deep belief network with restricted Boltzmann machines, *Neurocomputing*, 137, pp. 47–56.

Liu, P., Zheng, P. and Chen, Z. (2019) Deep learning with stacked denoising auto-encoder for short-term electric load forecasting, *Energies*, 12(12), p. 2445.

Mohammadi, M., Al-Fuqaha, A., Sorour, S. and Guizani, M. (2018) Deep learning for IoT big data and streaming analytics: A survey, *IEEE Communications Surveys & Tutorials*, 20(4), pp. 2923–2960.

Pandey, S. K. and Janghel, R. R. (2019) Recent deep learning techniques, challenges and its applications for medical healthcare system: A review, *Neural Processing Letters*, 50(2), pp. 1907–1935.

Peng, K., Jiao, R., Dong, J. and Pi, Y. (2019) A deep belief network-based health indicator construction and remaining useful life prediction using improved particle filter, *Neurocomputing*, 361, pp. 19–28.

Robinson, A. J., & Fallside, F. (1987). The utility driven dynamic error propagation network (Tech. Rep. No. CUED/F-INFENG/TR.1). Cambridge: Cambridge University Engineering Department.

Ruder, S. (2016) An overview of gradient descent optimization algorithms. *arXiv preprint arXiv:1609.04747.*

Sherstinsky, A. (2020) Fundamentals of recurrent neural network (RNN) and long short-term memory (LSTM) network, *Physica D: Nonlinear Phenomena*, 404, p. 132306.

Short, L. and Memory, T. (1995) "2 Constant Error BackProp," no. 1993, pp. 1–8.

Vokhmintcev, A., Melnikov, A., Timchenko, M., Kozko, A., Makovetskii, A. and Kober, A. (2018) Development of methods for selecting features using deep learning techniques based on auto-encoders. In *Applications of Digital Image Processing XLI*, Vol. 10752, p. 1075227. International Society for Optics and Photonics.

Werbos, P. J. (1988) Generalization of backpropagation with application to a recurrent gas market model, *Neural Networks*, 1(4), pp. 339–356.

Xu, W., Peng, H., Zeng, X., Zhou, F., Tian, X. and Peng, X. (2019) Deep belief network-based AR model for nonlinear time series forecasting, *Applied Soft Computing*, 77, pp. 605–621.

Zhang, L., Tan, J., Han, D. and Zhu, H. (2017) From machine learning to deep learning: progress in machine intelligence for rational drug discovery, *Drug Discovery Today*, 22(11), pp. 1680–1685.

Zhang, N., Ding, S. and Zhang, J. (2016) Multi-layer ELM-RBF for multi-label learning, *Applied Soft Computing*, 43, pp. 535–545.

Zhao, B., Lu, H., Chen, S., Liu, J. and Wu, D. (2017) Convolutional neural networks for time series classification, *Journal of Systems Engineering and Electronics*, 28(1), pp. 162–169.

9

Customer Analytics for Purchasing Behavior Prediction

Mitushi Agarwal, Priyanka Parashar, Aradhya Mathur, and Adwitiya Sinha
Jaypee Institute of Information Technology

CONTENTS

9.1 Introduction

A huge amount of data are generated every second for every action, every purchase, and every single click by the user that is now documented by the companies to enhance their already existing algorithms—therefore, there are lots of available data we will work upon and further want to generate various results. However, lots of these data are unstructured, sparse, and redundant from an analytical point of view, unless processed. It rarely provides any conclusive insights into user reactions by itself, and thus, a correct framework is required to make sense of this seemingly random data and reach a degree of comprehension. According to a study, it costs you five times more to accumulate new customers than it does to retain current customers, and thus the present customers are more likely to spend on the latest product of the corporate as compared to the new customers. This is often where customers' loyalty comes into action. Customer loyalty may well be a valuable concept because it allows us to take the chance of predicting the actions and behavior of individuals we trust. It's also an integral part of a business and helps in creating a corporate identity.

Therefore, once you understand the importance of customer value, you may want to induce your hands on all the information you'll be able to obtain about your customers to

DOI: 10.1201/9781003046431-9

assist you in choosing how best to serve them. However, not all customer data are equal. Balancing the information of your relevance or importance is the first task. Customer Loyalty Prediction is a model that was formulated to grasp the most important aspects of a customer's lifestyle. It is based on business analytics and acts as an intermediate between the customers and the merchants. The main target of this model is to seek out a balance between both the involved associates, that is, the customers and the merchants, by predicting the loyalty score of the customers. Customer holding back increases your customer's lifetime value and boosts your income. Although procuring new customers is critical, the company must emphasize holding current customers and ensuring reliable customers who will provide stable business activities. The predicted score of the purchasers categorizes the customers into various segments, which further helps us to research the purchasing behavior (i.e., products a customer will purchase supported his/her previous purchases) of the best and therefore a loyal group of customers, which are the main targets of this model. In today's hypercompetitive business environment, the marketing teams of the business stakeholders must know their customers inside out and maximize the return on every dollar spent. A marketer is anticipated to induce the foremost out of their budgets, growing the highest line through new customer acquisition while keeping expenses down and boosting the underside line through effective advertising. Predictive analytics can bring critical improvements to the present process, increasing the efficiency of selling efforts through segmentation.

Customer behavior analysis is an observational measure of how customers interact with your company. Customer segmentation is the beginning process within which the customers are first segmented to support their common characteristics and buying patterns. Then each segmented group is studied at various levels to look at how customers deal with the companies. It provides knowledge of the changing behavior of the customers. The model provides an idea of customers' past purchasing pattern and their likes. This study helps in determining the customers' perspective about the company and can confirm the same with the company's perspective about the customer and their regular purchasing pattern.

Customer Analytics is an idea to study the customer segmentation and predict the purchasing behavior of the buyers using Machine Learning Algorithm. What makes this particular model different is the fact that it targets the loyal and best customers of the corporate. This particular group of customers are the foremost frequent buyers with maximum purchases, and targeting this particular group of customers will be quite beneficial for the corporate because it will help with cost reduction because of an unwanted advertisement campaign; also, maintaining this range of customers can also help us to catch up with more customers, who could have similar range and buying pattern.

9.2 Literature Survey

Various existing surveys help in providing a better understanding of customer relationship management, strategies such as target marketing, different data analytics techniques, and various models that help in predictive analysis of the raw customer detail data. Harry Collins has summarized how various machine learning models can be created for predicting customer loyalty on a transactional dataset and how certain algorithms can be further used to improve the performance of the developed model. Sahar

F. Sabbeh has examined and studied the performance and accuracy of various strategies and classification models related to Artificial Intelligence on a media transmission dataset. This further helps in improving the efficiency of the model. This chapter also provides a better understanding of customer relationship management and how target marketing is a major strategy for acquiring profitable customers. It helps in customer retention as well as acquiring new customers. Nor Azura Md. Ghani provided an insight into the large-scale e-commerce platform and how to deal with that huge amount of data. It provides the approach of a clustering algorithm that categorizes the products into various clusters. It gives an idea to create a classification model on an online retail dataset, with the major help of the K-means clustering algorithm. Razieh Qiasi provides us with a better understanding of the importance of customer loyalty, the basis of targeting profitable customers, and how it is an essential strategy for organizations in e-commerce. It uses the RFM technique that helps in customer segmentation, along with various classification and data-mining techniques. This paper has been implemented on the grocery store data. Joseph P. Bigus has developed a Customer Relationship Management Analytics Framework that is a game-changer for the business. It is a type of framework that eliminates various stages that are required for developing an application and directly helps the creation of prepackaged solutions that can be easily located in various customer environments such as banks, industries, etc., without any customizations. Johnson Kampamba have tried to apply and explain the need for the STP strategy in the housing market. This paper shows the relevance of market research for decision making before investing. The importance of the need to focus on the demand for the product rather than developing resources based on assumptions. The major focus of the chapter was to identify a target market that might help in the elimination of the risks that are hindering the development of the business. Also, identify the customers' needs and wants by implementing the recommendations obtained from the research adds a greater value to the business. Michael Lynn helps us deeply understand how the segmentation, targeting, positioning, i.e., the STP strategy functions. This paper wants us to realize that the needs and wants of each individual are an important factor that is relevant for targeting the profitable segments and thus eliminating the competition, because eventually every other firm in the market will copy the strategy adopted by your own business and all the efforts will go in vain. Therefore, identifying those particular groups of customers who turn out to be profitable to your firm and then focusing only on them in a way that is cost-effective as well as satisfying will increase the sales rate of the firm and eventually help in capturing the market. CY Tsai works on a customer dataset that consists of the transactional history, past purchases, and the amount of money spent on each item. This paper majorly focuses on market segmentation, which is considered one of the important factors for customer relationship management. It allows the user to develop a clustering algorithm based on the products purchased by the customers, and the pattern and behavior of customers can be obtained in the form of clusters, and hence their relevance to one other can be determined from the same. Further application of the RFM technique helps in attaining the profitable customers from every cluster that is obtained by applying the described clustering algorithm. David Schmittlein has tried to identify whether using the segmentation technique could help in improving the estimation of future sales and also determine the segments of customers who are interested in buying the products. This can be done by obtaining the past purchase data for the customers, observing the pattern, and then formulating the algorithm to accurately forecast the sales. Another measure to improve the accuracy was by applying various kinds of algorithms and models and further classifying the customers into relevant segments based on their purchase rates, which allows target marketing and allows the firm to fulfill all the needs of the customers efficiently and effectively. Jiong Mu summarizes the importance of the need for the retention of the existing customer base and the emphasis on focusing on their needs to maintain the valuable relationships developed with them so far and improve the customer relationship management of the same. To do so, this paper implements algorithms that focus on future purchases and profitability. The outcome suggests the need to analyze all the variables to benefit the customer and fully understand their buying behavior pattern. M.N. Saroja has researched on various techniques that will help in the growth of the business and how analyzing the behavior of the customer is beneficial for the company. In today's world, every company wants to get ahead of their competitors; for doing so, it is very important to get to know their customers, their needs, their buying pattern, and the products they will be interested in. So, predictive analytics is one of the techniques that works well. In this research work, various tests have been performed by analyzing the customers on a different basis such as who buys more, who buys only in the sale period, and who prefer discounted products. Based on all these techniques, we can improve our sales by following different business strategies. Heiner Evanschitzky has explained how customer loyalty and loyalty programs are different and how they affect the company. The study focuses on two main points: the first one is how "customer loyalty" is different from "loyalty program," and the second focuses on how this loyalty program affects the customer, their behavior prediction, and what is the impact of the loyalty program on the company. As a result, it was found that the main role of loyalty programs is to don't let go of the customer from a particular brand and make sure if he/she buys and he/she spends more because of the loyalty offers. Pavels Goncarvos summarizes how data are not the problem nowadays. There are a lot of data available on customer behavior; our main challenge is how to utilize these data in an efficient manner for customer relationship management. Data analytics plays an important role in CRM. In this paper, the main aim is to analyze the different techniques for the CRM process. As a result, we found out that data analytics works well for customer relationship management and is very helpful for the industry in making decisions, and various data-mining techniques can be used for this purpose. Mark Camilleri tells that it might not be possible for a company to satisfy all of its customers, hence it divides and segments its customers based on their needs and targets the customers; accordingly they can segment customers into different categories based on their needs and preferences, thus creating different expertise for the customers that they might be receiving from other companies and being unique in understanding what the customers want and providing them with the experience that the customers want and be their regular customers. Tala Mirzaei suggests that customer relationship management is very important for any company for planning to target customers using this technique and provide a set layout that they can refer to and help them acquire new customers too.

9.3 Description and Experimentation

Customer Relationship Management is considered one of the most important tasks in any organization: some companies might as well adopt the strategy known as target marketing. The marketing team of such companies focuses on breaking down the market into various groups or segments and then targets the most profitable ones. The challenge that comes forward involves the effective utilization of the data present in the CRM processes and then further selection of appropriate techniques for the same.

Therefore, this research comes forward as an attempt to provide a solution to such a problem as follows:

- It begins by categorizing the customers into various groups, which helps the company in targeting the profitable segments, i.e., the loyal customer group.
- Further customer analysis helps us in receiving an insight into the customer's buying pattern, after which the products are segmented into various clusters followed by customer segmentation.
- The predictive analysis helps us further in training and testing our proposed model that predicts the purchasing behavior of the customer based on past purchases.

The online retail dataset contains details about the customers and their purchasing history over 10 months (Figures 9.1 and 9.2).

	InvoiceNo	StockCode	Description	Quantity	InvoiceDate	UnitPrice	CustomerID	Country
0	536365	85123A	WHITE HANGING HEART T-LIGHT HOLDER	6	12/1/2010 8:26	2.55	17850.0	United Kingdom
1	536365	71053	WHITE METAL LANTERN	6	12/1/2010 8:26	3.39	17850.0	United Kingdom
2	536365	84406B	CREAM CUPID HEARTS COAT HANGER	8	12/1/2010 8:26	2.75	17850.0	United Kingdom
3	536365	84029G	KNITTED UNION FLAG HOT WATER BOTTLE	6	12/1/2010 8:26	3.39	17850.0	United Kingdom
4	536365	84029E	RED WOOLLY HOTTIE WHITE HEART.	6	12/1/2010 8:26	3.39	17850.0	United Kingdom

FIGURE 9.1
The online retail dataset.

InvoiceNo	StockCode	Description	Quantity	UnitPrice	CustomerID	Country	Segment	Score	SegmentType
536367	84879	ASSORTED COLOUR BIR	32	1.69	13047	United Kingdom	344	11	Loyal Customers
536367	22745	POPPY'S PLAYHOUSE BE	6	2.1	13047	United Kingdom	344	11	Loyal Customers
536367	22748	POPPY'S PLAYHOUSE KI	6	2.1	13047	United Kingdom	344	11	Loyal Customers
536367	22749	FELTCRAFT PRINCESS CI	8	3.75	13047	United Kingdom	344	11	Loyal Customers
536367	22310	IVORY KNITTED MUG C	6	1.65	13047	United Kingdom	344	11	Loyal Customers
536367	84969	BOX OF 6 ASSORTED CC	6	4.25	13047	United Kingdom	344	11	Loyal Customers
536367	22623	BOX OF VINTAGE JIGSA	3	4.95	13047	United Kingdom	344	11	Loyal Customers
536367	22622	BOX OF VINTAGE ALPHA	2	9.95	13047	United Kingdom	344	11	Loyal Customers
536367	21754	HOME BUILDING BLOCI	3	5.95	13047	United Kingdom	344	11	Loyal Customers
536367	21755	LOVE BUILDING BLOCK	3	5.95	13047	United Kingdom	344	11	Loyal Customers
536367	21777	RECIPE BOX WITH META	4	7.95	13047	United Kingdom	344	11	Loyal Customers
536367	48187	DOORMAT NEW ENGLA	4	7.95	13047	United Kingdom	344	11	Loyal Customers
536368	22960	JAM MAKING SET WITH	6	4.25	13047	United Kingdom	344	11	Loyal Customers
536368	22913	RED COAT RACK PARIS F	3	4.95	13047	United Kingdom	344	11	Loyal Customers
536368	22912	YELLOW COAT RACK PA	3	4.95	13047	United Kingdom	344	11	Loyal Customers
536368	22914	BLUE COAT RACK PARIS	3	4.95	13047	United Kingdom	344	11	Loyal Customers
536369	21756	BATH BUILDING BLOCK	3	5.95	13047	United Kingdom	344	11	Loyal Customers
536370	22728	ALARM CLOCK BAKELIKI	24	3.75	12583	France	444	12	Best Customers
536370	22727	ALARM CLOCK BAKELIKI	24	3.75	12583	France	444	12	Best Customers
536370	22726	ALARM CLOCK BAKELIKI	12	3.75	12583	France	444	12	Best Customers

FIGURE 9.2
The extracted dataset of best, loyal, and potentially loyal customers along with RFM score and segment.

9.4 Implementation

The customer loyalty and purchasing behavior prediction model is based on a machine learning model. To implement this model using accurate algorithms, we have surveyed various steps. Figure 9.3 shows the flow of our research that summarizes the various steps performed.

9.4.1 Data Analysis

The first task was selecting an appropriate customer retail dataset that would help us in analyzing our problem, which was further followed by Data Cleaning and is the first critical step in any machine learning model; it helped in exploring the data for relevant features. It was observed that features like the amount of money spent, the number of transactions made by the buyer, the difference between the last and first purchases, types of products bought, etc. played an important role in developing a connection between the customer and the retailer. Formulation of a CRM system is an important part as it creates a relationship with customers that in turn creates loyalty and customer retention, which results in an increased profit for the business.

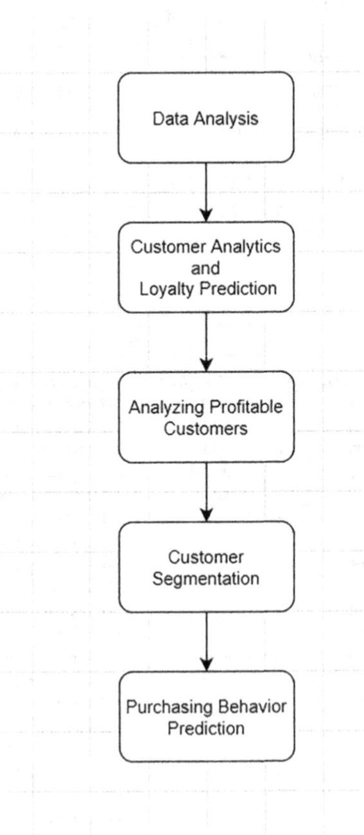

FIGURE 9.3
Flow diagram.

9.4.2 Customer Analytics and Loyalty Prediction

Customer analytics is a process in which customer behavior is analyzed and the information that we obtain from the analysis of customer behavior is used in making the viral decisions of the business. The segmentation done will influence sales, marketing, and decisions, and potentially the survival of an organization. The first step involves determining different categories of the customers by using the features extracted from the dataset, for instance, the customer id, invoice date, unit price, quantity, etc. To serve our purpose, the RFM segmentation model is applied.

RFM Segmentation Model: This model helps the marketers to focus on some specific group of customers, which are good for their business based on the behavior that is predicted based on their past transactions, which increases customer loyalty and higher rates of retention for the brand (Figure 9.4). RFM stands for:

- RECENCY (R): How many days since the last purchase.
- FREQUENCY (F): How many number of purchases.
- MONETARY VALUE (M): Total money spent on purchases.

These attributes have been created against each customer. This model is performed in the following three major steps:

- In building an RFM model, the very first step is to allocate R, F, and M values to every customer.
- In the second step, using Excel or any other tool, divide the list of customers into groups based on three dimensions, i.e., R, F, and M. We can also use K-means clustering, which divides the customers into groups with more similar characteristics.
- Lastly, to select the groups of shoppers to which specific styles of transmission are sent, support the RFM segments within which they're seen and further take out the foremost profitable segments.

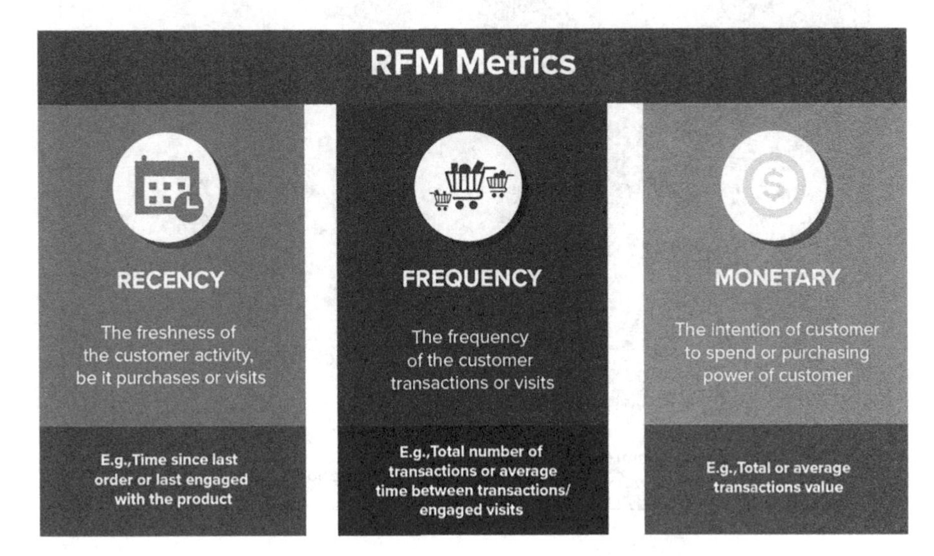

FIGURE 9.4
RFM metrics explaining the RFM in concise form.

The RFM analysis is a well-organized way because it includes objective and numeral results. It is a simple technique that is easy to implement and understand, and the output can also be easily explained to others. It can be performed without the requirement of data scientists or any other software tool.

9.4.3 Analyzing Profitable Customers

The RFM model provided us various categories of shoppers to appear upon and hence target our most profitable segments from the following:

- *Best Customers*: "444" (Highest frequency similarly as value with least recency)
- *Loyal Customers*: "344" (High frequency similarly as value with good recency)
- *Potential Loyalists*: "434" (High recency and value, average frequency)
- *Big Spenders*: "334" (High value but good recency and frequency values)
- *At-Risk Customers*: "244" (Customers shopping less often now who used to shop a lot)
- *Can't Lose Them*: "144" (Customers shopped way back who used to shop a lot)
- *Recent Customers*: "443" (Customers who recently started shopping for lots but with less monetary value)
- *Lost Cheap Customers*: "122" (Customers shopped way back but with less frequency and monetary value)

The score against each category defines the R-F-M metrics (Figure 9.5). For example, if we consider the Best Customer, this category contains customers who have all the R, F, and M values as 4, which means they have a recent transaction; they do it often and spend more

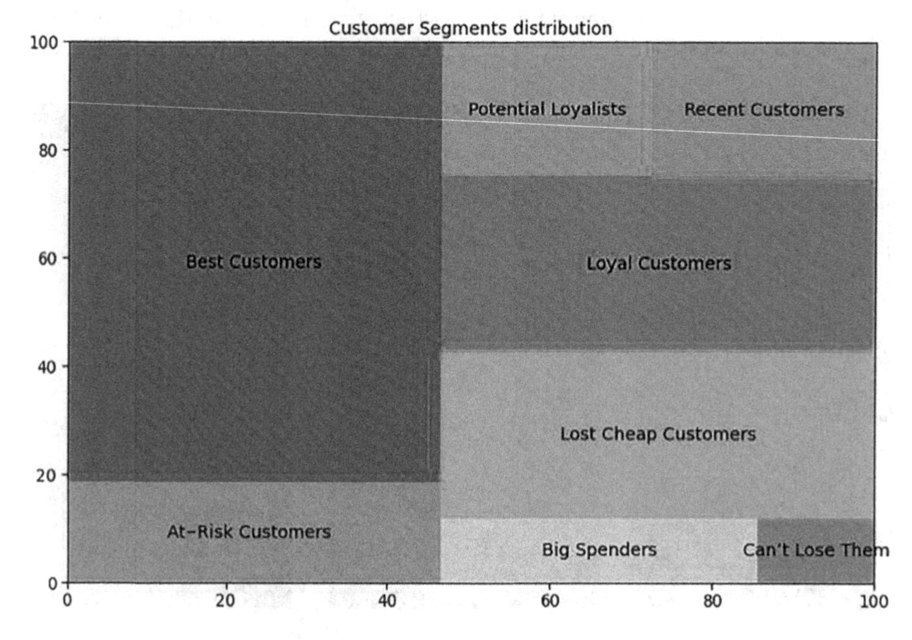

FIGURE 9.5
Pictorial representation of the various customer categories.

than other customers and similarly, followed by other categories described above. Now when viewed from the business perspective, the marketing team of any e-commerce business platform would mainly want to focus on the categories yielding the most profit. It was observed that segments like Best Customers, Loyal Customers, and Potential Loyalists clothed to be the foremost profitable segment among all. And hence the shoppers belonging to those particular segment blocks were extracted, i.e., the info against each customer ids with RFM score adequate to the values—"444," "344," and "434."

This dataset is now forwarded toward the customer segmentation part for predictive analysis of the data and eventually follow up with the results of behavior prediction.

9.4.4 Customer Segmentation

Customer segmentation is a process where we divide the buyer base of the industry into subgroups. We have to create the subgroups by using some characteristics such that the industry sells more products with less marketing costs. The segments formed should be such that the merchants can find the patterns that differentiate your customers. It is a matter of revealing information about a company's customer base, depending on their interactions with the business. In many cases, this interaction is in the form of their purchasing behavior and patterns. We search a number of ways within which this can be used. In this model, we are working on an online retail dataset to explore customer segmentation through an unsupervised learning model. Furthermore, we apply the association rule mining approach to search out interesting rules and patterns in this database. These customer segmentations, rules, and patterns will help in making interesting and useful decisions. This idea of segment-wise marketing will help the industry to sell more products with less marketing costs. Thus, the industry will make more profit. This is the reason why industries use customer segmentation nowadays. Customer segmentation is employed in other sectors like the retail sector, finance sector, and in customer relationship management-based products.

Industries nowadays use the **STP** approach to make the marketing strategy resistant. STP stands for Segmentation Targeting Positioning. The three stages are as follows:

- **Segmentation:** In this, we make segments of our consumer base using their personality characteristics. Once we've done the segmentation part, we will go on to the subsequent stage.
- **Targeting:** Here comes the part of the marketing teams. It evaluates the segments and tries to find out which type of product is suitable to which segments. The team performs this cycle for every segment, and finally, the team designs the customized products that are going to attract the purchasers of one or many segments. They're also going to select which product should be offered to which particular segment.
- **Positioning:** During this last stage, industries study the market options and what products they can offer to the consumers. The marketing team should bring up unique selling ideas. The team also tries to get to know how a specific segment responds to the brand, product, or service. This is often the way for industries to work out a way to best position their offering. Marketing and product teams of industry create a price proposition that explains how their offering is healthier than the other competitors. Lastly, the businesses start their campaign representing this value proposition such that the buyer base is happy about what they're getting.

Further, we have grouped the products into different classes. The *k*-means clustering method of sklearn is based on Euclidean distance. The silhouette score is used for defining the number of clusters that are best suited for our dataset.

9.4.4.1 k-Means/Centroid-Based Clustering

It arranges the data into nonhierarchical clusters. Most algorithms that are based on centroids are good but subtle to conditions and anomaly.

k-Means will randomly initiate centroids at random positions and steadily fit each data point to the closest centroid. Every data point signifies one customer, and the consumer closest to the same centroid is going to be within the same group. The centroid locations will be adjusted on their own based on the last nearest customer allotted to them. From this, it will learn to find other customers with the same characteristics on its own. Another method for finding the number of groups is the silhouette score. It considers both the intracluster and intercluster distances and return a score. The lesser the score, the more relevant the clusters formed. Silhouette is a technique used for the interpretation and validation of uniformity in clusters of data. The scores obtained in this were considered equivalent. Furthermore, as we see after five clusters, there are very few elements in many clusters. So, the dataset was segregated into five clusters. After the creation of product clusters, we move toward the creation of customer clusters or the customer categories, which is further followed by classification.

9.4.5 Purchasing Behavior Prediction

Predictive behavior modeling is applied to historical and transactional data to predict the long-term behavior of consumers. It typically wants to select the most effective marketing strategies to run on each group of consumers to spot which customers will like to change their spending level. The steps for the identical model have already been applied. This last step includes the classification of consumers into different classes to which they fit in.

Some classifiers were trained to categorize shoppers. The classifiers used are as follows:

- **Support Vector Machine Classifier:** The target of this classifier is to suit the given data and give back a hyperplane that divides our data, and when supplied with a few features, it returns the predicted class.
- **Logistic Regression:** This model is employed to predict the probability of a particular class or event occurring provided some previous data. It works with the binary data, where it is not sure either the event happens or not.
- **K-Nearest Neighbors:** This straightforward algorithm is based on the very simple procedure in which it saves all given cases and all the new cases are classified based on how similar is the measure.
- **Decision Tree:** It is an algorithm in which by learning some straightforward decision rules target value can be predicted. Used for classification problems.
- **Random Forest:** It is a classification algorithm that is an ensemble of the decision tree. Gives better prediction and accuracy than a decision tree.
- **Gradient Boosting Classifier:** A set of ML algorithms that we use to form a powerful predictive model by the combination of many weak models.

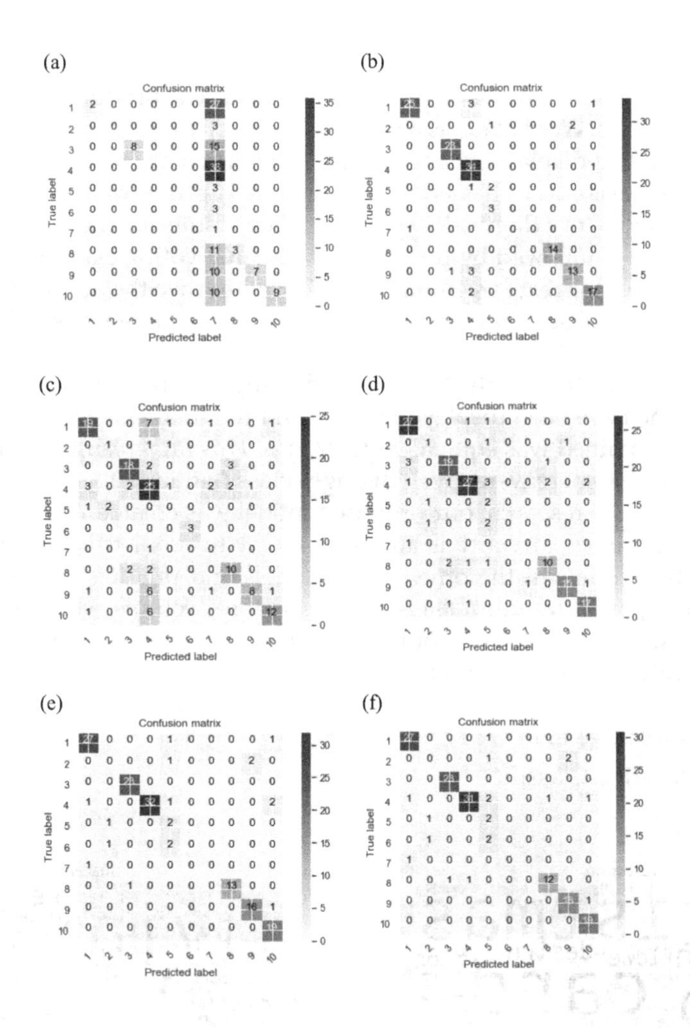

FIGURE 9.6
Confusion Matrix for (a) Support Vector Classifier (b) Logistic Regression (c) KNN (d) Decision Tree (e) Random Forest (f) Gradient Boosting.

We have used a confusion matrix to examine the comparison between predicted values and real values. The results of various classifiers can be combined to enhance our classification model (Figure 9.6). We have done this by using the voting classifier model. It combines various models into one single model that model is the most powerful model than any of the individual model alone. The entire analysis is based on the data of the first 10 months. To obtain the accuracy of our model, testing was performed over a 2-month data.

9.5 Result and Analysis

The distinct customer groups that were created in this research can help in a better understanding of the profitable customers and accordingly help the marketing team in making a proper plan of action for different segments. Accordingly, we have extracted our loyal and best customers and further applied the formulated machine learning model on them. The dataset consisted of 10 months of transactional data. The products were grouped into five different categories. Word cloud is a method of visually representing textual data. The highlighted or bold text displays its importance in a particular cluster (Figure 9.7).

First, the various classification models were applied to the training dataset that consisted of an 8-month history (Figure 9.8). The precision results for each classifier have been shown below, which is followed by the result of the final voting classifier that groups all the classifiers together to create a combined model to improve the accuracy.

From Figure 9.8, it is observed that the accuracy of the obtained model is equivalent to 88.51%.

The data of the prediction quality were tested using different classifiers; the precision results for the same have been displayed above, and the final test accuracy from the voting classifier thus obtained was equivalent to 82.41%. This final result obtained, therefore, shows that 82% of customers were given the right class (Figure 9.9).

In the last stage of the research, our proposed model was created using the voting classifier; this classifier comes forward as an efficient way to predict the output as it creates a single model that trains all the other weak models and predicts the output based on the majority voting of each output class, which is more accurate when compared to the accuracy of any single model. The voting classifier model here combines the following algorithms to create the required model:

- Random Forest Classifier
- Gradient Boosting Classifier

FIGURE 9.7
Word cloud of type of product in each cluster.

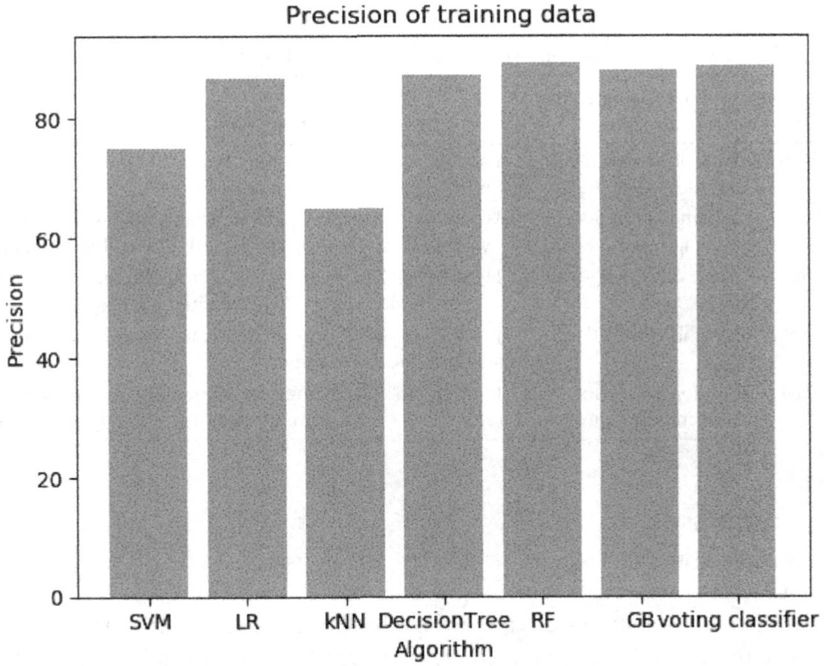

FIGURE 9.8
Precision of train data using different classifiers.

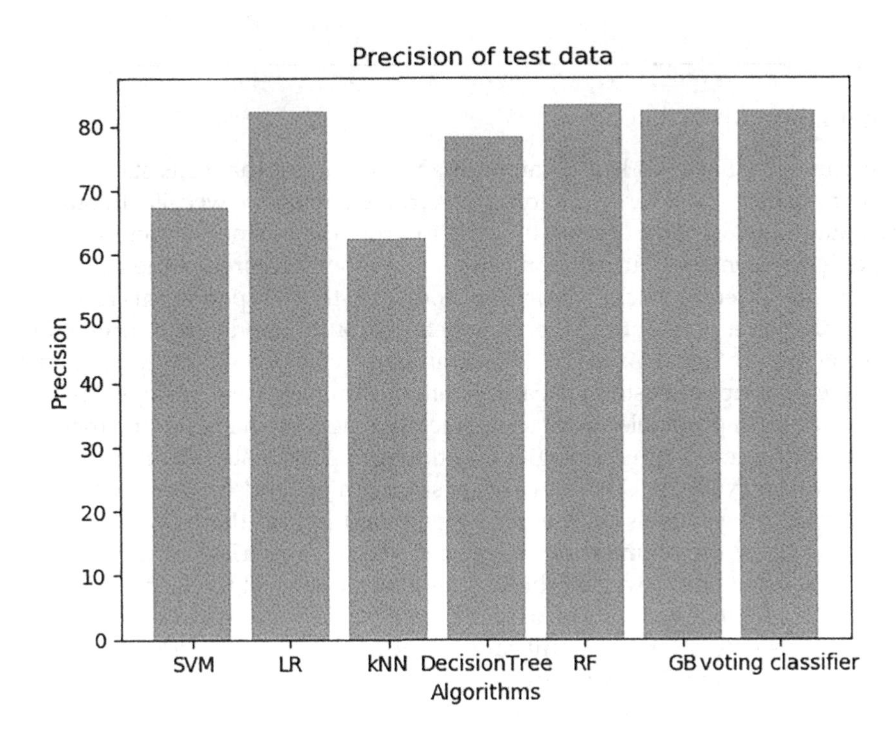

FIGURE 9.9
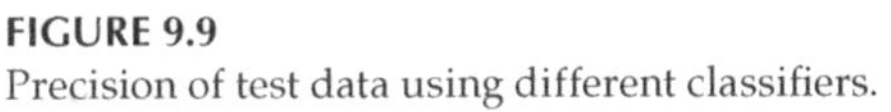
Precision of test data using different classifiers.

TABLE 9.1

Prediction of Purchasing Item for Top Five Loyal Customers Based on Their Previous Purchase

Serial No.	Customer Id	Purchasing Item Predicted	Past Items Frequently Purchased
1	14646	Pack of 12 woodland tissues	Pack of 12 woodland tissues, pack of 12 polka dot tissues, party candles, napkins, jam making set
2	18102	Cream heart holder, vintage memo board	Woodland whiteboard, cream heart holder, vintage memo board, vintage memo board, cream cardholder, blue owl soft toy
3	14911	Heart of wicker small	Heart of wicker small, heart of wicker large, pink candlestick, candleholder, heart of wicker small, carriage
4	17511	Lunchbox woodland	Jumbo bag strawberry, lunchbox woodland, lunch bag cars blue, chopsticks, coffee mug
5	16684	London bus coffee mug, ceramic storage jar	London bus coffee mug, ceramic storage jar, lunchbox woodland, save planet mug, ceramic storage jar, teacup and saucer

- Logistic Regression
- K-nearest Neighbor Classifier

The model thus obtained shows us that almost 83% of the customers were given the right class, therefore their purchasing items have been predicted accurately. The predicted values, i.e., the purchasing items for the top five customers have also been summarized in Table 9.1.

9.6 Conclusion

In this research, we have worked on an online retail dataset that consists of the transactional history of customers for 1 year for an electronic-commerce website. In our work, we see how data-mining techniques can be used to grow the business of online retailers by working on client-centered business and further how we can predict the behavior of our customers using different machine learning models. In the first part of this research, it has been found that data preparation, the interpretation of the model, and evaluation of the model are very important steps in the data-mining process. During this research, we came across different groups of customers based on their purchases. Accordingly, the company can decide their most profitable customers, which will help them in growing their business and choose effective marketing strategies for other groups of customers as well. According to this idea, we've extracted our loyal and best customers and further applied the various machine learning models on them. In the second half of the research, we grouped the products sold by the online store into five different categories. Further based on the amount of time spent on the website, how many times they visit the website, and the type of product they usually buy from the data of historical transactions of 10 months, we have classified our customers into 11 significant categories. After the formation of customer segments, for finding the first purchase of the customers, we trained several classifiers to categorize our customers into one among those 11 categories of consumers. Finally, based on the data of the last 2 months, the prediction quality was tested using different classifiers, the precision results for the same have been displayed, and the final test accuracy from the voting classifier thus obtained was equivalent to 82.41%. This final result obtained, therefore, shows that 82% of customers were given the right class, and the performance of the classifier, therefore, seems correct.

References

Bhavana, B., Reddy, K. S. P., Sailaja, P. and Raju, C. S. (2020) Machine learning model for predicting purchase nature of customer, *Sustainable Humanosphere*, 16(1), pp. 2113–2119.

Bigus, J. P., Chitnis, U., Deshpande, P. M., Kannan, R., Mohania, M. K., Negi, S. and White, B. F. (2009) CRM Analytics Framework. In COMAD.

Camilleri, M. A. (2018) Market segmentation, targeting and positioning. In Travel marketing, tourism economics and the airline product, pp. 69–83. Springer.

Evanschitzky, H., Ramaseshan, B., Woisetschläger, D. M., Richelsen, V., Blut, M. and Backhaus, C. (2012) Consequences of customer loyalty to the loyalty program and to the company, *Journal of the Academy of Marketing Science*, 40(5), pp. 625–638.

487Gončarovs, P. (2017) Data analytics in CRM processes: A literature review, *Information Technology and Management Science*, 20(1), pp. 103–108.

Kampamba, J. (2015) An analysis of the potential target market through the application of the STP principle/model, *Mediterranean Journal of Social Sciences*, 6(4), p. 324.

Lynn, M. (2011) Segmenting and targeting your market: Strategies and limitations.

Mathivanan, N. M. N., Ghani, N. A. M. and Janor, R. M. (2019) Analysis of K-Means Clustering Algorithm: A Case Study Using Large Scale E-Commerce Products. In 2019 IEEE Conference on Big Data and Analytics (ICBDA), pp. 1–4. IEEE.

Mirzaei, T. and Iyer, L. (2014) Application of predictive analytics in customer relationship management: A literature review and classification. In Proceedings of the Southern Association for Information Systems Conference, pp. 1–7.

Morwitz, V. G. and Schmittlein, D. (1992) Using segmentation to improve sales forecasts based on purchase intent: Which "intenders" actually buy?, *Journal of Marketing Research*, 29(4), pp. 391–405.

Mu, J., Xu, L., Duan, X. and Pu, H. (2013) Study on customer loyalty prediction based on RF algorithm, *JCP*, 8(8), pp. 2134–2138.

Qiasi, R., Baqeri-Dehnavi, M., Minaei-Bidgoli, B. and Amooee, G. (2012) Developing a model for measuring customer's loyalty and value with RFM technique and clustering algorithms, *The Journal of Mathematics and Computer Science*, 4(2), pp. 172–181.

Sabbeh, S. F. (2018) Machine-learning techniques for customer retention: A comparative study, *International Journal of Advanced Computer Science and Applications*, 9(2).

Tsai, C. Y. and Chiu, C. C. (2004) A purchase-based market segmentation methodology, *Expert Systems with Applications*, 27(2), pp. 265–276.

Van der Heijden, G., Collins, H. and Aslam, S. Predicting customer loyalty using various regression models, *Target*, 1(2), p. 3.

10

Discernment of Malaria-Infected Cells in the Blood Streak Images Using Advanced Learning Techniques

Megha Rathi, Chandna Gupta, Rachit Shukla, and Raja Raubins
Jaypee Institute of Information Technology

CONTENTS

10.1 Introduction

Caused by the *Plasmodium* parasite found in the female mosquitoes, Anopheles, malaria has become one of the most trivial and lethal diseases in the world. According to the reports published by the World Health Organization in 2018, an expected 219 million instances of malaria were recorded worldwide in 2017. Though these were fewer in comparison to the 239 million cases in 2010, no cogent improvements were made in the reduction and prevention of the disease. Among the cases recorded, most of them were found to be in the African region, which consisted of 92% of the total, followed by 5% cases in the South-East Asia zone, and 2% in the Eastern Mediterranean zone. Most of the hardships were carried by almost 15 countries in the African and Indian regions. The main cause of the spread in these regions is *Plasmodium falciparum*. There were an estimated 4,35,000 deaths because

DOI: 10.1201/9781003046431-10

of malaria in 2017, with children under 5 years old being the most vulnerable among the whole populace. Again, 80% of these deaths occurred in the African and Indian regions. Malaria is caused mainly by four types of parasites, namely:

1. ***Plasmodium falciparum***: *P. falciparum* is a unicellular protozoan parasite of humans, and hence the most carcinogenic form of *Plasmodium* that causes protozoa infection in the human body. The parasite is disseminated through the chomp of a feminine arthropod genus and causes the infection's most unsafe structure, the falciparum protozoa infection. It is responsible for around half of all the cases.
2. ***Plasmodium vivax***: *P. vivax* is a parasitic protozoa and human microbe. *P. vivax* is the most gradual and usually sensible reason for the recurrence of malaria. Contamination of vivax intestinal disease can prompt genuine ailment and passing, because of ordinary splenomegaly (an obsessively amplified spleen). *P. vivax* is exposed to the female Anopheles mosquito.
3. ***Plasmodium ovale***: *P. ovale* is a parasitic protozoa that causes Tertian intestinal sickness in humans. *P. ovale* is one of several such *Plasmodium* parasites that defile humans, including *P. falciparum* and *P. vivax*, responsible for most of the malaria diseases. This is unusual with both of these parasites, and considerably less risky than the falciparum.
4. ***Plasmodium malariae***: *P. malariae* is a parasitic protozoa that causes intestinal disease in humans. *P. malariae* is one of several such *Plasmodium* parasites that bother humans, including *P. falciparum* and *P. vivax*, responsible for most of the malaria diseases. Discovered worldwide, it causes alleged "benign malaria" and is not as harmful as *P. falciparum* and *P. vivax*.

Among the four listed above, the most prevalent parasite is *P. falciparum*. Once a person is bitten by a mosquito carrying the parasite, it spreads into their bloodstream and travels to the liver, which gets infected. Then the infection spreads to the red blood cells, a major component of the blood, where the parasites multiply and grow. The diagnosis of malaria can be done using various methods. The absolute initial step incorporates deriving suppositions from Clinical Diagnosis. This is primarily done from the symptoms shown by the patient, such as fever, migraines, muscle agony, chills, and perspiration. These assumptions must, however, be confirmed by performing approved laboratory tests. Some of the common tests are as follows:

1. **Microscopic Diagnosis:** The most widely recognized method for distinguishing a malaria parasite is by analyzing a patient's blood smear under a magnifying instrument. The specimen is recolored with Giemsa to render the parasite more conspicuous in the blood smear.
2. **Antigen Detection:** Antigen catch tests are equipped to quickly identify tiny parasites. Units for these tests are financially accessible. The two fundamental parasite antigens that are currently being utilized for tests are histidine-rich protein-2, which is only produced by *P. falciparum*, and the parasite lactate-dehydrogenase antigen, which is produced by all of the four *Plasmodium* species.
3. **Molecular Diagnosis:** Polymerase chain reaction is used to recognize parasite nucleic acids. To obtain results using this diagnosis takes an excessively long time, and hence, as a rule, this method is used as a confirmation test after the parasite has been identified through other methods.

4. **Serology:** This test recognizes antibodies against malaria parasites but does not span the prevailing infection but rather the prior vulnerability.
5. **Drug Resistance Test:** Drug resistance tests must be carried out specifically by research offices to assess the feebleness of antimalarial blends of parasites gathered from a specific patient.

All these procedures depend largely on the skills of the pathologist conducting the tests and the conditions of the clinic, such as light exposure and the quality of equipment used. Automation and computerization of the process will reduce the cost, time, and labor involved while also improving efficiency. Our aim in this work is to improve the accuracy to the highest levels.

10.2 Related Work

For the last decade, the computerization of the malaria detection process has attracted many researchers. Preprocessing of the acquired blood smear images is the foremost task in this process. The pictures should be tidied up first as they contain numerous high-recurrence commotions such as Poisson Noise, Salt and Pepper Noise, Gaussian Noise, and Speckle Noise. Channels, for example, middle channels, Gaussian channels, and Weiner channels are utilized to remove such commotions. Cecilia Di Ruberto (2002) proposed a morphological method to deal with a section of the blood smear pictures by using granulometry to assess the size of the red platelets and to smother and smoothen certain regions. Adedeji Olugboja and Zenghuhi Wang (2017) compared different machine learning classifiers and found that Fine Gaussian SVM and Subspace KNN have the best performance, with an overall accuracy of 82% and 86.3%, respectively. Hanung Adi Nugroho et al. (2015) put forth a method to obtain parasite cells using *K*-implies bunching calculation and morphological reproduction. The images were first preprocessed using the contrast stretching method, and the noise was removed using a median filter and then segmentation was applied. They used Artificial Neural Networks (ANNs) for classification. Ahmed Elmubarak Bashir et al. (2017) converted the acquired images from RGB to Grayscale to reduce processing time and used the square median filter to remove high-frequency noise. They also used ANNs to classify the images as infected or noninfected. Pattanaik et al. (2017) proposed a three-step solution. They utilized the Kalman channel to expel the noise from the pictures and additionally used Reinhard's technique for shading Normalization for preprocessing. Softmax Classifier was used for arrangement purposes. They also used three different datasets for training their model with accuracies of 96.71%, 95.10%, and 94.13%, respectively. Penas et al. (2017) used a saturation channel to give a fruitful distinction between the parasite and the nonparasite regions. A series of morphological transformations were used to clean up the images. Connected Component Analysis was used for segmentation, and Convolutional Neural Networks (CNNs) were used for classification. This process gave a training veracity of 94% and a validation veracity of 87.6%. The method proposed by Shipra Saraswat et al. (2017) take advantage of median filtering for noise removal and histogram equalization to enhance the image. Morphological operations were used to differentiate cells from each other and HSV segmentation was applied. Dhanya Bibin et al. (2017) worked on the Deep Belief Networks for the classification of parasitic and nonparasitic images. Ishan R. Dave (2017) proposed a method to detect all the life stages of *P. vivax* and leukocytes. The author first segmented the image into foreground and background, which was followed by the conversion of the images into HSV color space. In yet another novel work, authors Paluck et al. (2019) detected irregularities in blood samples using WBC differential count. Another study by Ekam et al. (2019) used a deep learning model for the analysis and segmentation of brain tumors. Rajat et al. (2020), in their research work, employed learning techniques to uncover mutated motifs in DNA sequences.

10.3 Basic Methodology

In this work, we employed a strategy that includes four fundamental strides to distinguish the malarial parasite from the blood smear pictures. The initial step is to tidy up the pictures obtained from the dataset. This includes expelling high-recurrence clamor and upgrading the complexity of pictures. To separate the picture into regions of intrigue and nonareas of intrigue, picture division is applied using watershed segmentation. The following and most significant advance is the component extraction method, which will assist the classifier in distinguishing between the tainted and parasitized pictures. This is followed by the last advance, which is the grouping of the pictures using CNNs.

10.3.1 Preprocessing

Preprocessing of images is the initial move toward this undertaking since we can't compose a special calculation for every condition under which a picture is taken, and hence, when we obtain a picture, we will, in general, consider it as a structure that would allow for a general calculation to unravel. Fundamentally, preprocessing is expected to convert a picture into a two-fold/grayscale to recreate it (because various pictures are of various measurements), and to lessen the clamor, with the aim that procedures on these pictures can be easily executed. Figure 10.1 presents the preprocessing steps.

10.3.1.1 Noise Removal and Filtering

For all image-processing tasks, we first need to import the NumPy as well as the cv2 module in Python, which will make available the functionalities required to apply on our

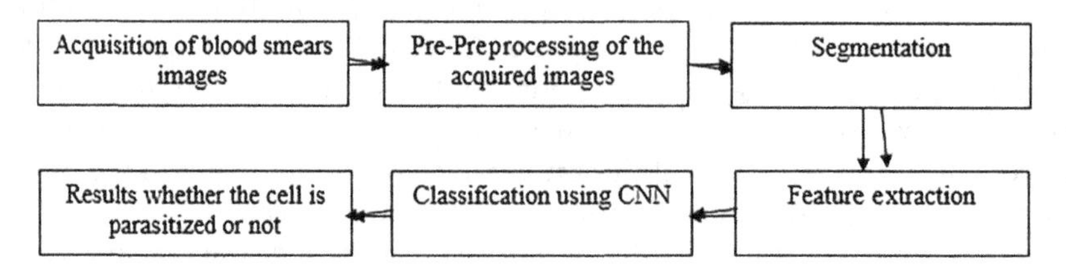

FIGURE 10.1
Flowchart representing methodology.

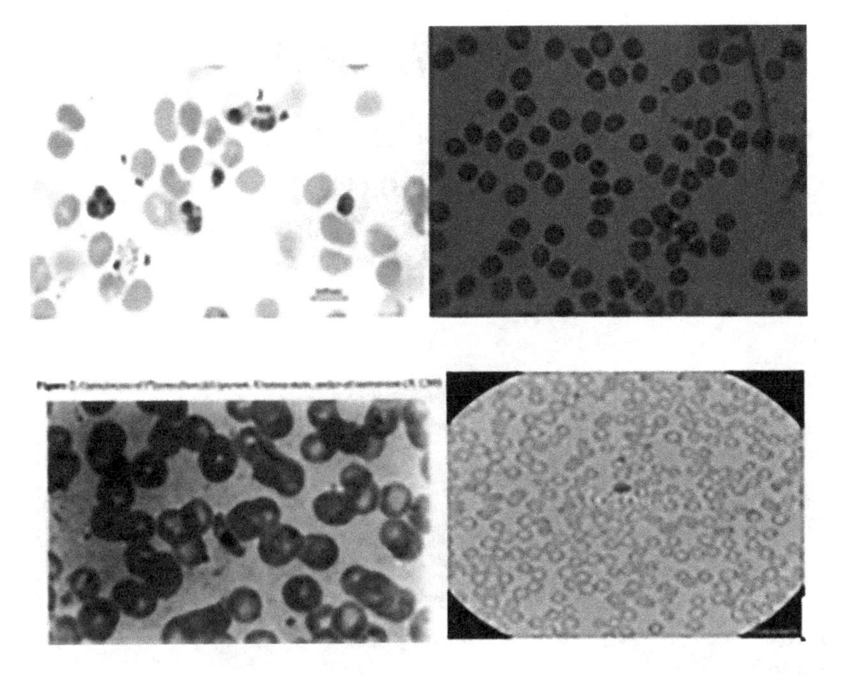

FIGURE 10.2
Diverseness in the different blood samples collected.

original image. For example, to read the original image, simply call the imread function in the cv2 module. As we can easily distinguish by seeing the blood smear image, there is some noise in the original image. That's why the denoising technique has become an important step to apply to sample images. To denoise, different algorithms have been developed, such as a median filter, Gaussian filter, Wiener filter, etc. As in our work, we have used the Cv2.fastNlMeansDenoisingColored () function provided by the cv2 module to denoise the colored image different blood sample diversity is shown in Figure 10.2.

```
Pseudo code:
noised_image=img.imread(r'''file_path''') de_noised image=cv2.fastNIMeans
DenoisingColored(noised_image,None,par1,par2,par3,par4) plt.show()
```

10.3.1.2 RGB to Gray/Binary Conversion

After denoising, our next step is to convert images from RGB scale to grayscale. To convert to gray, we simply call the cvtColor function, which permits us to convert an image from the current color space to another. This function receives the original image as the first input and the color code as the second input. Since we want to convert our original image to gray color space, we use the code COLOR_BGR2GRAY Figures 10.3 and 10.4 show RGB and Grayscale images.

```
Pseudo code:
Imag=cv2.imread(r'''file path''',0)
Gray=cv2.cvtcolr(Imag,cv2.COLOR_BGR2 GRAY
Cv2.imshow('RGB SCALED IMAGE',Imag) Cv2.imshow('GRAY SCALED IMAGE',Grey1)
```

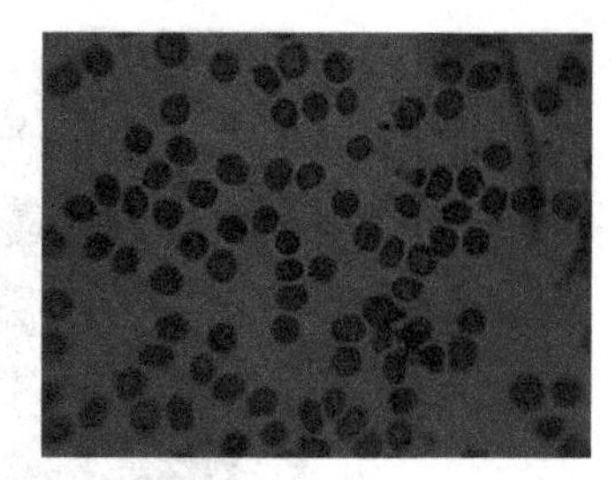

FIGURE 10.3
RGB color space.

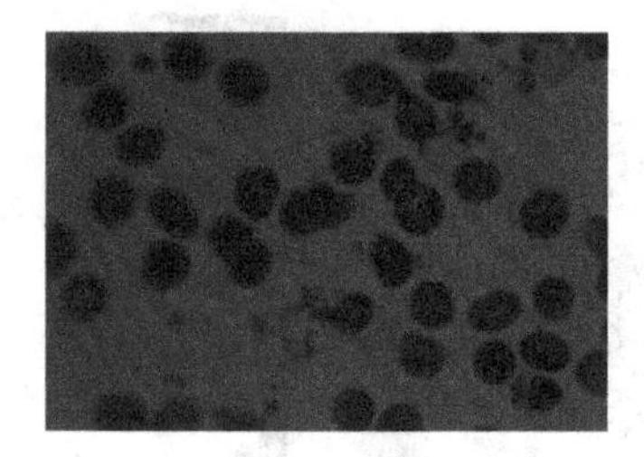

FIGURE 10.4
Grayscale color space.

10.3.2 Segmentation

Picture segmentation is the division of a picture into regions or classes that relate to various items or parts of articles. Every pixel in a picture is dispensed to one of some of these classes.

The principal objective of the division is to unmistakably separate between the article and the foundation in a picture. Essentially, we divide and separate our pictures into two significant parts. For example,

a. Region of interest
b. Nonarea of interest

Steps included in picture segmentation:

A masked picture is essentially a picture in which a portion of the pixel power esteems is zero, and others are nonzero.

10.3.2.1 Watershed Segmentation

Segmented image is shown in Figure 10.5. From the masked picture, we understood that there were still some coterminous issues and we used the watershed division to take care of this issue. Watershed is a region-based technique that makes use of image morphology. Basically works by extracting background from the foreground image and then applying markers for the detection of boundaries.

10.3.3 Feature Extraction

The central purpose of this step is to extract the main features from the preprocessed images. These are the features that will help us to distinguish between the parasitized

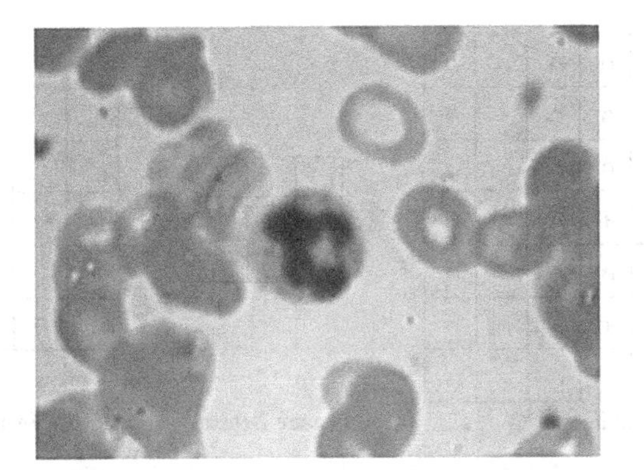

FIGURE 10.5
Segmentation applied on masked image.

cell and the nonparasitized cell. There are two approaches to feature extraction: Feature Extraction and Feature Selection. The shape and size of the infected cells are some of the features that help the classifier give us the desired results.

10.3.4 Classification

A simple neural network comprises an input layer, where the input of any image, text, or entity is fed in, and a concealed layer, where counterfeit neurons take in a lot of weighted data sources and produce a yield through an activation function, which can be a sigmoid or a logistic function, a ReLu or a rectified linear unit function, or a hyperbolic tangent function. For the sake of simplicity, we go with ReLu as it deals with the vanishing gradient problem, while sigmoids have slow convergence, which is not of much benefit to the activation layer.

After the image-processing part, where the RGB is converted to binary/grayscale using the cv2 module in Python and then using watershed segmentation, we arrive at the part where we put CNN to use.

Our work is divided into four steps, which finally will provide an output to confirm whether our model is accurate. The four basic steps are as follows:

10.3.4.1 Convolution

The first step that we apply to the input image is convolution, as CNN deals with computer vision, i.e., images unlike the newly developed RNNs, i.e., Recurrent Neural Networks. It is one of the categories of ANNs usually used for speech identification and natural language processing (NLP) programs. RNNs are equipped to identify data's sequential characteristics and use patterns to anticipate the next likely scenario, or the output according to the input fed in.

We create as many feature maps as we can so that we get our first proper convolutional layer, which can also be known as a convolution matrix (Figure 10.7).

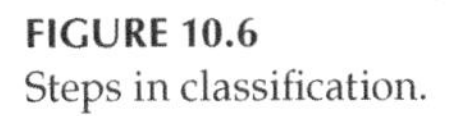

FIGURE 10.6
Steps in classification.

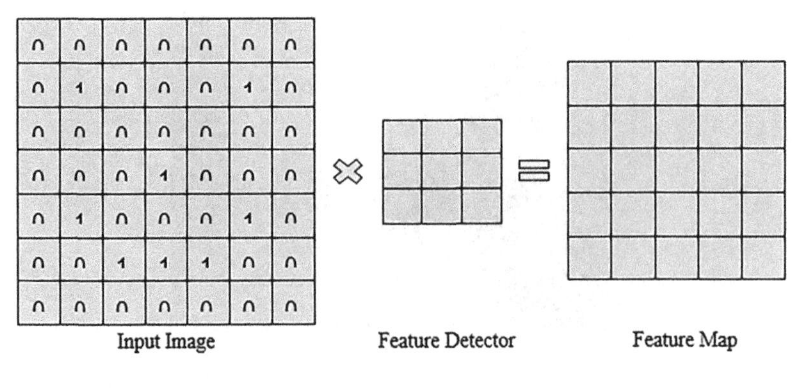

FIGURE 10.7
Convolutional matrix.

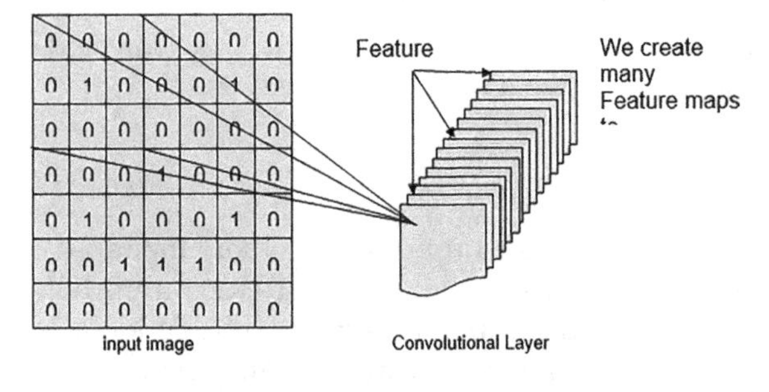

FIGURE 10.8
Working of convolution layer.

Figure 10.8 shows how it looks virtually.

Now, the reason why we're applying the rectifier is that we want to increase nonlinearity in our image or our network, and our CNN and our rectifier act as that filter or act as the function that breaks up the linearity; and the reason why we want to increase nonlinearity in our network is that images themselves are excessively nonlinear, particularly if you're recognizing different objects next to each other or in the background.

10.3.4.2 Max Pooling

The second step is max pooling. Max pooling is a sample-based Discretization algorithm used to minimize and break down the key features of an image. The intent is to undersample the input representation (image, hidden-layer output matrix, etc.), abbreviating its extensity and granting it for presumptions to be made about highlights accommodated in the sublocales, making it one of the most helpful and important techniques for identifying the main features in the input entity.

It is finished by administering a maximum channel or highlight extractor to (as a rule) uncover subdistricts of the underlying portrayal of any picture or lattice.

Suppose we have a 4×4 framework speaking to our first info. Suppose, also, that we have a 2×2 channel that we'll pass over our information utilizing the convolution system. We have a walk of 2 and won't cover districts of the information framework.

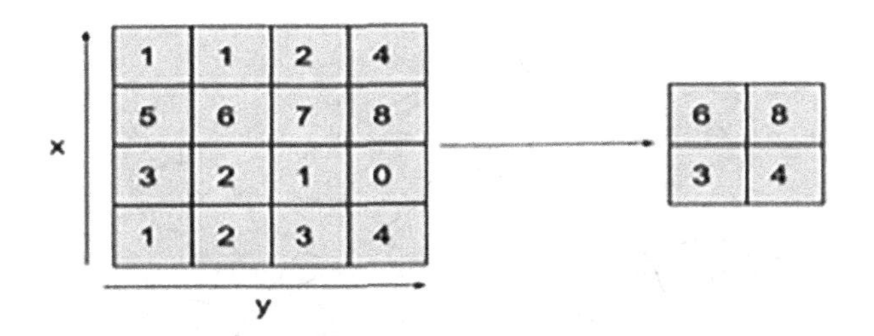

FIGURE 10.9
Max pool with 2×2 filters and stride.

For every one of the areas spoken to by the channel, we will take the maximum of that district and make another yield lattice, where every component is the maximum of a locale in the first information.

10.3.4.3 Flattening

Flattening is very basic and small, but it is very important and needs to be attended to. And so what we are going to do is take the pooled feature map and flatten it into a column. The numbers are taken row by row and put into one long column. And the reason for that is because we want to later input this into an ANN for further handling, which is a normal ANN and the final step in our approach.

10.3.4.4 Full Connection

Now, the full connection step is where we input the flattened column into the input layer of the artificial neural network that we have defined for ourselves. We define our classifier and assign hidden layers and output layers to get the metrics or the result.

```
classifier. Add (Dense (activation = 'relu', units = 128)) classifier.
Add (Dense (activation = 'sigmoid', units = 1))
```

This is one way of writing the code for the ANN.

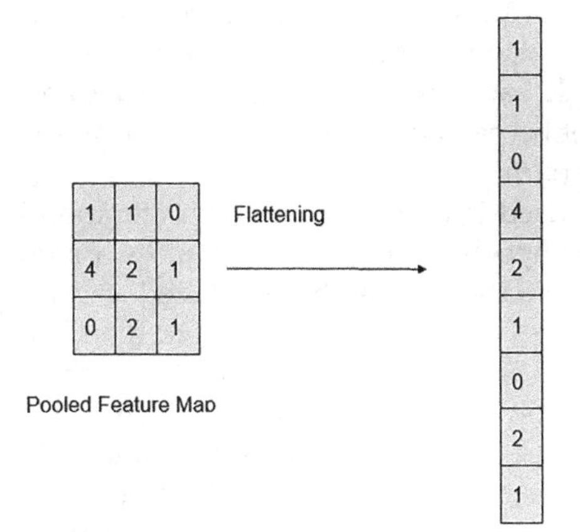

FIGURE 10.10
Flattening.

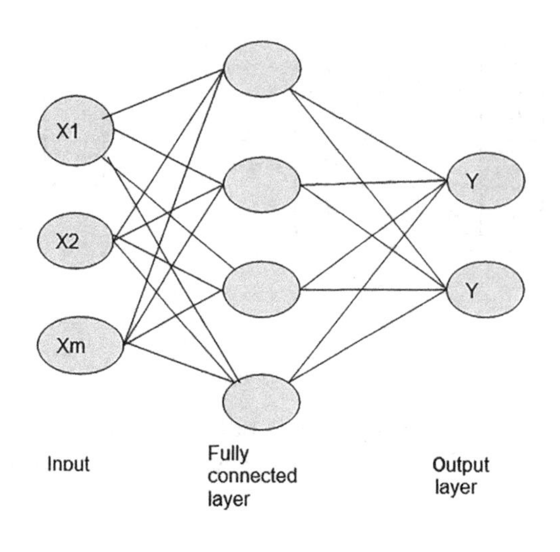

FIGURE 10.11
Full connection layer.

10.4 Results

In our study, the number of epochs is 25, with a batch size of 250 being used on a dataset of 1,364 images. Since one epoch is too big to feed to the computer at once, we divide it into smaller batches to ease up the efficiency and fasten the process of training our model. We are dealing with val_loss and loss, which are two different entities in nature. Val_loss is the value of the cost function for our cross-validation data, and loss is the value that we are dealing with for the cost function in our data. Also, val_acc is the accuracy that has been computed on the validation set, which is data that have not been seen by the computer. As in our training and test data, we are trying to predict how accurate our data are; so, to make our model estimate and generalize new data, validation accuracy is looked upon because we use the validation split. An increase in the validation accuracy determines how good one's model is, therefore we pick the model with the highest validation accuracy of the given epoch (Figure 10.12: Metrics 1 and Figure 10.13: Metrics 2). Also, there is a term called batch size, which is also important and used in metrics. The instances of training samples present in a single batch are termed as batch size, and meanwhile, batches used as a whole dataset cannot be passed as neural nets at once, so we divide the dataset into several batches or sets or parts.

One other term that is used is known as iteration. Iteration is the instances of batches required to complete one epoch. Our study introduces a method of detecting malaria-infected cells from a large dataset of both parasitized and nonparasitized blood smear images. The method involves enhancing the acquired blood smear images through various image-processing techniques, extracting the distinguishing features through feature extraction methods, and applying CNNs for classification and final results. A total of 1,364 (approx. 80,000 cells) pictures were utilized to prepare and test the classifier [15]. The proposed software yielded an accuracy of 89.60% in the detection of the *Plasmodium* parasite. This computerized system will not only provide better results than the manual methods but can also help in less-developed areas where proper medication and testing facilities are scarce.

```
Epoch 1/25
250/250 [==============================] - 419s 2s/step - loss: 0.6863 - acc: 0.5426 - val_loss:
0.6734 - val_acc: 0.5740
Epoch 2/25
250/250 [==============================] - 336s 1s/step - loss: 0.6370 - acc: 0.6342 - val_loss:
0.6050 - val_acc: 0.6790
Epoch 3/25
250/250 [==============================] - 297s 1s/step - loss: 0.5880 - acc: 0.6897 - val_loss:
0.5425 - val_acc: 0.7450
Epoch 4/25
250/250 [==============================] - 366s 1s/step - loss: 0.5408 - acc: 0.7328 - val_loss:
0.4894 - val_acc: 0.7615
Epoch 5/25
250/250 [==============================] - 382s 2s/step - loss: 0.4910 - acc: 0.7665 - val_loss:
0.4439 - val_acc: 0.8025
Epoch 6/25
250/250 [==============================] - 374s 1s/step - loss: 0.4568 - acc: 0.7817 - val_loss:
0.4361 - val_acc: 0.8040
Epoch 7/25
250/250 [==============================] - 368s 1s/step - loss: 0.4195 - acc: 0.8073 - val_loss:
0.3997 - val_acc: 0.8190
Epoch 8/25
250/250 [==============================] - 329s 1s/step - loss: 0.3903 - acc: 0.8257 - val_loss:
0.3941 - val_acc: 0.8285
Epoch 9/25
250/250 [==============================] - 292s 1s/step - loss: 0.3736 - acc: 0.8313 - val_loss:
0.3952 - val_acc: 0.8150
Epoch 10/25
250/250 [==============================] - 341s 1s/step - loss: 0.3375 - acc: 0.8515 - val_loss:
0.3730 - val_acc: 0.8290
Epoch 11/25
250/250 [==============================] - 322s 1s/step - loss: 0.3209 - acc: 0.8592 - val_loss:
0.3351 - val_acc: 0.8655
Epoch 12/25
250/250 [==============================] - 376s 2s/step - loss: 0.3039 - acc: 0.8694 - val_loss:
0.3505 - val_acc: 0.8450
```

FIGURE 10.12
Metrics 1: Accuracy of epoch.

10.5 Conclusion

Our study introduces a method of detecting malaria-infected cells from a large dataset of both parasitized and nonparasitized blood smear images. The method involves enhancing the acquired blood smear images through various image-processing techniques, removing the noise from images, extracting the distinguishing features through feature extraction methods, and applying CNNs for classification and final results. Convolutional Networks are used because they can adequately catch the Spatial and Temporal conditions in a picture through the use of pertinent channels. The construction lays out chief fitting to the pictures because of the depletion in the number of parameters included and reclaimable of loads. A total of 1,364 (approx. 80,000 cells) pictures were utilized to prepare and test the classifier. The proposed software yielded an accuracy of 94.18% in the detection of the *Plasmodium* parasite. This computerized system will not only provide better results than the manual ways but can also help in less-developed areas where proper medication and testing facilities are scarce.

```
Epoch 13/25
250/250 [==============================] - 359s 1s/step - loss: 0.2836 - acc: 0.8728 - val_loss:
0.3549 - val_acc: 0.8525
Epoch 14/25
250/250 [==============================] - 296s 1s/step - loss: 0.2681 - acc: 0.8848 - val_loss:
0.4089 - val_acc: 0.8355
Epoch 15/25
250/250 [==============================] - 308s 1s/step - loss: 0.2423 - acc: 0.8998 - val_loss:
0.3490 - val_acc: 0.8585
Epoch 16/25
250/250 [==============================] - 302s 1s/step - loss: 0.2330 - acc: 0.9010 - val_loss:
0.3456 - val_acc: 0.8670
Epoch 17/25
250/250 [==============================] - 351s 1s/step - loss: 0.2221 - acc: 0.9083 - val_loss:
0.4071 - val_acc: 0.8525
Epoch 18/25
250/250 [==============================] - 326s 1s/step - loss: 0.2037 - acc: 0.9164 - val_loss:
0.3288 - val_acc: 0.8625
Epoch 19/25
250/250 [==============================] - 300s 1s/step - loss: 0.1913 - acc: 0.9189 - val_loss:
0.3656 - val_acc: 0.8710
Epoch 20/25
250/250 [==============================] - 299s 1s/step - loss: 0.1857 - acc: 0.9250 - val_loss:
0.3292 - val_acc: 0.8730
Epoch 21/25
250/250 [==============================] - 293s 1s/step - loss: 0.1675 - acc: 0.9325 - val_loss:
0.3302 - val_acc: 0.8745
Epoch 22/25
250/250 [==============================] - 296s 1s/step - loss: 0.1667 - acc: 0.9323 - val_loss:
0.3543 - val_acc: 0.8720
Epoch 23/25
250/250 [==============================] - 300s 1s/step - loss: 0.1621 - acc: 0.9339 - val_loss:
0.3367 - val_acc: 0.8705
Epoch 24/25
250/250 [==============================] - 295s 1s/step - loss: 0.1476 - acc: 0.9381 - val_loss:
0.3798 - val_acc: 0.8815
Epoch 25/25
250/250 [==============================] - 294s 1s/step - loss: 0.1444 - acc: 0.9418 - val_loss:
0.3633 - val_acc: 0.8960
```

FIGURE 10.13
Metrics 2: Accuracy of epoch.

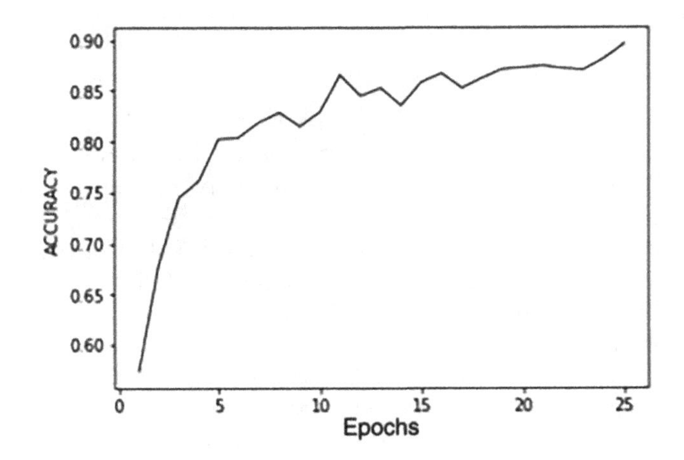

FIGURE 10.14
Epoch versus Accuracy.

10.6 Future Scope

In further work in the project, we can apply Big Data algorithms to accommodate the large datasets that will be used if the system is used in real-world applications. Also, other Machine Learning and Deep Learning algorithms can be used and combined to improve the accuracy of the system.

References

Bashir, A., Mustafa, Z. A., Abdelhameid, I. and Ibrahem, R. (2017) Detection of malaria parasites using digital image processing. In 2017 International Conference on Communication, Control, Computing and Electronics Engineering (ICCCCEE), pp. 1–5. IEEE.

Bibin, D., Nair, M. S. and Punitha, P. (2017) Malaria parasite detection from peripheral blood smear images using deep belief networks, *IEEE Access*, 5, pp. 9099–9108.

Chahal, E. S., Haritosh, A., Gupta, A., Gupta, K., Sinha, A. (2019) "Deep learning model for brain tumor segmentation & analysis," IEEE 3rd International Conference on Recent Developments in Control, Automation & Power Engineering (RDCAPE), pp. 378–383.

Dave, I. R. (2017) Image analysis for malaria parasite detection from microscopic images of thick blood smear. In 2017 International Conference on Wireless Communications, Signal Processing and Networking (WiSPNET), pp. 1303–1307. IEEE.

Deep, P., Shukla, S., Pandey, E., Sinha, A. (2019) "Detection of Abnormalities in Blood Sample using WBC Differential Count," Springer 9th International Conference on Soft Computing for Problem Solving (SocProS), AISC Springer, vol. 1, pp. 227–240.

Di Ruberto, C., Dempster, A., Khan, S. and Jarra, B. (2002) Analysis of infected blood cell images using morphological operators, *Image and Vision Computing*, 20(2), pp. 133–146.

Kumar, G. and Bhatia, P. K. (2014) A detailed review of feature extraction in image processing systems. In 2014 Fourth international conference on advanced computing & communication technologies, pp. 5–12. IEEE.

Kumar, N. and Nachamai, M. (2012) Noise removal and filtering techniques used in medical images, *Indian Journal of Computer Science and Engineering*, 3(1), pp. 146–153.

Makler, M. T., Palmer, C. J. and Ager, A. L. (1998) A review of practical techniques for the diagnosis of malaria, *Annals of Tropical Medicine and Parasitology*, 92(4), pp. 419–434.

Nugroho, H.A. "P. vivax (malaria) infected human blood smears" obtained from https://data.broadinstitute.org/bbbc/BBBC041/.

Nugroho, H. A., Akbar, S. A. and Murhandarwati, E. E. H. (2015) Feature extraction and classification for detecting malaria parasites in thin blood smear. In 2015 2nd International Conference on Information Technology, Computer, and Electrical Engineering (ICITACEE), pp. 197–201. IEEE.

Olugboja, A. and Wang, Z. (2017) Malaria parasite detection using different machine learning classifiers. In 2017 International Conference on Machine Learning and Cybernetics (ICMLC), Vol. 1, pp. 246–250. IEEE.

Parashar, R., Goel, M., Sharma, N., Jain, A., Sinha, A., Biswas, P. (2020) "Discovering mutated motifs in DNA Sequences: A comparative analysis," In: Bansal, P., Tushir, M., Balas, V., Srivastava, R. (Eds.) Proceedings of International Conference on Artificial Intelligence and Applications (ICAIA 2020). Advances in Intelligent Systems and Computing, Springer, Singapore, volume 1164, pp. 257–269.

Pattanaik, P. A., Swarnkar, T. and Sheet, D. (2017) Object detection technique for malaria parasite in thin blood smear images. In 2017 IEEE International Conference on Bioinformatics and Biomedicine (BIBM), pp. 2120–2123. IEEE.
Peñas, K. E. D., Rivera, P. T. and Naval, P. C. (2017) Malaria parasite detection and species identification on thin blood smear using a convolutional neural network. In 2017 IEEE/ACM International Conference on Connected Health: Applications, Systems and Engineering Technologies (CHASE), pp. 1–6. IEEE.
Saraswat, S., Awasthi, U. and Faujdar, N. (2017) Malarial parasites detection in RBC using image processing, In 2017 6th International Conference on Reliability, Infocom Technologies and Optimization (ICRITO), pp. 599–603. IEEE.
Tangpukdee, N., Duangdee, C., Wilairatana, P. and Krudsood, S. (2009) Malaria diagnosis: a brief review, *The Korean Journal of Parasitology*, 47(2), p. 93.
World Health Organization Report (2018), available at https://www.who.int/malaria/publications/world-malaria-report-2018/en

11

Handwritten Text Recognition with IoT Devices

Omkareshwar Tripathi and Mayank Tiwari
G L Bajaj Institute of Technology and Management

Anish Gupta
Chandigarh University

CONTENTS

11.1 Introduction

Understanding data has become easier these days with the advancement of technology. Several techniques can be used to understand, manipulate, and analyze data. These techniques involve IoT, Machine Learning, Data Science, Deep Learning, and much more. IoT in simple words can be defined as the system of computing devices that are interconnected over the Internet. These devices collect and share data and can also transfer data over the network without any external interference. The IoT generates humongous data from the sensors attached to the machine or from the words we speak daily in front of these IoT devices (smart speakers, e.g., Alexa, Google Home, etc.). An IoT device comes in all shapes and sizes, from a Smartphone to self-driving cars. IoT has grown over the years and different industries are trying to use this in the best way possible. There has been a significant rise in the number of businesses that use IoT technologies, i.e., from 13% (2014) to 25% (today). With a greater number of people having easy access to technology, the number of IoT-connected devices worldwide is projected to be around 43 billion by 2023.

DOI: 10.1201/9781003046431-11

Data Science is a multifaceted field that provides different techniques, such as data mining, and uses different visualization techniques that help in gaining some knowledge about the data. Machine Learning provides different algorithms/techniques for understanding and visualizing data. Therefore, the amount of data generated by the IoT devices requires filtering and analysis, which can easily be done by applying Data Science and Machine Learning.

And the question that arises here is, why is the IoT important? The answer is quite simple: the Internet is all over the world and the number of devices connected and using the services of the Internet are enormous. IoT devices are being widely used in many different industries, for example, they can be used to monitor one's health condition, used in agriculture, etc. Earlier, the local computer system with sensors attached to them was used and was controlled by an embedded module. IoT devices can also be used to sense the surroundings in many ways such as monitoring pressure, viscosity, temperature, etc. IoT allows us to transmit the data into the cloud at a component level. And coming back to Machine Learning, it is a research field at the intersection of statistics, artificial intelligence (AI), and computer science. The systems built on Machine Learning algorithm learn from the data provided, i.e., data that is derived from past experiences, and after learning the trend and behavior of past data, the machine learning model predicts about the future trend. Machine learning is subcategorized into three types:

11.1.1 Supervised Learning

In supervised learning, the machine learning model is provided a dataset that is used to train the model. Once the model gets trained, future trends can be predicted when new data are given as input. The predictions made by the model are based on the learning from the dataset provided. Supervised learning deals with labeled data and the computational complexity of supervised learning is high. Supervised learning makes use of regression and classification methods for building predictive models. The algorithms are Linear Regression, Logistic Regression, Decision Tree, Naive Bayes, Support Vector Machine, K-Nearest Neighbor, and many more (Figure 11.1).

11.1.2 Unsupervised Learning

Unsupervised learning deals with unlabeled data. In unsupervised learning, the machine learning model learns by observing patterns in the data. Based on these observations and patterns, the model creates clusters in it. What it cannot do is label the cluster or it cannot

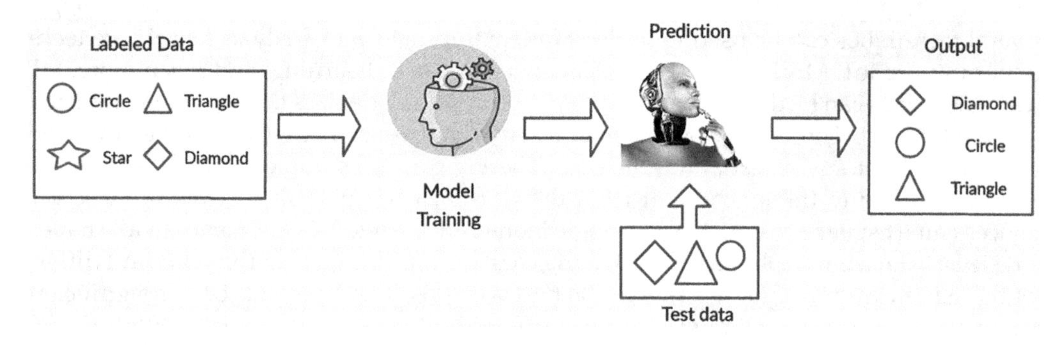

FIGURE 11.1
Supervised learning.

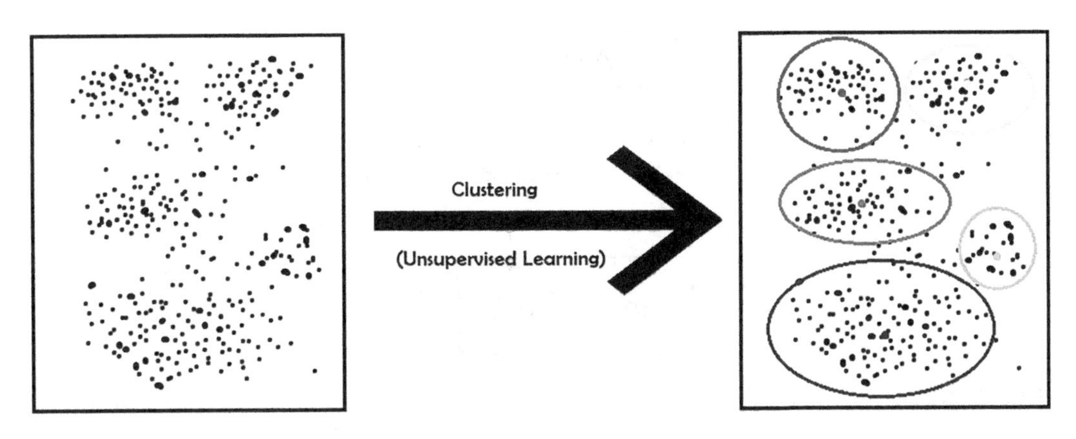

FIGURE 11.2
Unsupervised leaning using clustering.

TABLE 11.1
Difference between Two Types of Learning

Supervised Learning	Unsupervised Learning
1. Deals with labeled data.	1. Deals with unlabeled data.
2. Model is trained using the labeled data.	2. Model learns by observing patterns in the data.
3. Computational complexity is high.	3. Computational complexity is low.
4. Uses regression and classification methods for prediction.	4. Uses clustering methods for predicting new data.

say, for example, whether it is a group of cats. When we pass the image of a cat or a dog to the model, it will add the cat's images to the cluster of cat and the dog images to the not-cat cluster. In unsupervised learning, the clustering algorithms that are used are Hierarchical clustering, *k*-means clustering, and Principal Component Analysis. Unsupervised learning also includes association rule-mining problems. The computational complexity of unsupervised learning is low.

- **Clustering:** For unsupervised learning, clustering plays a vital role. It usually finds patterns or similarities in a collection of unclassified data. Clustering algorithms will execute data and find natural clusters (groups) if they are present in existing data. Modification can be done in an algorithm for identifying the number of clusters present in the data (Figure 11.2).

The difference between supervised learning and unsupervised learning is summarized in Table 11.1.

11.1.3 Reinforcement Learning

The reinforcement learning agent learns by interacting with the environment and finds what the best outcome is. A point is rewarded or penalized for a correct or a wrong answer and the model trains itself based on the points rewarded for a correct answer. And after training, it is ready to predict the new data when given to it. The problem of Split Delivery

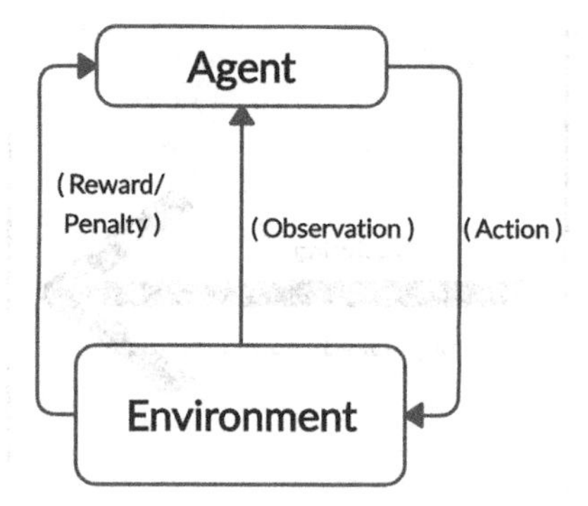

FIGURE 11.3
Working of a reinforcement agent.

Vehicle Routing can be solved by using reinforcement learning. It is different from supervised learning in a way that in supervised learning, the model is trained with data that also contains the output, whereas, in reinforcement learning, the model/agent decides on its own what to do to perform the given task. Reinforcement learning is bound to learn from experience in the absence of training data (Figure 11.3).

The use of IoT devices has rapidly increased over the years. IoT devices are very useful in our day-to-day lives. IoT is surely the next frontier in the race for its share in the market with wireless networks and revolutionary computing capabilities. There are numerous applications of the IoT. IoT is used in different sectors or fields such as Healthcare, Industrial Automation, Smart Cities, Agriculture, Home Applications, Cars, and many more. The applications of IoT in all the above-mentioned fields are explained as follows:

- **Healthcare**: IoT has extensive applicability in the Healthcare sector. IoT provides unexampled advantages, which could improve the efficiency and the quality of treatments. This involves various health-monitoring devices (wearables), which can perform real-time monitoring of an individual. Devices can be in the form of fitness bands or watches or some other kind of device, which is capable of monitoring the heart rate, blood pressure, glucometer, etc., which will help to understand the status of an individual in a better way. Data that are collected from the IoT devices (monitoring devices) can be analyzed and used for the future in predicting the state of any individual at an early stage. From monitoring patients' health to tracking real-time location (by tagging the sensors with IoT devices) of medical equipment such as nebulizers, wheelchairs, and other monitoring equipment, IoT devices are of great use to hospitals or the healthcare sector.
- **Industrial Automation**: In the industry of automation, the IoT is proving to be a game-changer. To the automation industry, IoT plays an important role as IoT devices are effective, affordable, and responsive in creating system architectures. IoT-based solutions add value to industrial equipment.
- **Smart Cities**: With the enhancement in technologies, smart cities are in buzz these days. In layman's terms, a smart city is a city where technologies are used to improve the lives of the citizens and the businesses that inhabit it. IoT for a

smart city can be used in various areas such as monitoring traffic in a particular area, Street Lighting, Waste Management, Smart Parking, Public Safety, etc. IoT provides various opportunities for cities to use data to cut pollution, keep citizens safe, and manage traffics.

- **Smart Home Applications**: Do you wonder if you could control the lights, air-conditioner, and other electrical appliances straight away from your Smartphone? Well, because of IoT this is all possible. This can help in saving energy as it will help to turn off the appliances after a certain amount of time. IoT for a smart home can also be used for security purposes, where any intrusion will be informed on your Smartphone.
- **Agriculture**: IoT in agriculture can be termed as smart farming, where devices can be used to monitor the crop by using various sensors for humidity, light, temperature, and acidity of the soil, etc. Smart farming, when compared to the conventional approach, turns out to be better than the latter. Farmers can monitor the field and crops from anywhere with the help of IoT devices.
- **Cars**: IoT is effectively used in the automobile industry nowadays. It is changing the user requirement and style of living. Automobile industry is effective in using IoT applications in their products, for example, in-car navigation, telematics, and entertainment. Rear-view cameras and proximity sensors are used for blind-spot detection and assist in easier parking.

Another term associated with the data is Deep Learning. Deep learning is a subset of machine learning in AI, which is capable of learning from unlabeled or unstructured data. Netflix movie recommendation system, Facebook friend's suggestion, or songs recognized by apps uses the concept of Deep Learning. Deep Learning uses multiple layers between the input and output layers. Each mathematical manipulation is considered a layer. Complex sensing and recognition tasks can be performed by implementing deep learning with IoT. This could bring a generation of applications that can initiate a new realm of interactions between humans and their physical surroundings.

IoT is a wonderful technology and has countless applications and advantages. It is a boon to society by making our day-to-day lives easier and smoother. Machine Learning and IoT integration can be of great use for further development in certain areas, and in coming years, we can expect that security and privacy might be resolved.

Data are generated heavily regularly, and therefore it is important to analyze these data, and by implementing various data analysis and visualization techniques, we can make the best out of these data. IoT devices generating data and by implementing Machine Learning and Deep Learning techniques properly, a great contribution can be made to the required sectors (Figure 11.4).

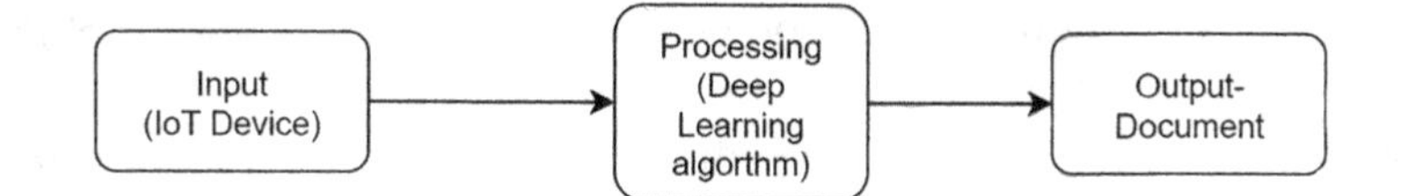

FIGURE 11.4
Analytical process.

11.2 Related Work

IoT is a great field as it provides plenty of areas to work in. And the areas or sectors to work in just significantly increases when IoT combines its application with Machine Learning and Deep Learning. IoT has great applications and some of the works that have been conducted in this sector are as follows:

- Mano et al. (2016) mentioned in their work how IoT devices can be used to identify patient identity and emotion that will enhance healthy smart homes. To help patients and elderly people in an in-house Healthcare context, patient images, and emotional identification can be used. His work also discusses that previous or current studies in this area do not use patient images to monitor patients.
- Su et al. (2018) proposed a lightweight approach to detect DDoS (distributed denial-of-service) malware in the IoT environments. First, extraction of one-channel gray-scale images from binaries is performed, and then, a convolutional neural network (CNN) (lightweight) is used to classify IoT malware families. The outcomes tender that the proposed system attains an accuracy of 94.0% for the classification of good ware and DDoS malware, and an accuracy of 81.8% for the classification of good ware and the two main malware families.
- The concept of musical chairs for real-time image identification using IoT devices was used by Hadidi et al. (2018). This musical chair is an approach by which Deep Neural Network models are distributed across multiple devices. It alters the available computing devices at runtime and accommodates the underlying dynamics of the IoT network. The musical chair is applied to two known image identification models (Alex Net and VGG 16) and then to the Raspberry PIS network. This approach demonstrates that it (the collaboration of IoT devices with the musical chair) achieves the same real-time performance without the additional cost of maintaining the server.
- An IoT-based image identification system was used by Tseng et al. (2017) to design a multicultural service to enhance the command of image identification. Statistical histogram-based *k*-means clustering was the basic technique used in this approach for image segmentation. However, it is a time-consuming method. Therefore, to obtain the PDF of pixels and to divide these weights to equal intensity, the use of statistical histograms has been proposed. *k*-Means clustering, histogram, and *k*-means were combined to overcome the high computational cost for clustering. The results showed that the recommended method is more reliable than *k*-means clustering because it is ten times rapid than *k*-means clustering.
- In their work, Diro and Chilamkurti (2018) used intensive learning to detect attacks on the social IoT. The performance of traditional machine learning approaches and deep learning models is compared and distributed attack detection is evaluated against a centralized detection system. This experiment shows that using an intensive learning model, a distributed attack detection system is superior to a centralized detection system.
- To identify a healthy brain, an ischemic stroke, or a hemorrhagic stroke, an IoT framework with a CNN was presented by Dourado Jr. et al. (2019) for analysis of strokes from CT-scanned images. This approach enables automation of the diagnosis procedure because it can capture information that becomes unwanted from the human eye.

- Meidan et al. implemented various machine learning algorithms on network traffic data to identify IoT devices that are connected to a network. From nine diverse IoT devices and personal computers and Smartphones, data were collected to prepare and estimate the classifier. A multiple-stage Meta classifier is trained using supervised learning. The classifier can differentiate between traffic generated using IoT and non-IoT devices in the first stage. In the second phase, each IoT device is associated with a specific IoT device class. The analysis accuracy obtained by the model is 99.28%.
- Because of wastage and suboptimal prices, around 50% of the farm produce never reaches the end consumer. In their work, Shenoy and Pingle (2016) provided the solution for transport cost, predictability of prices based on past data, and current market conditions, minimizing the agents between the farmers and the end consumers by using an IoT-based solution.
- To provide advanced and intelligent services to the IoT that integrate into the network with various devices, such as user privacy and address attacks (e.g., spoofing attacks, denial of jamming and service attacks, and many more) have to protect. Xiao et al. (2018) reviewed IoT security solutions, which are based on machine learning techniques, including supervised learning, unusable learning, and reinforcement teaching. The challenges that require attention to implement the above-mentioned machine learning technology-based security schemes in IoT systems are discussed in this chapter.
- An IoT-based plant-infected leaf segmentation and identification method is proposed. It is based on the following algorithms, namely, *k*-means clustering, a fusion of super-pixel clustering, and PHOG. First, the super-pixel clustering algorithm splits the color diseased leaf image into a few compact super-pixels. Then, *k*-means clustering is used to segment the lesion image from each super-pixel. And finally, the PHOG attributes are obtained from the three color components of each segmented bruise image and its grayscale image and combine the four PHOG descriptors as a vector. A possible solution for plant diseased leaf image segmentation and recognition has been provided by Zhang et al. (2018).

11.3 Observations

It has been observed that in the coming years, IoT will play an integral part in one's day-to-day life as well as in various sectors such as Healthcare, building smart cities, industrial advancement, etc. IoT is a great deal in automation and people are adopting smart devices faster than ever. The IoT devices generate a lot of data daily, and applying different machine learning algorithms to the generated data can help in making these IoT devices smarter than before. The number of IoT devices over the years has increased exponentially. There were 15.41 billion IoT-connected devices in 2015 and it increased to 26.66 billion IoT-connected devices in 2019 and is expected that there will be 75 IoT devices by the end of 2025. Also, 127 devices are connected to the Internet per second. The above statistics clearly states that IoT certainly has a future. The researchers aimed at exploring different sectors or areas where IoT can be implemented and how it can be used for the betterment of society.

Different devices are being built in the Healthcare sector to monitor the individual's health and any unusual symptoms can be detected at an early stage. With time these devices are also getting better and giving results with better accuracy. IoT can be integrated with Deep Learning to give better results in the field of image recognition. CNNs can be used with IoT devices for better results.

Hadidi et al. (2018) used the concept of Deep Neural Network with IoT for image recognition. And to improve the success rate for image recognition using IoT devices and Deep Learning can be used together by altering the learning rate or the hyperparameter tuning or by increasing layers in the Deep Learning algorithms. IoT is the future and will surely do a great deed for humankind.

11.4 Open Challenges

Coupled with the advent of deep learning, IoT devices are getting better and efficient. By incorporating image recognition techniques with IoT devices the scope of IoT becomes even broader. Image processing is an integral part of an IoT device to work more efficiently and can solve many of our problems and to incorporate image processing with these devices there are many steps involved. Image processing involves many layers of proceeding. These proceedings are image accusation, preprocessing and analysis, image segmentation, and classification. Now to replace the use of paper with these techniques, we first need to address the problem with the authenticity of the document. Not to perform all these operations consecutively needs fast processing at a low cost, so that it is available to a larger audience. An IoT device can perform these operations. IoT devices are reliable and due to their modularity, we can add external equipment to authenticate the data such as biometrics, e-signature, watermarking, etc.

The greatest challenge in this area is to automate the whole process so that everything becomes faster and more reliable and to do that we need lots and lots of data to train models and make them more reliable. To build such a device requires high computational power. But to get that power at a low cost is very difficult. So, hardware and software must work simultaneously with each other. Both hardware and software should work in a chain and they should communicate with each other. The whole process should be very efficient and accurate. There are various aspects related to this problem. The main concern is of accuracy and reliability of the device. To get better accuracy, we need lots and lots of data, and for reliability, we need good hardware. With the current advancement in mobile technology, many of the challenges can be addressed. But proper implementation is required: everything should work in a chain and proper communication should be available between hardware and software.

11.5 Proposed Solutions

The use of IoT devices is increasing day by day. In many fields, a sustainable environmental approach can also be achieved with the use of IoT devices. There is a significant improvement in the technology of writing pads and many other similar products. Coupled

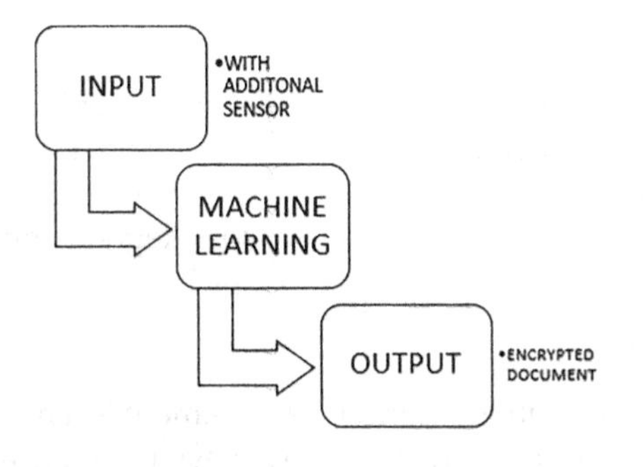

FIGURE 11.5
Working of proposed model.

with the advancement in machine learning, these writing pads can be used in place of paper in many fields.

Many image recognition algorithms, such as CNNs, can be used with these types of IoT devices. Algorithms such as natural language processing can be used to convert speech into text and they can be used to replace paper at various levels.

To authenticate the data, many techniques, such as biometric watermarking or e-signature, can be used to authenticate the data. These techniques can be used to encrypt the document to maintain its authenticity and used in place of paper. Figure 11.5 outlines the proposed framework incorporating the aforementioned aspects.

11.5.1 Input

For input, we can use many devices, such as digital writing pads, touchscreen displays, and so on. Now, the choice of the input device depends on various factors.

These factors are as follows:

- Choice of model
- Accuracy and reliability
- Ease of use
- Familiar interface

The input is very important as it will directly interact with the person of interest. Now, it is also important to consider how the input device will process the given input. It is very important for the input device to natively process the input as per the model requirements. If the input device requires an additional step to process the input as per requirement, then because of the additional step, the required processing power increases. Our main concern while designing the whole model should be that it requires less processing power so that it can work efficiently even on low-processing power devices. The lesser the processing, the better the model will perform, so it is very important to take processing as a concern while designing a model. Another factor to be considered is set of input which further can be processed on variety of individuals. The input device should be accurate and precise. It should sense inputs for fast and the scope of error should be very small. An input device should be capable enough to take inputs without any errors in it so it can be processed further.

11.5.2 Additional Sensors

Additional sensors with input devices are required to provide security to the input device. The addition of additional sensors can also increase the performance of the input device. These additional sensors can be of various types, for example, a biometric sensor, light sensor, scanner etc. The choice of additional sensors depends on the model.

11.5.3 Machine Learning

Now, to make this device possible, the main requirement is a machine learning model that can recognize the given input. There are many ways through which a machine learning model can be constructed, for example, a sci-kit learn model that can identify an image or a CNN model. The main preference should be on building an efficient and accurate model. Nowadays, deep learning models are performing well. Building a deep learning model for the device will help to make a better device. The focus while building the model should be more on performance. There are many parameters to a deep learning model. The parameters should be tuned in such a way that they perform well after deployment. These parameters are called hyperparameters, which help in tuning the model to perform and give better results on data that it has never seen. There are many ways to build and tune these parameters. The choice of the deep learning model influences the whole model, so it is better to try different models with various parameters. Many open-source libraries can be used to build this model. But while choosing the library, the user should keep in mind that the library should remain open source under good governance. There are many ways to build a deep learning neural network model from scratch, but choosing a library makes it easier to build a Deep Neural Network.

11.5.4 Output

The output from this device can be used in many ways. It can be directly stored in the database; it can also be used for further reference. But the output should also contain the real input from the user in the form of an image with the digital signature or with the biometric watermark. There are several ways to encrypt the document, so the choice depends on how the model is made and for what purpose it will be used.

11.5.5 Dataset

The proposed model uses two different datasets (English and Hindi) for the recognition of the characters, which are given as the input to the device. These datasets are stored in a local computer. The datasets are extracted from Kaggle (English Characters) &UCI Machine Learning Repository (Hindi Characters). EMNIST data are used for the English character dataset. The dataset contains all the English alphabets. The images are 32×32 pixel black and white images. For the Hindi (Devanagari) character dataset, the image resolution is 32×32 and the actual character is cantered within 28×28 pixels, padding of 2 pixels is added on all four sides of an actual character (Figures 11.6 and 11.7).

FIGURE 11.6
Hindi character dataset.

FIGURE 11.7
English character dataset.

FIGURE 11.8
Input device.

11.5.6 Implementation

The experimental setup for the proposed system is shown in Figure 11.8. For this device, a Raspberry Pi combined with touch input is being used. Touch input is programmed in such a way that it takes all the input from the user and converts it into a JPEG file and then that JPG file is then processed in the future to recognize what is written in that file. A CNN model is being used to recognize that the text Raspberry Pi is trained with that model. When the user gives input to the touch sensor that input is converted into a JPEG file. After converting input into a JPEG file that file is then stored in the database of the device. Now, after this, the JPG file is then used to recognize the text input given by the user. To recognize text from the JPG file, CNN model is being used. Now what happens is that model breaks down the image and recognizes what is written in the document by matching it to the dataset with which we have already trained it with. To recognize different scripts with this model, we can change the dataset, which is being used under the hood.

11.6 Conclusion

We have seen tremendous growth in machine learning in the past few years. Right now, the whole world is trying to develop a sustainable environmental approach, and through this study, we can use these techniques of machine learning and IoT devices to replace papers at places where it is least required. Just by incorporating these devices in an institution the amount of paper that is being used can be dropped at various levels. These techniques are

```
/62 [==============================] - 3s 48ms/step - loss: 0.0367 - acc: 0.9909 - val_loss: 0.0209 - val_acc: 0.9910
och 497/500
/62 [==============================] - 3s 48ms/step - loss: 0.0305 - acc: 0.9914 - val_loss: 0.0220 - val_acc: 0.9910
och 498/500
/62 [==============================] - 3s 48ms/step - loss: 0.0465 - acc: 0.9864 - val_loss: 0.0236 - val_acc: 0.9910
och 499/500
/62 [==============================] - 3s 48ms/step - loss: 0.0294 - acc: 0.9919 - val_loss: 0.0235 - val_acc: 0.9895
och 500/500
/62 [==============================] - 3s 48ms/step - loss: 0.0344 - acc: 0.9909 - val_loss: 0.0206 - val_acc: 0.9940
NFO] evaluating network...
              precision    recall  f1-score   support

           1       0.98      0.98      0.98        49
          10       1.00      1.00      1.00        46
          11       1.00      0.96      0.98        53
          12       1.00      1.00      1.00        59
           2       0.99      1.00      0.99        68
           3       0.98      1.00      0.99        51
           4       1.00      1.00      1.00        55
           5       1.00      0.98      0.99        57
           6       0.98      1.00      0.99        57
           7       1.00      1.00      1.00        55
           8       1.00      1.00      1.00        61
           9       1.00      1.00      1.00        52

    accuracy                           0.99       663
   macro avg       0.99      0.99      0.99       663
ighted avg       0.99      0.99      0.99       663

NFO] serializing network and label binarizer...

ase) C:\Users\omkar\Desktop\vovels digit recogination\Python_Hindi_char>python predict.py --image output/013_01.jpg --model output/smallvggnet.model --label-bin output/smallvggnet_lb.pickle --width 28 --hei
28
ing TensorFlow backend.
NFO] loading network and label binarizer...
RNING:tensorflow:From C:\Users\omkar\Anaconda3\lib\site-packages\keras\backend\tensorflow_backend.py:517: The name tf.placeholder is deprecated. Please use tf.compat.v1.placeholder instead.

RNING:tensorflow:From C:\Users\omkar\Anaconda3\lib\site-packages\keras\backend\tensorflow_backend.py:4138: The name tf.random_uniform is deprecated. Please use tf.random.uniform instead.

RNING:tensorflow:From C:\Users\omkar\Anaconda3\lib\site-packages\keras\backend\tensorflow_backend.py:245: The name tf.get_default_graph is deprecated. Please use tf.compat.v1.get_default_graph instead.

RNING:tensorflow:From C:\Users\omkar\Anaconda3\lib\site-packages\keras\backend\tensorflow_backend.py:174: The name tf.get_default_session is deprecated. Please use tf.compat.v1.get_default_session instead.

RNING:tensorflow:From C:\Users\omkar\Anaconda3\lib\site-packages\keras\backend\tensorflow_backend.py:181: The name tf.ConfigProto is deprecated. Please use tf.compat.v1.ConfigProto instead.

19-11-23 14:39:12.745612: I tensorflow/core/platform/cpu_feature_guard.cc:142] Your CPU supports instructions that this TensorFlow binary was not compiled to use: AVX2
19-11-23 14:39:12.752940: I tensorflow/stream_executor/platform/default/dso_loader.cc:42] Successfully opened dynamic library nvcuda.dll
19-11-23 14:39:13.417225: I tensorflow/core/common_runtime/gpu/gpu_device.cc:1640] Found device 0 with properties:
```

FIGURE 11.9
Performance of the model.

cost-effective and also reliable. With these techniques, the process that requires so many people can be automated. A better, efficient, and environmentally friendly ecosystem can be made with the use of machine learning and IoT devices (Figure 11.9).

References

Diro, A. A., & Chilamkurti, N., 2018, 82, 761–768, Distributed attack detection scheme using deep learning approach for Internet of Things. *Future Generation Computer Systems.*

Dourado Jr, C. M., da Silva, S. P. P., da Nóbrega, R. V. M., Barros, A. C. D. S., Reboucas Filho, P. P., & de Albuquerque, V. H. C., 2019, 152, 25–39, Deep learning IoT system for online stroke detection in skull computed tomography images. *Computer Networks.*

Hadidi, R., Cao, J., Woodward, M., Ryoo, M. S., & Kim, H. 2018, 1, Real-time image recognition using collaborative IoT devices. In *Proceedings of the 1st on Reproducible Quality-Efficient Systems Tournament on Co-designing Pareto-efficient Deep Learning*, ACM digital library.

Mano, L. Y., Faiçal, B. S., Nakamura, L. H., Gomes, P. H., Libralon, G. L., Meneguete, R. I., et al., 2016, 89, 178–190. Exploiting IoT technologies for enhancing Health Smart Homes through patient identification and emotion recognition. *Computer Communications.*

Pandey, P. S., 2017, 1–5. Machine learning and IoT for prediction and detection of stress. In *2017 17th International Conference on Computational Science and Its Applications (ICCSA).* IEEE.

Qi, X., & Liu, C., 2018, 367–372. Enabling deep learning on IoT edge: Approaches and evaluation. In *2018 IEEE/ACM Symposium on Edge Computing (SEC).* IEEE.

Shanthamallu, U. S., Spanias, A., Tepedelenlioglu, C., & Stanley, M., 2017, 1–8. A brief survey of machine learning methods and their sensor and IoT applications. In *2017 8th International Conference on Information, Intelligence, Systems & Applications (IISA) IEEE.*

Shenoy, J., & Pingle, Y., 2016, 1456–1458. IoT in agriculture. In *2016 3rd International Conference on Computing for Sustainable Global Development (INDIACom).* IEEE.

Su, J., Vasconcellos, V. D., Prasad, S., Daniele, S., Feng, Y., & Sakurai, K. 2018, 2, 664–669. Lightweight classification of IoT malware based on image recognition. In *2018 IEEE 42nd Annual Computer Software and Applications Conference (COMPSAC)*. IEEE.

Tseng, H. T., Hwang, H. G., Hsu, W. Y., Chou, P. C., & Chang, I., 2017, 9(7), 125. IoT-based image recognition system for smart home-delivered meal services. *Symmetry*.

Xiao, L., Wan, X., Lu, X., Zhang, Y., & Wu, D., 2018, pp 35(5), 41–49. IoT security techniques based on machine learning: How do IoT devices use AI to enhance security? *IEEE Signal Processing Magazine*.

Zhang, S., Wang, H., Huang, W., & You, Z., 2018, 157, 866–872. Plant diseased leaf segmentation and recognition by fusion of superpixel, K-means, and PHOG. *Optik*.

12

Circadian Rhythm and Lifestyle Diseases

Sakshi Singh and Priyadarshini
Jaypee Institute of Information Technology

CONTENTS

12.1 Introduction

Two and a half billion years ago, life evolved on earth following a 24-hour rhythm, and all organisms developed mechanisms to predict the everyday variations in environmental factors including light and temperature fluctuations. Organisms ranging from cyanobacteria, fungi, plants, to animals, including human beings, all developed biological clocks, known as the circadian clocks (CCs) to anticipate the 24-hour day, to demonstrate

DOI: 10.1201/9781003046431-12

endogenic rhythmic cycles of 24 hours to acquire the right physiological and behavioral characteristics corresponding to apposite time frame every single day (Panda, 2016).

The expression "circadian" has been acquired from Latin origins, wherein "circa" means "around," and "diem" means "day," implying around the day. Circadian rhythms (CRs) are different from diurnal rhythms owing to their endogenous generation within the organism and the fact that they perpetuate even in the nonappearance of external time signals. Even though CRs are endogenous yet they show entrainment, i.e., they are still synchronized by the external environment in such a manner that the period of the biological clock matches with the earth's period of rotation, which is 24 hours.

Any mental, physical, or behavioral changes in the body, which follow a daily cycle of about 24 hours displaying endogenous oscillations, entrainable to zeitgebers, can be said to follow a CR, the most common example being the 24-hour sleep and wake cycles (Panda, 2016; Gerhart-Hines & Lazar, 2015).

Over a hundred million years, these rhythms have evolved to coordinate metabolism by temporarily dividing contrasting metabolic processes such as anabolism and catabolism, and by anticipating repetitive feeding–fasting cycles to maximize metabolic competence.

12.2 Characteristic Features of Mammalian CC

Section 12.2.1 highlights the significant characteristics and architecture of methods and platforms.

12.2.1 Architecture of the Biological Clock

The suprachiasmatic nuclei (SCN) and the peripheral clocks (PCs) are the two main constituents of the mammalian circadian system. First, the central pacemaker, referred to as the master clock, is found within the two SCN, situated over the optic chiasmin the hypothalamus. Each SCN comprises about 10,000 heterogeneous neurons, each of which autonomously generates CRs. SCN obtains light inputs from the ipRGCs found inside theretina. Second, there are several smaller clocks all over the body in peripheral tissues, comprising organs such as lungs, liver, skeletal muscles, fibroblasts, and cardiovascular, hepatic, pancreatic, adipose, and gastrointestinal tissues (Hastings et al., 2003). Hence, the mammalian circadian system comprises the master pacemaker, the hypothalamic SCN situated in the brain, which serves the function of entrainment of the PCs spread across the body, i.e., simply synchronizing them (Patel, Velingkaar & Kondratov, 2014; Hardeland, 2016).

This combination of clocks cooperatively regulates a broad range of metabolic targets, including glucocorticoids (Marcheva et al., 2013), AMPK, the chief controller of cellular energy homeostasis (Jenwitheesuk et al., 2014), rate-restricting steps in the production of fatty acids and cholesterol (Xydous, Prombona & Sourlingas, 2014) and hepatic cyclic AMP (cAMP) response element-binding (CREB) to regulate gluconeogenesis (Zhang et al., 2010). As a net result, a rhythmic pattern is observed across a span of 24 hours in several metabolic processes, namely insulin sensitivity, fat oxidation, cholesterol production, insulin secretion, and ATP expenditure (Marcheva et al., 2013; Gerhart-Hines & Lazar, 2015).

Regulation of other processes such as the emission of cortisol and melatonin hormones, oxidative stress response (Patel, Velingkaar & Kondratov, 2014), and maintaining the temperature of the body are all regulated by the master clock SCN, which depends on the hypothalamic nuclei, vasoactive intestinal peptide (VIP), and pineal gland for maintaining these processes (Jenwitheesuk et al., 2014; Hardeland, 2016). Metabolism control by the central pacemaker is believed to be carried out through diffusible factors, mainly cortisol, melatonin, and synaptic prognostication, also through the autonomic nervous system (Mohawk, Green & Takahashi, 2012).

The CC depends on cellular cues and lightweight feedback to sync with solar time and oscillates over 24 hours. The clock obtains regular signals from external sources supplying sunlight and darkness to control biological time. The peripheral body tissues combine these clock signals with environmental influences and behavioral factors (for example, exposure to the sun, sleep, physical activity, and feeding) and their independent patterns to rhythmically control metabolism (Barclay, Tsang & Oster, 2012). Consequently, the natural physiology, behavior, and cellular biochemical transfers within a life form are all essentially regulated by the biological clock.

12.2.2 SCN: Anatomy and Molecular Oscillations

Bilaterally symmetrical and paired, the SCN consists of closely compacted, short-diameter neurons situated laterally from the third ventricle, above the optic chiasm. Researchers agree upon subdividing the SCN into two major subparts: a ventral region referred to as the "core" and a dorsal region known as the "shell" (Antle et al., 2009; Golombek & Rosenstein 2010). It is believed that SCN core neurons combine external signals, together with light signals obtained from the retinohypothalamic tract (RHT), thalamus as well as midbrain structures, including the raphe nucleus (Morin & Allen, 2006). Core neurons use gamma-aminobutyric acid and VIP, or gastrin-releasing peptide (GRP), to transmit these signals to the remaining SCN. The amplitudes of gene translation patterns and neuronal action are comparatively small and high in the core and shell neurons, respectively (Yan & Okamura, 2002). Core and shell outputs of SCN neurons primarily migrate to certain hypothalamic zones, such as the subparaventricular zone (sPVZ) and the dorsal medial hypothalamus (DMH). Projections from these hypothalamic relay nuclei are redirected all over the central nervous system and endocrine system (Dibner, Schibler & Albrecht, 2010).

The SCN is composed of heterogeneous subpopulations of neurons. These individual constituent neurons of the SCN, upon being isolated in vitro (Webb et al., 2009), function as independent cell-autonomous oscillators, whereas they work naturally in an integrated paired manner in vivo.

12.2.3 Molecular Mechanism of Core Circadian Genes

BMAL-1 and CLOCK along with Period-1, Period-2, CRY-1, and CRY-2 are considered to be the core genes of the autoregulatory transcription and translation feedback mechanism underlying the mammalian circadian regulation. BMAL-1 and CLOCK genes encode for transcriptional activators, which direct the expression of the period gene (Per-1 and Per-2) and Cryptochrome (CRY-1 and CRY-2), protein products of which successively feedback to repress CLOCK and BMAL-1 (Reppert and Weaver, 2002).

The CLOCK and BMAL-1 proteins, which are bHLH-PAS transcription factors, form a heterodimer with each other (Gekakis et al., 1998). The CLOCK–BMAL-1 complex coheres to regulatory elements that contain Enhancer boxes of DNA driving in the rhythmic transcription of the PERIOD gene (Per-1, Per-2, and Per-3) and CRYPTOCHROME gene (Cry-1 and Cry-2) into repressor proteins (Shearman et al., 2000). These genes (Period and Cryptochrome) are activated in the mouse by CLOCK and BMAL-1 during the day time, their RNAs are transcribed in the afternoon, and their proteins (Per and Cry) undergo translation and accumulate by the late afternoon or dusk (Lee et al., 2001).

Cry and Per are interrelated, and also associated with the serine-threonine kinases (Gallego & Virshup, 2007), and transfer back inside the cell nucleus upon nightfall (Lowrey & Takahashi, 2011). As their levels inside the nucleus increase during the night, they can directly interact with the CLOCK–BMAL-1 and inhibit the activation potential of CLOCK–BMAL-1, which thereby turns off their transcription. Because of the comparatively small half-lives of Per and Cry protein products, they become the targets of E3 ligase complexes and consequently undergo breakdown by the proteasome (Preußner & Heyd, 2016). The Cry and Per eventually disappear by the end of the night and BMAL-1 and CLOCK can stimulate a novel round of transcription by the next morning.

12.3 Hormonal Mechanism of Sleep–Wake Cycle

The sleep–wake (SW) cycle is the ultimate observable circadian pattern. Sleep plays a significant role in recovering brain capacity, shutting off external stimuli and off-line processing of data gained while awake, promoting the plasticity of cortical improvements affecting thinking, consolidating memory, and disappearance, and stimulating the newly discovered glymphatic network responsible for getting rid of brain metabolites (Musiek & Holtzman, 2016).

NREM (nonrapid eye movement) and REM (rapid eye movement) are the two definite phases of sleep, on grounds of typical polysomnography signals, which vary regularly during the night. The wake time determines the sum of Slow Wave Sleep (SWS) in the circadian period. The circadian constituent of the sleep predisposition mechanism is probably regulated by the SCN using the indirect stimulations of the sleep promotion center located in the preoptic ventrolateral nucleus (Deurveilheret & Semba 2005).

In diurnal organisms such as humans, melatonin is an essential regulator of physiological sleep. The rapid rise in sleep tendency at night time in humans typically happens 2 hours past the beginning of endogenous melatonin secretion (Lavie, 1997; Zisapel, 2007); in fact, the nocturnal melatonin period transmits information about the night length to the brain and different organs, along with the SCN itself. For both normal and blind subjects, the circadian melatonin cycle is directly linked to the sleep pattern (Zisapel, 2001).

The retina perceives light from the retinohypothalamic pathway, which is different from the optic nerve. If no light signals are transmitted to the SCN, often seen in fully blind people with no eyes present, no entrainment occurs, resulting in their duration remaining unmatched to the day's 24-hour cycle.

Endogenous melatonin synthesis and secretion are regulated by the SCN and take place during the dark (quiescent) period of the electrical activity of the SCN. In the light period, melatonin production is stopped and rises in the evening to hit the peak at midnight. Daylight has two impacts on the secretion of pineal melatonin: daylight–dark cycles synchronize pattern (due to CR entrainment) and intense night light exposure rapidly lowers serotonin N-acetyltransferase (SNAT) action, thereby inhibiting melatonin release (Albrecht, 2012) and rapidly decreasing blood and CSF levels of melatonin. A clear and accurate measure of the endogenous circadian phase is the dim light melatonin onset (DLMO), which is the initial hike in the melatonin production during the early part of the night under dim light conditions (Zisapel, 2001).

Daytime administration of melatonin (when endogenously absent) in human beings contributes to the development of exhaustion and drowsiness (Gorfineet al., 2006). Essentially, melatonin is not a sedative: in nocturnal species, melatonin is correlated with awakening cycles, not sleeping, and in humans, its sleep-inducing effects become vital around 2 hours after intake, close to the circadian pattern at night (Zisapel, 2007).

12.4 CR and Sleep-Related Disorders

A wide range of circadian phenotypes or chronotypes can be found in the human population, with larks (early category) being one extreme category and owls (late category) the other. The chronotype of a person is determined using certain genes and mutations of these genes can clarify the genetic variations. Researchers have recently established variants in several loci linked to the clock gene predominantly PER homologs 2 and 3, behind the common public, being typically more active, involved, and alert in the morning (Hu et al., 2016; Jones et al., 2016). Various chronotypes may change sleep cycles to match both their CC and communal commitments.

Derangement with the outside light and dark sequences contributes to critically disturbed SW rhythms, persistent lethargy, and fatigue. The reason for this may be the fundamental defects in the clock apparatus, contributing to non-24-hour rhythms or shortcomings in the input mechanisms and entrainment organization resulting in a disarrayed circadian pattern. For instance, core clock machinery defects include late or advanced sleep phase disorders, for example, Familial advanced sleep phase syndrome is related to mutations in PER 2 (Toh et al., 2001) whereas Familial delayed sleep phase syndrome (FDSPS) is related to CK 1 Delta mutation (Xu et al., 2005). Mutations in CRY-1 have lately been related to FDSPS with a relatively soaring prevalence of 0.60% in the population, hence disturbing sleep patterns in significant percentage of people (Patke et al., 2017).

On account of fast or slow molecular clocks, the time frame identified as ideal by the clock for sleep is changed about the ambient light and dark cycle, which results in acute misalignment. Additionally, there are also rather common situations where input pathways are deficient. Low light rates in the nursing home setting contribute to circadian cycle disturbance (Most et al., 2010), and individuals with serious sight impairment attributable to either hereditary factors or injuries, lack light signals to the CC leading to acute misalignment (Sack et al., 1992). For such cases, behavioral patterns enforced by treatment or feeding can help mitigate this disturbance, but PCs that are desynchronized display absence of entrainment, which is evident as in somnolence and disturbed sleep.

12.4.1 Intrinsic CRSW Disorders

This section illustrates the intrinsic factors of CRSW disorders (CRSWDs), which include DSWPD, ASWPD, N24SWD, and ISWRD, respectively.

12.4.1.1 Delayed SW-Phase Disorder

Delayed SW-phase disorder (DSWPD) presents itself as a postponement of the chief sleep episode in comparison to the ideal timing of the individual or the timing required to meet family, educational, or occupational demands. Patients experience severe difficulties in falling asleep at bedtimes deemed standard among their peers, as well as with waking up at the mandatory or preferred hours, but the quality of sleep is typically reported as standard when the person sleeps at delayed hours.

12.4.1.2 Advanced SW-Phase Disorder

Advanced SW-phase disorder (ASWPD) is distinguished by the advancement of the main sleep incident in comparison to the expected or necessary SW periods for the patient. Patients with ASWPD experience severe difficulties in remaining awake throughout the evening hours, and often fall asleep before finishing job tasks and social or family responsibilities. Therefore, wake time is also undesirable and before time, and is found atypical according to peers.

12.4.1.3 Non-Twenty-Four-Hour SW Rhythm Disorder

Non-twenty-four hour SW rhythm disorder (N24SWD) is considered as a patient's struggle to acclimatize to the light–dark period and clock cycle of 24 hours. Thus, patients display SW cycles that indicate gradual postponement or advancement, based on the duration of their endogenous CRs. The duration of high sleep predisposition slowly changes over the asymptomatic phase, so that patients feel hypersomnolence in the daytime and insomniac throughout the night. Most N24SWD patients are entirely blind, although this condition often exists among the sighted as well.

12.4.1.4 Irregular SW Rhythm Disorder

In irregular SW rhythm disorder (ISWRD), patients do not have a distinct CR of SW activity. Therefore, in addition to extreme sleepiness and extended daytime sleep episodes, the affected people undergo lengthy intervals of wakefulness during the usual nocturnal sleep cycle. Sleep is broken, and mostly inadequate. In patients with neurodevelopmental or neurodegenerative diseases, ISWRD is more frequently found and may present specific difficulties to the caregivers (Jagannath et al., 2017).

12.4.2 Extrinsic or Environmentally Influenced CRSWDs

This section illustrates the extrinsic factors, which include jet lag, and shift work.

12.4.2.1 Jet Lag

Jet lag refers to either sleepiness or alertness that comes about at inappropriate hours of the day concerning the regional time, occurring after frequent travels beyond multiple time zones. The investigative criteria for jet lag include complaints of sleeplessness or prolonged daytime sleep correlated with transmeridian plane travel through more than onetime zone, related disruption of activities during the day, general uneasiness, or gastrointestinal disruption within 2 days after the flight (Arendt, 2018).

The quick transition over different time zones results in some disparity or loss of contemporaneousness between the actions of the endogenous rhythm-producing machinery and the domestic socioenvironmental time cues. The SCN gradually acclimatizes to sudden variations in time signals (Buijs et al., 2016).

The central clock changes on average roughly 1 hour per day without harm and takes 24 hours to complete the transition from every hour of time zone adjustment. The rate of acclimatization differs significantly in individuals, with time zone adjustment being retarded toward the east at large. The alteration rate mostly varies during the span of adaptation, generally more rapidly in the early stages than in the ending stages. Furthermore, it is normal for voyagers across multiple time zones to acclimatize in the "incorrect" phase, for example, by holding up by 16 hours as a replacement for advancing 8 hours, i.e., antidromic instead of orthodromic adaptation (Takahashi et al., 2001). There is dissociation among peripheral oscillators, as well as between peripheral oscillators and SCN, with various mechanisms responding to adaptation at varying rates (Dibner & Schibler, 2015). Hence, endogenous rhythms, apart from being out of synchrony with the surroundings (exterior desynchronization), are also out of harmony with each other (core desynchronization). In a phenomenon known as re-entrainment by "partition," certain rhythms make headway while others impede. Adaptation to these phase shifts gets harder as we age based on currently unknown grounds (Moline et al., 1992).

12.4.2.2 Shift Work

Shift work is a concept that includes a vast range of untypical work shift arrangements ranging from intermittent overnight phone-call services, rotating shifts, and continuous, regular night shifts. This can also refer to schedules that require waking up early from nocturnal sleep. The diversity of job-timing patterns makes it extremely tricky to generalize about shift work (Sack et al., 2007).

Shift work usually includes working beyond the normal time bracket of about 7 am–11 pm like at night, during the late evening, even work commencing before 6 am. The shifts may be regular, gradually changing, frequently switching from night to day (rotating backward), or day to night (rotating forward), even without any specific trend (Lunn et al., 2017). Field research focusing on real LAN exposure has not excessively been conducted, although a recent study indicates that it may vary from 50 to 100 photopic lux, with certain exposures higher than 200 lux (Hunter & Figueiro, 2017).

12.4.3 Therapeutic Options for CRSWDs

Since light is the main zeitgeber for the pacemaker of the SCN, it has been seen to have some effectiveness in bright light therapies along with cognitive-behavioral therapies, such as planned outdoor activity, which stimulate the endogenous rhythms (Atkinson et al., 2007; Zee, Attarian & Videnovic, 2013). For many years, melatonin cycle has been known to be regulated by the CC and hence may be of use to alter the clock phase by operating through the melatonin receptors found in the SCN neurons and several other populations of cells in the body. Melatonin is tested as a potential chronotherapeutic medication and results are excellent for particular circadian disorders (Mundey et al., 2005).

In the case of aged patients, melatonin, known by the trade name Circadin, is prescribed to relieve insomnia (Lemoine et al., 2012), and overcome jet lag. In the year 2014, the US FDA gave assent to Tasimelteon for the treatment of N24SWD in individuals who suffered total visual impairment (Lockley et al., 2015). Agomelatine, a potent melatonin agonist, is used for the management of disorders related to depression (Kennedy et al., 2006).

Solt and his colleagues were able to isolate a novel REV-ERBa receptor agonist that was successful in modulating sleep and metabolism in mouse models (Sato et al., 2004). KL044 is a potent small-molecule cryptochrome stabilizer that can regulate the activity of the CC through its action on the Cry gene (Lee et al., 2015).

12.5 Circadian Regulation of Metabolism

Eubacterial, fungal, and plant studies (Paranjpe & Sharma, 2005) indicate that CCs provide an evolutionary benefit by facilitating effective fuel accession and preservation at the correct period of the day. The CC in mammals regulates anabolic and catabolic mechanisms in peripheral tissues together with synchronizing feed–fast and SW cycles.

A change in perception of CC activity occurred about two decades ago with the findings that the CLOCK–BMAL-1 heterodimer oscillates not just inside the SCN master clock neurons, but even inside peripheral cells and the fact that isolated cells show circadian oscillations ex vivo upon integration with serum, cyclase agonists, or temperature adjustments. In peripheral organs, particularly the liver, there are more cyclic genes than in the hypothalamus itself (Zhang et al., 2014).

The perception of the molecular association between CCs and metabolism was changed when it was observed that the homozygous Clock-mutant mouse became vulnerable to obesity (Turek et al., 2005). Apart from sleep disturbances and circadian behavior, these mice acquired early-life hyperphagia, which further resulted in hyperleptinemia-, hyperlipoproteinemia-, and hypoinsulinemia-induced high blood sugar, which is characteristic of endocrine pancreatic malfunction (Marcheva et al., 2010). The knocking out of BMAL-1 or Per-2 and double knocking out of Cry1 and Cry2 makes the mice vulnerable to obesity and metabolic diseases (Barclay et al., 2013).

Similarly, in human trials, circadian misalignment raises the rates of glucose, insulin, and triglycerides (Scheer et al., 2009) and reduces energy expenditure (McHill et al., 2014). Besides, damaging the activity of Clock in mice suppresses gluconeogenesis, while knocking out the BMAL-1 gene completely stops it (Rudic et al., 2004). Conditional removal of the CLOCK gene is responsible for diabetes. Disruption of the CLOCK gene in islets of Langerhans in the pancreas contributes to transcriptome-wide differences in gene expression, participating in the assembly of synaptic vesicles, growth, and development within these cells.

The association between circadian oscillators in peripheral tissues and metabolic disorders was further verified by the abolition of the molecular clocks restricted to skeletal muscles, smooth muscles, adipose tissue, pancreas, myeloid cells, liver, and arcuate nucleus AgRP neurons. Following these results, polymorphisms in many clock genes were related to obesity or symptoms of certain metabolic disorders and cardiovascular effects in humans. Observational findings in limited population samples found that polymorphisms observed in the CLOCK genes were correlated with susceptibility to obesity, type 2 diabetes (T2D), and hypertension are linked with two SNPs in the BMAL-1 gene. Also, several studies have identified a relation between CRY2 polymorphisms and high fasting glucose. The link between REV-ERBα polymorphisms and obesity in men has also been identified (Maury, 2019). Recently, Volodymyr Petrenkoa and his colleagues discovered that alpha and beta islet cells isolated from the pancreas of T2D patients have clocks with decreased circadian amplitudes and less potential for in vitro synchronization relative to their nondiabetic equivalents. Regulation of time series profiles of insulin and glucagon production was disturbed for T2D subjects, suggesting that islet cell-autonomous clocks

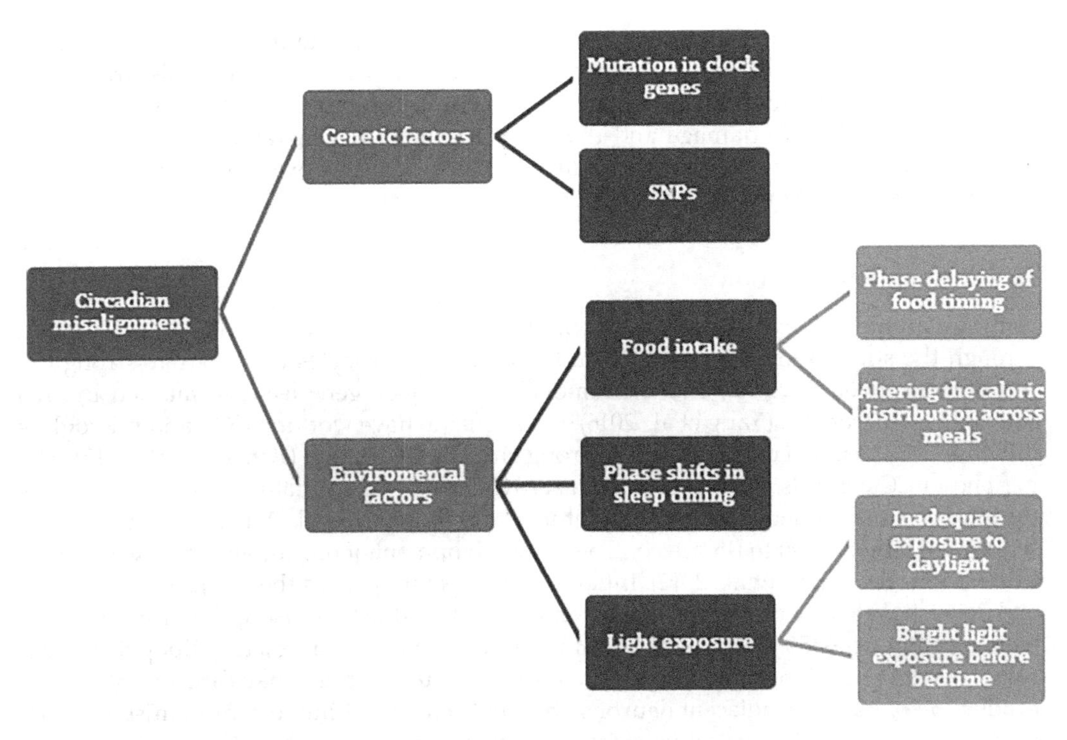

FIGURE 12.1
Probable factors contributing to circadian misalignment.

have a partial role to play in T2D development. This finding highlights an association between the CCs and T2D, indicating that clock regulators carry promise as future therapeutic agents for diabetes (Petrenko et al., 2020) (Figure 12.1).

12.6 Circadian Disruption and Neurodegeneration

Most brain cells, including neurons, astrocytes, and microglia, exhibit CC genes, which show oscillations in cell culture. Transcriptomic analyses of cerebellar and brain stem extracts from mice indicate that there are circadian oscillations in hundreds of genes. Mice without Bmal1 experience alarming neurological phenotypes, including extreme spontaneous astrogliosis, enhanced oxidative disruption, synaptic degeneration, reduced brain functional integration, and disrupted thinking (Leng et al., 2019).

12.6.1 Oxidative Stress

Oxidative tension in flies results in circadian disturbance and loss of sleep as seen in aging (Koh et al., 2006). Conversely, disturbance of the circadian mechanism by Bmal1 deletion induces elevated oxidative stress in several mice organs (Kondratov et al., 2006), including the brain. Bmal1 activates redox-related gene transcription in the brain, including Nqo1 and Aldh212. As reactive oxygen species are generated as by-products of increased neuronal activity in the brain, the CC can help to temporarily coordinate the expression of redox protection genes with diurnal variations in brain metabolism. Therefore, disturbance of the usual circadian process in neurological disease settings may make the brain more susceptible to oxidative damage and thus facilitate neurodegeneration. Consequently, a reduced Bmal1 expression aggravates neuronal loss induced by oxidative stress in in vitro and in vivo models (Musiek et al., 2013).

12.6.2 Neuroinflammation

Neuroinflammation is a significant contributor to neurodegeneration, mostly propagated through the stimulation of astrocytes and microglia. Astrocytes demonstrate strong CC function (Prolo et al., 2005), and the removal of the clock gene leads to marked in vivo activation of astrocytes (Yang et al., 2016). Microglia do have working CCs, and microglia's inflammatory reaction demonstrates a strong circadian variation (Fonken et al., 2015). The peripheral CC controls the response of macrophages to inflammatory stimuli, as well as the movement of monocytes to inflammatory areas (Nguyen et al., 2013). Rev-Erbα, a transcriptional target direct to Bmal1, controls the development of pro-inflammatory cytokines in macrophages (Gibbs et al., 2012). Inflammation also influences the clock, as both Bmal1 and Rev-erbα rates are highly inhibited in macrophages due to transcriptional suppression by microRNA miR-155(Curtis et al., 2015) in reaction to the inflammatory lipopolysaccharide. Therefore, local inflammation in the hippocampus or cortex may directly inhibit the production of Bmal1 in adjacent neurons and glial cells, resulting in compromised Bmal1-mediated production of oxidative stress response and making these cells vulnerable to neurodegeneration.

12.6.3 Neurodegenerative Disorders

Neurodegenerative disorder patients usually undergo CR disruptions (CRD) far more acutely than usual age-related CRD. For example, patients have been observed to be more alert or energetic throughout the night relative to daytime, and often the 24-hour SW cycle is reversed or lost. Importantly, research indicates that disturbance of circadian functions may be early signs of neurodegeneration and may also be a contributing factor in the progression of neurodegenerative disorders in stable adults over the age of 60 years (Musiek & Holtzman, 2016).

Given the varying pathogenesis and symptomatic variability within different neurodegenerative disorders, disturbances in circadian cycles, or near 24-hour cyclic variations in a variety of physiological and behavioral processes are observed in the cases of Parkinson's disease, Huntington's disease, and Alzheimer's disorder (Jagannath et al., 2017).

12.6.4 AD and Related Dementias

Patients with mild to extreme AD were found to have significantly more serious circadian disruptions relative to healthy adults of the same age group. These included greater fragmentation, dampened amplitude, and phase delay relative to the more common advanced circadian phase correlated with normal aging (Videnovi et al., 2014). It was proposed that the rise in behavioral and neuropsychiatric symptoms of AD patients at the time of sunset, known as "sundowning", is perhaps because of the phase delay of temperature and hormone cycles (Canevelli et al., 2016). Normally, healthy elderly individuals experience age-related ASWPD, whereas CRSWD found most prominently in the case of AD patients is ISWRD. ISWRD is based on the deficiency of consistent 24-hour SW cycles, typically with long wakefulness periods throughout the night followed by intermittent sleep bursts throughout the day that may get worse in case of acute AD (Auger et al., 2015). DNA methylation mutations of clock genes cause loss of cognition and behavioral alterations in people with AD (Leng et al., 2019).

12.6.5 Parkinson's Disease

Symptoms of PD in both motor and nonmotor forms reveal disturbances in their normal 24-hour oscillations. CR disruption in patients with PD is characterized by a decrease in CR intensity but no statistically meaningful difference in circadian phases, contrary to patients with ADRD (Hood & Amir, 2017). SW problems as a whole are the most severe nonmotor manifestations in patients with PD, affecting up to 80% of patients with PD. Also, five of the six studies investigating circadian characteristics in PD patients recorded either extreme daytime sleepiness (EDS) or variations in sleep timing (Leng et al., 2019). Continuous treatment with levodopa can weaken CR activity. There were reduced BMAL-1 and RORα in animals with 6-OHDA-induced PD.

12.7 Role of CR in Cancer

Failure of the normal CR may contribute to an elevated risk of different types of cancer. Drosophila's CR gene TIMELESS is stimulated in cancer of the nasopharynx, and its amplified expression is correlated with reduced overall survival. Alternatively, excessive expression of the gene TIMELESS contributed as an inhibitor of cisplatin-induced apoptosis, which in turn triggered the Wnt signaling pathway (Maiese, 2017).

Signals of Wnt and β-catenin will inhibit autophagy (Masia et al., 2013), apoptosis, impact sensory modalities (Maiese, 2017), and contribute to the proliferation of stem cells (Maiese, 2015). However, these pathways encourage angiogenesis (Chen et al., 2015) and contribute to tumor development (Jia et al., 2017), which may be in parallel with clock gene proliferation pathways such as TIMELESS. Disruption of circadian cycles through shift work on a clinical basis indicates that these tasks can often raise the likelihood of contracting cancer. Long-term, frequent night shift work for female nurses has an elevated chance of breast cancer (Wegrzyn et al., 2017). Increased circadian gene expression hClock can also result in the development of tumors, like the spread of cancerous tumors of the bowel, by an increase in gene expression associated with angiogenesis (Wang et al., 2017).

12.8 Conclusion

In brief, the mammalian CC and the SW cycle are multilayered, integrated processes that have many pathways that affect brain activity and neurodegeneration. Despite considerable strides in discovering the fundamental processes regulating the CC, and the neuronal structure of sleep, our awareness about how such structures influence the aging of the brain and the development of neurodegenerative disorders is still quite limited. A more thorough understanding of the process by which particular neurodegenerative diseases and pathogenic proteins influence circadian and sleep processes is required. The role played by circadian gene products in aging and neurodegeneration apart from sleep and circadian processes is required to explain the underlying mechanisms in depth.

References

Albrecht, U. (2012). Timing to perfection: the biology of central and peripheral circadian clocks. *Neuron, 74*(2), 246–260. doi:10.1016/j.neuron.2012.04.006.

Antle, M. C., Smith, V. M., Sterniczuk, R., Yamakawa, G. R., & Rakai, B. D. (2009). Physiological responses of the circadian clock to acute light exposure at night. *Reviews in Endocrine and Metabolic Disorders, 10*(4), 279–291. doi:10.1007/s11154-009-9116-6.

Arendt, J. (2018). Approaches to the pharmacological management of jet lag. *Drugs, 78*(14), 1419–1431. doi:10.1007/s40265-018-0973-8.

Atkinson, G., Edwards, B., Reilly, T., & Waterhouse, J. (2006). Exercise as a synchroniser of human circadian rhythms: An update and discussion of the methodological problems. *European Journal of Applied Physiology, 99*(4), 331–341. doi:10.1007/s00421-006-0361-z.

Auger, R. R., Burgess, H. J., Emens, J. S., Deriy, L. V., Thomas, S. M., & Sharkey, K. M. (2015). Clinical practice guideline for the treatment of intrinsic circadian rhythm sleep-wake disorders: Advanced sleep-wake phase disorder (ASWPD), delayed sleep-wake phase disorder (DSWPD), non-24-hour sleep-wake rhythm disorder (N24SWD), and irregular sleep-wake rhythm disorder (ISWRD). An update for 2015. *Journal of Clinical Sleep Medicine, 11*(10), 1199–1236. doi:10.5664/jcsm.5100.

Barclay, J. L., Shostak, A., Leliavski, A., Tsang, A. H., Jöhren, O., Müller-Fielitz, H., Landgraf, D., Naujokat, N., van der Horst, G. T., & Oster, H. (2013). High-fat diet-induced hyperinsulinemia and tissue-specific insulin resistance in cry-deficient mice. *American Journal of Physiology-Endocrinology and Metabolism, 304*(10). doi:10.1152/ajpendo.00512.2012.

Barclay, J. L., Tsang, A. H., & Oster, H. (2012). Interaction of central and peripheral clocks in physiological regulation. *Progress in Brain Research the Neurobiology of Circadian Timing,* 163–181. doi:10.1016/b978-0-444-59427-3.00030-7.

Buijs, F. N., León-Mercado, L., Guzmán-Ruiz, M., Guerrero-Vargas, N. N., Romo-Nava, F., & Buijs, R. M. (2016). The circadian system: a regulatory feedback network of periphery and brain. *Physiology, 31*(3), 170–181. doi:10.1152/physiol.00037.2015.

Canevelli, M., Valletta, M., Trebbastoni, A., Sarli, G., D'Antonio, F., Tariciotti, L., de Lena, C., & Bruno, G. (2016). Sundowning in dementia: Clinical relevance, pathophysiological determinants, and therapeutic approaches. *Frontiers in Medicine, 3.* doi:10.3389/fmed.2016.00073.

Curtis, A. M., Fagundes, C. T., Yang, G., Palsson-McDermott, E. M., Wochal, P., McGettrick, A. F., Foley, N. H., Early, J. O., Chen, L., Zhang, H., Xue, C., Geiger, S. S., Hokamp, K., Reilly, M. P., Coogan, A. N., Vigorito, E., FitzGerald, G. A., & O'Neill, L. A. (2015). Circadian control of innate immunity in macrophages by Mir-155 targeting BMAL1. *Proceedings of the National Academy of Sciences, 112*(23), 7231–7236. doi:10.1073/pnas.1501327112.

Debebe, A., Medina, V., Chen, C.-Y., Mahajan, I. M., Jia, C., Fu, D., He, L., Zeng, N., Stiles, B. W., Chen, C.-L., Wang, M., Aggarwal, K.-R., Peng, Z., Huang, J., Chen, J., Li, M., Dong, T., Atkins, S., Borok, Z., … Stiles, B. L. (2017). Wnt/β-catenin activation and macrophage induction during liver cancer development following steatosis. *Oncogene, 36*(43), 6020–6029. doi:10.1038/onc.2017.207.

Deurveilher, S., & Semba, K. (2005). Indirect projections from the suprachiasmatic nucleus to major arousal-promoting cell groups in rat: implications for the circadian control of behavioural state. *Neuroscience, 130*(1), 165–183. doi:10.1016/j.neuroscience.2004.08.030.

Dibner, C., & Schibler, U. (2015). Circadian timing of metabolism in animal models and humans. *Journal of Internal Medicine, 277*(5), 513–527. doi:10.1111/joim.12347.

Dibner, C., Schibler, U., & Albrecht, U. (2010). The mammalian circadian timing system: organization and coordination of central and peripheral clocks. *Annual Review of Physiology, 72*(1), 517–549. doi:10.1146/annurev-physiol-021909-135821.

Fonken, L. K., Frank, M. G., Kitt, M. M., Barrientos, R. M., Watkins, L. R., & Maier, S. F. (2015). Microglia inflammatory responses are controlled by an intrinsic circadian clock. *Brain, Behavior, and Immunity, 45*, 171–179. doi:10.1016/j.bbi.2014.11.009.

Gallego, M., & Virshup, D. M. (2007). Post-translational modifications regulate the ticking of the circadian clock. *Nature Reviews Molecular Cell Biology, 8*(2), 139–148. doi:10.1038/nrm2106.

Gekakis, N., Staknis, D., Nguyen, H. B., Davis, F. C., Wilsbacher, L. D., King, D. P., Takahashi, J. S., & Weitz, C. J. (1998). Role of the clock protein in the mammalian circadian mechanism. *Science, 280*(5369), 1564–1569. doi:10.1126/science.280.5369.1564.

Gerhart-Hines, Z., & Lazar, M. A. (2015). Circadian metabolism in the light of evolution. *Endocrine Reviews, 36*(3), 289–304. doi:10.1210/er.2015-1007.

Golombek, D. A., & Rosenstein, R. E. (2010). Physiology of Circadian Entrainment. *Physiological Reviews, 90*(3), 1063–1102. doi:10.1152/physrev.00009.2009.

Gorfine, T., Assaf, Y., Goshen-Gottstein, Y., Yeshurun, Y., & Zisapel, N. (2006). Sleep-anticipating effects of melatonin in the human brain. *NeuroImage, 31*(1), 410–418. doi:10.1016/j.neuroimage.2005.11.024

Hardeland, R. (2016). Melatonin and the pathologies of weakened or dysregulated circadian oscillators. *Journal of Pineal Research, 62*(1). doi:10.1111/jpi.12377.

Hastings, M. H., Reddy, A. B., & Maywood, E. S. (2003). A clockwork web: Circadian timing in brain and periphery, in health and disease. *Nature Reviews Neuroscience, 4*(8), 649–661. doi:10.1038/nrn1177.

Hood, S., & Amir, S. (2017). Neurodegeneration and the circadian clock. *Frontiers in Aging Neuroscience, 9*. doi:10.3389/fnagi.2017.00170.

Hu, Y., Shmygelska, A., Tran, D., Eriksson, N., Tung, J. Y., & Hinds, D. A. (2016). GWAS of 89,283 individuals identifies genetic variants associated with self-reporting of being a morning person. *Nature Communications, 7*(1), 10448. doi:10.1038/ncomms10448.

Hunter, C. M., & Figueiro, M. G. (2017). Measuring light at night and melatonin levels in shift workers: A review of the literature. *Biological Research For Nursing, 19*(4), 365–374. doi:10.1177/1099800417714069.

Jagannath, A., Taylor, L., Wakaf, Z., Vasudevan, S. R., & Foster, R. G. (2017). The genetics of circadian rhythms, sleep and health. *Human Molecular Genetics, 26*(R2), R128–R138. doi:10.1093/hmg/ddx240.

Jenwitheesuk, A., Nopparat, C., Mukda, S., Wongchitrat, P., & Govitrapong, P. (2014). Melatonin regulates aging and neurodegeneration through energy metabolism, epigenetics, autophagy and circadian rhythm pathways. *International Journal of Molecular Sciences, 15*(9), 16848–16884. doi:10.3390/ijms150916848.

Jiang, L., Yin, M., Wei, X., Liu, J., Wang, X., Niu, C., Kang, X., Xu, J., Zhou, Z., Sun, S., Wang, X., Zheng, X., Duan, S., Yao, K., Qian, R., Sun, N., Chen, A., Wang, R., Zhang, J., … Meng, D. (2015). Bach1 represses Wnt/β-catenin signaling and angiogenesis. *Circulation Research, 117*(4), 364–375. doi:10.1161/circresaha.115.306829.

Jones, S. E., Tyrrell, J., Wood, A. R., Beaumont, R. N., Ruth, K. S., Tuke, M. A., et al. (2016). Genome-wide association analyses in 128,266 individuals identifies new morningness and sleep duration loci. *PLoS Genetics, 12*(8), e1006125. doi:10.1371/journal.pgen.1006125.

Kennedy, S. H., & Emsley, R. (2006). Placebo-controlled trial of agomelatine in the treatment of major depressive disorder. *European Neuropsychopharmacology, 16*(2), 93–100. doi:10.1016/j.euroneuro.2005.09.002.

Koh, K., Evans, J. M., Hendricks, J. C., & Sehgal, A. (2006). A drosophila model for age-associated changes in sleep:wake cycles. *Proceedings of the National Academy of Sciences, 103*(37), 13843–13847. doi:10.1073/pnas.0605903103

Kondratov, R. V., Kondratova, A. A., Gorbacheva, V. Y., Vykhovanets, O. V., & Antoch, M. P. (2006). Early aging and age-related pathologies in mice deficient in BMAL1, the core Component of the circadian clock. *Genes & Development, 20*(14), 1868–1873. doi:10.1101/gad.1432206

Lavie, P. (1997). Melatonin: Role in gating nocturnal rise in sleep propensity. *Journal of Biological Rhythms, 12*(6), 657–665. doi:10.1177/074873049701200622.

Lee, C., Etchegaray, J., Cagampang, F. R., Loudon, A. S., & Reppert, S. M. (2001). Posttranslational mechanisms regulate the mammalian circadian clock. *Cell, 107*(7), 855–867. doi:10.1016/s0092-8674(01)00610-9.

Lee, J. W., Hirota, T., Kumar, A., Kim, N., Irle, S., & Kay, S. A. (2015). Development of small-molecule cryptochrome stabilizer derivatives as modulators of the circadian clock. *ChemMedChem, 10*(9), 1489–1497. doi:10.1002/cmdc.201500260.

Lemoine, P., & Zisapel, N. (2012). Prolonged-release formulation of melatonin (Circadin) for the treatment of insomnia. *Expert Opinion on Pharmacotherapy, 13*(6), 895–905. doi:10.1517/14656566.2012.667076.

Leng, Y., Musiek, E. S., Hu, K., Cappuccio, F. P., & Yaffe, K. (2019). Association between circadian rhythms and Neurodegenerative Diseases. *The Lancet Neurology, 18*(3), 307–318. doi:10.1016/s1474-4422(18)30461-7.

Lockley, S. W., Dressman, M. A., Licamele, L., Xiao, C., Fisher, D. M., Flynn-Evans, E. E., Hull, J. T., Torres, R., Lavedan, C., & Polymeropoulos, M. H. (2015). Tasimelteon for non-24-hour sleep–wake disorder in totally blind people (set and reset): Two multicentre, randomised, double-masked, placebo-controlled phase 3 trials. *The Lancet, 386*(10005), 1754–1764. doi:10.1016/s0140-6736(15)60031-9.

Lowrey, P. L., & Takahashi, J. S. (2011). Genetics of circadian rhythms in mammalian model organisms. *The Genetics of Circadian Rhythms Advances in Genetics*, 175–230. doi:10.1016/b978-0-12-387690-4.00006-4.

Lunn, R. M., Blask, D. E., Coogan, A. N., et al. (2017). Health consequences of electric lighting practices in the modern world: a report on the National Toxicology Program's workshop on shift work at night, artificial light at night, and circadian disruption. *Science of the Total Environment, 607–608*, 1073–1084. doi:10.1016/j.scitotenv.2017.07.056.

Maiese, K. (2015). SIRT1 and Stem Cells: In the forefront with cardiovascular disease, neurodegeneration and cancer. *World Journal of Stem Cells, 7*(2), 235. doi:10.4252/wjsc.v7.i2.235.

Maiese, K. (2017). Moving to the rhythm with clock (circadian) genes, autophagy, mTOR, and SIRT1 in degenerative disease and cancer. *Current Neurovascular Research, 14*(3). doi:10.2174/1567202614666170718092010.

Marcheva, B., Ramsey, K. M., Buhr, E. D., Kobayashi, Y., Su, H., Ko, C. H., Ivanova, G., Omura, C., Mo, S., Vitaterna, M. H., Lopez, J. P., Philipson, L. H., Bradfield, C. A., Crosby, S. D., JeBailey, L., Wang, X., Takahashi, J. S., & Bass, J. (2010). Disruption of the clock components clock and BMAL1 leads to hypoinsulinaemia and diabetes. *Nature, 466*(7306), 627–631. doi:10.1038/nature09253.

Marcheva, B., Ramsey, K. M., Peek, C. B., Affinati, A., Maury, E., & Bass, J. (2013). Circadian clocks and metabolism. *Handbook of Experimental Pharmacology*, (217), 127–155. doi: 10.1007/978-3-642-25950-0_6.

Maury, E. (2019). Off the clock: From circadian disruption to metabolic disease. *International Journal of Molecular Sciences, 20*(7), 1597. doi:10.3390/ijms20071597.

McHill, A. W., Melanson, E. L., Higgins, J., Connick, E., Moehlman, T. M., Stothard, E. R., & Wright, K. P. (2014). Impact of circadian misalignment on Energy Metabolism during simulated night-shift work. *Proceedings of the National Academy of Sciences, 111*(48), 17302–17307. doi:10.1073/pnas.1412021111.

Mohawk, J. A., Green, C. B., & Takahashi, J. S. (2012). Central and peripheral circadian clocks in mammals. *Annual Review of Neuroscience, 35*(1), 445–462. doi:10.1146/annurev-neuro-060909-153128.

Moline, M. L., Pollak, C. P., Monk, T. H., Lester, L. S., Wagner, D. R., Zendell, S. M., Graeber, R. C., Salter, C. A., & Hirsch, E. (1992). Age-related differences in recovery from simulated Jet Lag. *Sleep, 15*(1), 28–40. doi:10.1093/sleep/15.1.28.

Morin, L., & Allen, C. (2006). The circadian visual system, 2005. *Brain Research Reviews, 51*(1), 1–60. doi:10.1016/j.brainresrev.2005.08.003.
Most, E. I., Scheltens, P., & Someren, E. J. (2010). Prevention of depression and sleep disturbances in elderly with memory-problems by activation of the biological clock with light - a randomized clinical trial. *Trials, 11*(1). doi:10.1186/1745-6215-11-19.
Mundey, K., Benloucif, S., Harsanyi, K., Dubocovich, M. L., & Zee, P. C. (2005). Phase-dependent treatment of delayed sleep phase syndrome with melatonin. *Sleep, 28*(10), 1271–1278. doi:10.1093/sleep/28.10.1271.
Musiek, E. S., & Holtzman, D. M. (2016). Mechanisms linking circadian clocks, sleep, and neurodegeneration. *Science, 354*(6315), 1004–1008. doi:10.1126/science.aah4968.
Musiek, E. S., Lim, M. M., Yang, G., Bauer, A. Q., Qi, L., Lee, Y., Roh, J. H., Ortiz-Gonzalez, X., Dearborn, J. T., Culver, J. P., Herzog, E. D., Hogenesch, J. B., Wozniak, D. F., Dikranian, K., Giasson, B. I., Weaver, D. R., Holtzman, D. M., & FitzGerald, G. A. (2013). Circadian clock proteins regulate neuronal redox homeostasis and neurodegeneration. *Journal of Clinical Investigation, 123*(12), 5389–5400. doi:10.1172/jci70317.
Nguyen, K. D., Fentress, S. J., Qiu, Y., Yun, K., Cox, J. S., & Chawla, A. (2013). Circadian gene BMAL1 regulates diurnal oscillations of LY6C hi inflammatory monocytes. *Science, 341*(6153), 1483–1488. doi:10.1126/science.1240636.
Ortiz-Masiá, D., Cosín-Roger, J., Calatayud, S., Hernández, C., Alós, R., Hinojosa, J., Apostolova, N., Alvarez, A., & Barrachina, M. D. (2013). Hypoxic macrophages impair autophagy in epithelial cells through WNT1: Relevance in IBD. *Mucosal Immunology, 7*(4), 929–938. doi:10.1038/mi.2013.108
Panda, S. (2016). Circadian physiology of metabolism. *Science (New York, N.Y.), 354*(6315), 1008–1015. doi: 10.1126/science.aah4967.
Paranjpe, D. A., & Sharma, V. K. (2005). Evolution of temporal order in living organisms. *Journal of Circadian Rhythms, 3*, 7. doi:10.1186/1740-3391-3-7.
Patel, S. A., Velingkaar, N. S., & Kondratov, R. V. (2014). Transcriptional control of antioxidant defense by the circadian clock. *Antioxidants & Redox Signaling, 20*(18), 2997–3006. doi:10.1089/ars.2013.5671.
Patke, A., Murphy, P. J., Onat, O. E., Krieger, A. C., Özçelik, T., Campbell, S. S., & Young, M. W. (2017). Mutation of the human circadian clock gene cry1 in familial delayed sleep phase disorder. *Cell, 169*(2), 203–215.e13. doi:10.1016/j.cell.2017.03.027.
Petrenko, V., Gandasi, N. R., Sage, D., Tengholm, A., Barg, S., & Dibner, C. (2020). In pancreatic islets from type 2 diabetes patients, the dampened circadian oscillators lead to reduced insulin and glucagon exocytosis. *Proceedings of the National Academy of Sciences of the United States of America, 117*(5), 2484–2495. doi:10.1073/pnas.1916539117.
Preußner, M., & Heyd, F. (2016). Post-transcriptional control of the mammalian circadian clock: implications for health and disease. *Pflügers Archiv - European Journal of Physiology, 468*(6), 983–991. doi:10.1007/s00424-016-1820-y.
Prolo, L. M. (2005). Circadian rhythm generation and entrainment in astrocytes. *Journal of Neuroscience, 25*(2), 404–408. doi:10.1523/jneurosci.4133-04.2005.
Reppert, S. M., & Weaver, D. R. (2002). Coordination of circadian timing in mammals. *Nature, 418*(6901), 935–941. doi:10.1038/nature00965.
Sack, R. L., Auckley, D., Auger, R. R., Carskadon, M. A., Wright, K. P., Vitiello, M. V., et al. (2007). Circadian rhythm sleep disorders: part I, basic principles, shift work and jet lag disorders. *Sleep, 30*(11), 1460–1483. doi:10.1093/sleep/30.11.1460.
Sato, T. K., Panda, S., Miraglia, L. J., Reyes, T. M., Rudic, R. D., McNamara, P., Naik, K. A., FitzGerald, G. A., Kay, S. A., & Hogenesch, J. B. (2004). A functional genomics strategy reveals Rora as a component of the mammalian circadian clock. *Neuron, 43*(4), 527–537. doi:10.1016/j.neuron.2004.07.018.
Scheer, F. A., Hilton, M. F., Mantzoros, C. S., & Shea, S. A. (2009). Adverse metabolic and cardiovascular consequences of circadian misalignment. *Proceedings of the National Academy of Sciences, 106*(11), 4453–4458. doi:10.1073/pnas.0808180106.

Shearman, L. P., Sriram, S., Weaver, D. R., Maywood, E. S., Chaves, I., Zheng, B., et al. (2000). Interacting molecular loops in the mammalian circadian clock. *Science (New York, N.Y.), 288*(5468), 1013–1019. doi:10.1126/science.288.5468.1013.

Takahashi, T., Sasaki, M., Itoh, H., Yamadera, W., Ozone, M., Obuchi, K., Matsunaga, N., Sano, H., & Hayashida, K.-I. (2001). Re-entrainment of the circadian rhythms of plasma melatonin in an 11-H eastward bound flight. *Psychiatry and Clinical Neurosciences, 55*(3), 275–276. doi:10.1046/j.1440-1819.2001.00857.x.

Toh, K. L., Jones, C. R., He, Y., Eide, E. J., Hinz, W. A., Virshup, D. M., Ptácek, L. J., & Fu, Y. H. (2001). An hPer2 phosphorylation site mutation in familial advanced sleep phase syndrome. *Science (New York, N.Y.), 291*(5506), 1040–1043. doi:10.1126/science.1057499.

Turek, F. W., Joshu, C., Kohsaka, A., Lin, E., Ivanova, G., McDearmon, E., Laposky, A., Losee-Olson, S., Easton, A., Jensen, D. R., Eckel, R. H., Takahashi, J. S., & Bass, J. (2005). Obesity and metabolic syndrome in circadian clock mutant mice. *Science, 308*(5724), 1043–1045. doi:10.1126/science.1108750.

Wang, Y., Kuang, Z., Yu, X., Ruhn, K. A., Kubo, M., & Hooper, L. V. (2017). The intestinal microbiota regulates body composition through NFIL3 and the circadian clock. *Science, 357*(6354), 912–916. doi:10.1126/science.aan0677.

Webb, A. B., Angelo, N., Huettner, J. E., & Herzog, E. D. (2009). Intrinsic, nondeterministic circadian rhythm generation in identified mammalian neurons. *Proceedings of the National* Academy of Sciences, 106(38), 16493–16498. doi:10.1073/pnas.0902768106.

Wegrzyn, L. R., Tamimi, R. M., Rosner, B. A., Brown, S. B., Stevens, R. G., Eliassen, A. H., Laden, F., Willett, W. C., Hankinson, S. E., & Schernhammer, E. S. (2017). Rotating night-shift work and the risk of breast cancer in the nurses' health studies. *American Journal of Epidemiology, 186*(5), 532–540. doi:10.1093/aje/kwx140.

Xu, Y., Padiath, Q. S., Shapiro, R. E., Jones, C. R., Wu, S. C., Saigoh, N., et al. (2005). Functional consequences of a CKIdelta mutation causing familial advanced sleep phase syndrome. *Nature, 434*(7033), 640–644. doi:10.1038/nature03453.

Xydous, M., Prombona, A., & Sourlingas, T. (2014). The role of H3K4me3 and H3K9/14ac in the induction by dexamethasone of Per1 and Sgk1, two glucococorticoid early response genes that mediate the effects of acute stress in mammals. *Biochimica Et Biophysica Acta (BBA) - Gene Regulatory Mechanisms, 1839*(9), 866–872. doi:10.1016/j.bbagrm.2014.07.011.

Yan, L., & Okamura, H. (2002). Gradients in the circadian expression ofPer1andPer2genes in the rat suprachiasmatic nucleus. *European Journal of Neuroscience, 15*(7), 1153–1162. doi:10.1046/j.1460-9568.2002.01955.

Yang, G., Chen, L., Grant, G. R., Paschos, G., Song, W.-L., Musiek, E. S., Lee, V., McLoughlin, S. C., Grosser, T., Cotsarelis, G., & FitzGerald, G. A. (2016). Timing of expression of the core clock gene BMAL1 influences its effects on aging and survival. *Science Translational Medicine, 8*(324). doi:10.1126/scitranslmed.aad3305.

Zee, P. C., Attarian, H., & Videnovic, A. (2013). Circadian rhythm abnormalities. *CONTINUUM: Lifelong Learning in Neurology, 19*(1), 132–147. doi:10.1212/01.con.0000427209.21177.aa.

Zhang, E. E., Liu, Y., Dentin, R., Pongsawakul, P. Y., Liu, A. C., Hirota, T., et al. (2010). Cryptochrome mediates circadian regulation of cAMP signaling and hepatic gluconeogenesis. *Nature Medicine, 16*(10), 1152–1156. doi:10.1038/nm.2214.

Zhang, R., Lahens, N. F., Ballance, H. I., Hughes, M. E., & Hogenesch, J. B. (2014). A circadian gene expression atlas in mammals: Implications for biology and medicine. *Proceedings of the National Academy of Sciences, 111*(45), 16219–16224. doi:10.1073/pnas.1408886111.

Zisapel, N. (2001). Circadian rhythm sleep disorders. *CNS Drugs, 15*(4), 311–328. doi:10.2165/00023210-200115040-00005.

Zisapel, N. (2007). Sleep and sleep disturbances: Biological Basis and clinical implications. *Cellular and Molecular Life Sciences, 64*(10), 1174–1186. doi:10.1007/s00018-007-6529-9.

13

Deep Learning for Automated Disease Detection

Neeraj Yadav and Somya Tripathi
JSS Academy of Technical Education

Ekam Singh Chahal
Jaypee Institute of Information Technology

Suruchi Sabherwal
JSS Academy of Technical Education

CONTENTS

13.1 Introduction: Background and Driving Forces

Deep learning (DL) is an artificial intelligence subfield concerned with algorithms that emulate the function and structure of the human brain artificially named Artificial Neural Networks (ANN). DL is heavily used in the realm of healthcare in diverse dimensions, thus increasing the possibility of diagnosis for most diseases that had remained unidentified in the past (Alwan, 2011; Lakomkin et al., 2019). This chapter highlights the various datasets available in the medical science domain and their usage in biomedical research, giving their characteristics, to what extent these datasets have been used for disease detection and the opportunities for further exploration, with the challenges faced during implementation. The Electronic Health Record (EHR) dataset utilized for biomedical research, which comprises some discharge summaries and progress reports of various patients, is one of the datasets studied in this chapter. The second dataset is image data (Chollet, 2016), the third is skin cancer data, the fourth is cardiac MRI segmentation data, and the fifth is lung cancer detection data. Some of the research areas for which the DL techniques mentioned in this chapter can be useful include biomedical research, skin cancer detection, lung cancer detection, etc.

DOI: 10.1201/9781003046431-13

A person with a recent cardiac arrest or an aberrant heart rhythm is an appropriate example. Due to this reason, CHA2DS2-VASc may be acquired from as marginal as 25 positive labels. Every year, a massive amount of unlabeled sensor data points are created, containing valuable information such as heart rhythm resting and changes in heart rate. It is associated with health conditions as disparate as heart failure, atrial fibrillation, sleep apnea, diabetes, and irritable bowel syndrome, and is generated by wearables, such as Apple Watch and Fitbit. There are several challenges to the use of heart rate-related sensors from a medical perspective. First, there are chances of the presence of significant errors in sensors. Second, to sustain battery life, the calculation rate varies. Third, the use of wearables in an outpatient setting, day-to-day behaviors, such as stress, walking, exercise, drinking coffee, or consuming alcohol, can confuse basic heuristics.

In the United States, skin cancer is the most common disease. In this chapter, we have used the ISIC Database with 1,280 images for melanoma detection. Every year, it kills more than 10,000 people in the United States. A person infected with the disease has a 5-year cure rate of 98% if treated in the first stage. If cancer spreads to nearby cells, the cure rate degrades to 63%, and the cure rate plunges to a marginal value of 17% if found in the last stage. Since melanoma spreads inimically, monitoring it and discovering skin lesions have become very important (Esteva et al., 2015). Experiments include traditional models such as support vector machines (SVMs) and ANNs. The human heart is a machine that could continuously work for centuries without any failure. One of the simplest ways for determining how well the heart functions is to calculate its ejection fraction, which means that after the heart relaxes in its diastole, it fills with blood, estimating the amount with which it transfers the blood when it contracts in its systole. The initial step in calculating this metric is to *segment* (delineate the area of) the ventricles in cardiac pictures. Deep neural networks (NNs) have achieved substantial accuracy at recognizing patterns from data that have both noise and complexity. These networks have also achieved low error rates in applications such as automatically detecting diabetic retinopathy from input pictures, detecting skin cancer from mobile phone cameras, and initiating electronic medical data comprising a person's health problems (Esteva et al., 2017; Tison et al., 2017; He et al., 2015, 2016a, 2016b; McManus et al., 2013). We have summarized the existing research going on in this area in the following paragraph.

A novel and interesting work has been proposed by Joseph et al. on object detection approach termed as "You Only Look Once" (He et al., 2016a, 2016b; Ren et al., 2015; Redmon et al., 2016). This approach frames object detection as a regression problem with the objective of separating bounding boxes spatially and associating class probabilities. In this approach, a single NN forecasts bounding boxes and associated probability in one evaluation. In previous approaches, for detecting lung cancer, Kisgsley et al. used a 3D-deep CNN to develop a model for automated diagnosis (Kuan et al., 2017). In this process, they have taken CT scan images provided by Kaggle. Also, they have utilized an additional dataset, The Lung Nodule Analysis dataset (2016). Using both datasets, they have trained the framework in two stages. The framework comprises four separate NNs: nodule detector, malignancy detector, nodule extractor, nodule classifier, and patient classifier. These NNs work in a pipeline fashion, classifying the probability of a patient as having cancer or not. They have presented the performance results of each component of the overall framework in the concluding section. A novel four-stage pipelined-based model for nodule detection in lung has been done by Rafael et al. (2021). This is a data-driven process with a flexible and configurable four-stage pipeline wherein the first stage identifies the nodules. The second stage reidentifies them from CT scans of a patient, the third stage quantifies their growth, and the fourth stage predicts their malignancy. They have utilized deep CNN for this model. The paper concludes by giving the present solution and future directions. In another similar research, to ease the burden of radiologists in spending hours studying CT scan images to detect lung cancer in patients (Shi, 2018), researchers implemented DL algorithms for accurate diagnosis of lung cancer using Lung Image database consortium (LIDC). They compared the performances of two 2.5D CNN model, two 3D CNN models, and one 3D CNN with a spatial transformer network module for detecting lung nodules. Results indicate that 2.5D-1 outperforms 2.5D-2. Also, 3D models perform better than 2.5D models. Haofu et al. have worked on detecting skin disease using disease-targeted characteristics and lesion-targeted characteristics (Liao et al., 2016). After implementation, authors have achieved a mean average precision of 0.70 in case of lesion-targeted datasets compared to other types.

13.2 Dataset Description

In this chapter, we have explored a total of six datasets. First, the EHR dataset includes progress notes and discharge summaries collected from three institutions: Partners HealthCare, Beth Israel Deaconess Medical Center, and University of Pittsburgh Medical Center. The organizers personally annotated 826 medical records, which served as a reference for all three tasks in the process. We have used a total of 349 annotated medical notes for the training process. The remaining 477 annotated medical notes were used for the testing process and for evaluating the performance of the participating system. This innovative hybrid approach, which blends rule-based techniques with machine learning (ML), is utilized to properly recognize medical individuals and their arguments.

Images were gathered for creating training datasets to generate manually annotated datasets capable of extracting clinical entities such as medical problems, tests, and therapies. Phase images of bacterial microcolonies were collected from bacterial cells. Images of marked nuclei, which are fluorescent by nature, and cytoplasm phase images were taken from the dataset of mammalian cells. To classify each pixel of the training dataset sample as either cellular interior, boundary, or background, image was applied. Annotation of a microscope picture with a maximum of 500 nuclei, 300 bacteria, or 100 mammalian cells was abundant in producing a training dataset that led to a high-performance Conv-Net; the annotation job takes around 2–4 hours. A set of representative images for each distinctive class was created by taking a small sample patch around each annotated pixel. We observed that the system performs better at run time if sampling of patches is done, as it conserves the GPU memory. We applied random down sampling on the collection of images so that different classes have an equivalent number of representations in the collection.

International Skin Imaging Collaboration (ISIC) and Dermnet's are the two most common public datasets containing images of melanoma. ISIC dataset has unbalanced data with 1,031 benign images and only 249 malignant images (Li et al., 2020; Liao et al., 2016). Dermnet's dataset has 23,000 images, out of which 500–2,500 images belong to various classes of skin disease (23 classes), 1,000 images belong to the melanoma class, and the remaining 22,000 images create a very heavy and unbalanced dataset. The study done in this chapter focuses only on the ISIC dataset. The dataset from the Acropolis Convention Center contains only 243 physician-segmented images. Because of the tiny dataset, it would be impossible to presume generalization to unknown objects! It is a common practice to apply affine transformations to data, such as zooms, random rotations, shears, and translations. Furthermore, we also used elastic deformations, which locally compresses and stretches the image.

Such augmentations prevent the network from memorizing only the training instances and help the network realize that the RV is rigid, with the shape of a crescent object that may grow in several directions (Redmon et al., 2016; Ren et al., 2015). We created the transformations on the fly in our training framework so that the network, in every iteration, has a new set of random transformations. By standardizing the background pixel intensities to lie between 0 and 1, we note that only 5.1% of the pixels are part of the RV cavity throughout the whole dataset. While building the loss functions, we observed that the unweighted average performed best.

The dataset used has patients' CT scan images with their cancer status from the dataset Lung Module Analysis 2016 (LUNA16), which gives module annotations. It has its issues, as this dataset does not include a patient's cancer status (Kuan et al., 2017). Therefore, we have used both datasets to train the framework in the following two stages:

- The Lung Module Analysis 2016 (LUNA16) dataset contains 888 axial CT scan pictures of chest cavities from different individuals acquired straight from the LIDC/IDRI database.
- A total of 2,101 axial CT scan images of the chest cavities of various patients are present in the Kaggle Data Science Bowl 2017 (KDSB17) dataset.

Electronic medical records contain important data about patients. One can obtain data from various sources, including empirically from laboratories or through different wearables, which indirectly give continuous patient health data. To acquire a trained DNN that can reliably predict a given patient's diagnoses, we obtained patient data from wearables; thus, integrating both the data will offer us a dataset with a lot of data about the patient.

13.3 Data Processing and Techniques

This section explains the various approaches and techniques used for each dataset (ML approach, Hybrid Natural Language Processing (NLP)-based approach for assertion classification, and concept extraction) described in this chapter with a short description of the experiments conducted.

13.3.1 Electronic Health Record

An exploratory analysis of ML-based methods to specify the existing medical terms and document a new hybrid abstraction technique which shows better performance. The approaches followed in the EHR dataset include the following:

- ML-based approaches for EHR: The role of the EHR task includes the recognition of medical entity borders and the allocation of their linguistic forms (problem, evaluation, or treatment). Every word is then assigned a label with the help of a multiclass classifier. In this processing, we have investigated various ML techniques and feature kinds.

- ML algorithms: In EHR, SVM, and Conditional Random Fields (CRF) are the two ML algorithms that are most commonly applied. A polynomial of degree two-based kernel function, a double-sized context window, and a pairwise (one-against-one) multiclassification approach based on the tasks identified by EHR are used in this processing.
- Types of features:
  - NLP systems' lexical and semantic information: Mostly structured definitions and forms of semantics recognized by NLP systems (Dai and Le, 2015). For this task a total of three different NLP systems were applied: (i) KnowledgeMap; (ii) MedLEE; and (iii) a Dictionary-based Semantic Tagger, which uses public and private terminologies and processes them for evaluations, medical conditions, evaluations, and cure.
  - Discourse Information: Parts of the clinical notes and observation archives are collected by personalized software built for the task details.
- Experiment: For training, we applied a five-fold cross-validation method. We found out that test data of one-fold was the best-tailored parameter for individual ML algorithms. We started with a benchmark approach that essentially requires bag-of-word features, and then added additional feature types and recorded their related outcomes to assess the effects of various feature types.
- Assertion classification using ML-based methods: The function of assertion classification is to classify a medical problem into one of the six possible labels (Present, Absent, Conditional, Hypothetical, Possible, and Not associated with the patient). We developed an SVM-based multiclass classifier using the LIBLINEAR package, which works very efficiently on enormous datasets. We structurally studied the contributions of every individual feature type using the 349 training notes for the assertion task.
- Concept extraction and assertion classification using a hybrid NLP system: A new hybrid approach was built for clinical concept extraction and to classify the assertion. The hybrid system has four components: (i) the CRF based on the EHR module that utilizes optimized features sets and parameters as shown by the training dataset; (ii) a post-processing approach that utilizes heuristic-based guidelines to rectify potential errors and increase performance; (iii) a combination module that maximizes multiple classifier's results; and (iv) SVM-based assertion classifier.
- Post-processing module: There are many rules to rectify false negatives, like the system-missed abbreviations. Some rules are used to evaluate a semantically ambiguous entity's correct type depending on its meaning. A medicine-related word, for instance, is usually classified as "Treatment."
- Combination module: The combination module functions in the following ways:
  - It produces the *intersection* of all the medical entities extracted from three different classifiers. The result constitutes only the medical entities extracted by these three classifiers.
  - A *union* operation is conducted between the step 1 intersection and the output from the fourth classifier, indicating the addition of any entity obtained by either the first step or the fourth classifier. Figure 13.1 gives a diagrammatic overview of the MedNet system with multiple modules.

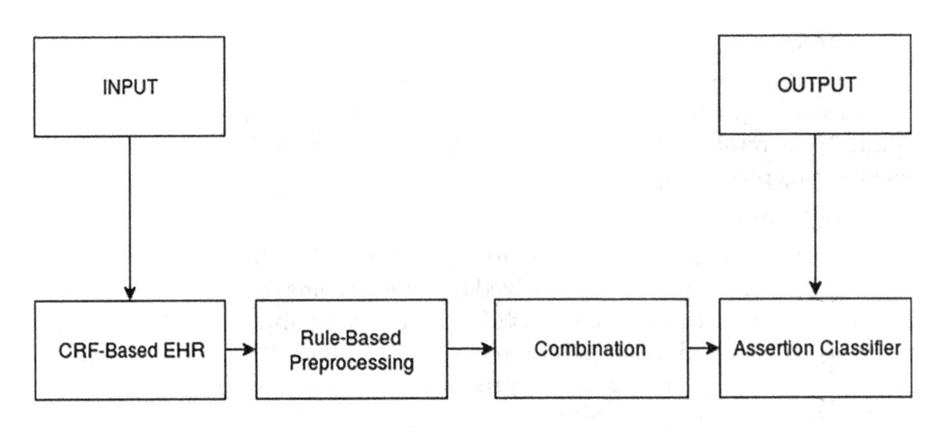

FIGURE 13.1
Overview of MedNet system.

13.3.2 Image Data

Robustness to illumination and microscope camera variability is one of the concerns about the efficiency of a trained ConvNet (CNN). This involves the following stages:

- Training data normalization
    - First, we normalize every training image before using them to train a ConvNet by dividing it by the average pixel value of that image.
    - For normalizing, we have used an average filter to measure the normalized image's local average, and then it is used to subtract it from the normalized image.
    - After sampling a given image, data augmentation methods such as random rotation (0, 90, 180, or 270 degrees) and random reflection on the image before using our training dataset.
    - After data normalization and data augmentation, the dataset contained a total of ~200,000–400,000 patches.
    - The training dataset associated with the mammalian cytoplasm consisted of two distinct channels: (i) an image and (ii) a phase image.
- Segmenting image
    - We have passed the whole image to a trained ConvNet, which generates a pixel-wise classification, i.e., the segmentation mask for that input image. It is achieved with the help of a fully convolutional implementation of ConvNets, which can directly work on a whole image. For improving segmentation accuracy, we have used model parallelism (averaging of five trained network results).
    - ConvNet classification predictions were accurate. However, we discovered that creating binary segmentation masks still required further refining. For example, for pictures containing mammalian cells, classification predictions from ConvNets, when combined with a nuclear marker, direct the active contour algorithm, improving the segmentation process.

13.3.3 Skin Cancer

There are only 249 images of Malignant Melanoma in the ISIC dataset that makes this dataset highly unbalanced (Li et al., 2020).

- Preprocessing: In preprocessing, first, we cleaned the raw images followed by segmenting the skin lesion. After this, we reshaped the images to suitable dimensions according to the NN's input layer. Figure 13.2 represents several pictures derived from the ISIC Database.
- Image processing: This consists of gathering the raw data set from the ISIC Database. Images in the dataset contained various forms of noise. Some issues with the raw data included:

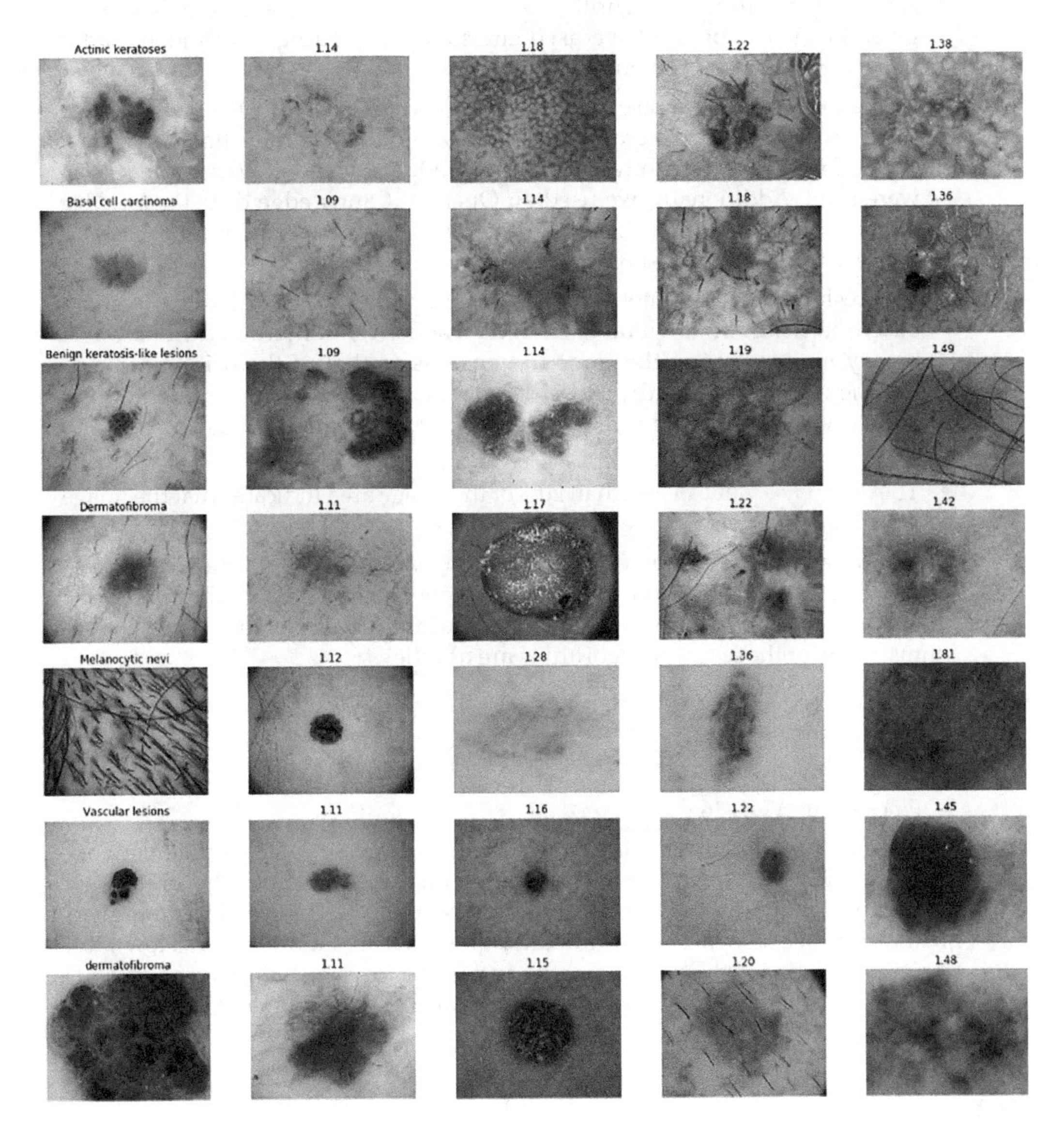

FIGURE 13.2
Raw images of skin cancer. (Derived from ISIC Database, 2016.)

- Vignetting that existed in some images; the existence of bright-colored Band-Aids in a few images.
- The dataset was imbalanced with a ratio of 1:4 for malignant and benign lesions. Models often get biased because of an imbalanced dataset; to overcome this, we constructed a balanced subset of the training and test dataset for both training and evaluation phases. The largest training dataset consisted of 346 images in total.
- After importing each image, we applied a light Gaussian kernel blur to smooth the edges, reducing the outline of any light-colored and thin hair in the provided image.
- After this, we used an algorithm to decide whether vignetting exists in the given image or not. We have used circle crop vignetting which identified if there was a need to crop out the corners.
- In the majority of the cases, moles/lesions were located in the center of the image, which meant the cropping process around the corners had a marginal effect. To increase the contrast of the image, OpenCV and grayscale conversion were used. Additionally, we used an OpenCV Canny edge detector for edge detection and contours. Only the contours having the maximum areas and above a predefined threshold were kept.
- A threshold filter was constructed by computing a normalized threshold based on mean pixel value of images. The newly constructed filter was a binary matrix having the same dimensions as that of the original image. We then used a manually labeled filter to compute lesion segmentation's accuracy and compared the newly generated filter's pixel areas with labeled masks.
- The difference value of less than 20% of the image area indicates that the image is segmented correctly, else it is incorrectly segmented.

- Image formatting (NN): For correct resolution, we converted all the images into square-shaped images. As the images were now fully processed, they were in a suitable matrix form and were utilized as an input for different learning algorithms. Some of the learning algorithms are as follows:
  - Image formatting: Keras CNN
  - Logistic regression
  - NN
  - Fine-tuning VGG-16
- Transfer learning: Transfer learning is used on a Convolutional Neural Network (CNN) model. In the transfer learning process, we can apply the pretrained network weights directly to a new problem statement. Transfer learning gives great results because these pretrained weights consist of more generic, low-level features like edges which act as an important factor while classifying various image types. Figure 13.3 shows the performance of errors generated from various models.

13.3.4 Cardiac MRI Segmentation

For finishing the training and delivery within 4 weeks, we implemented the U-Net model.

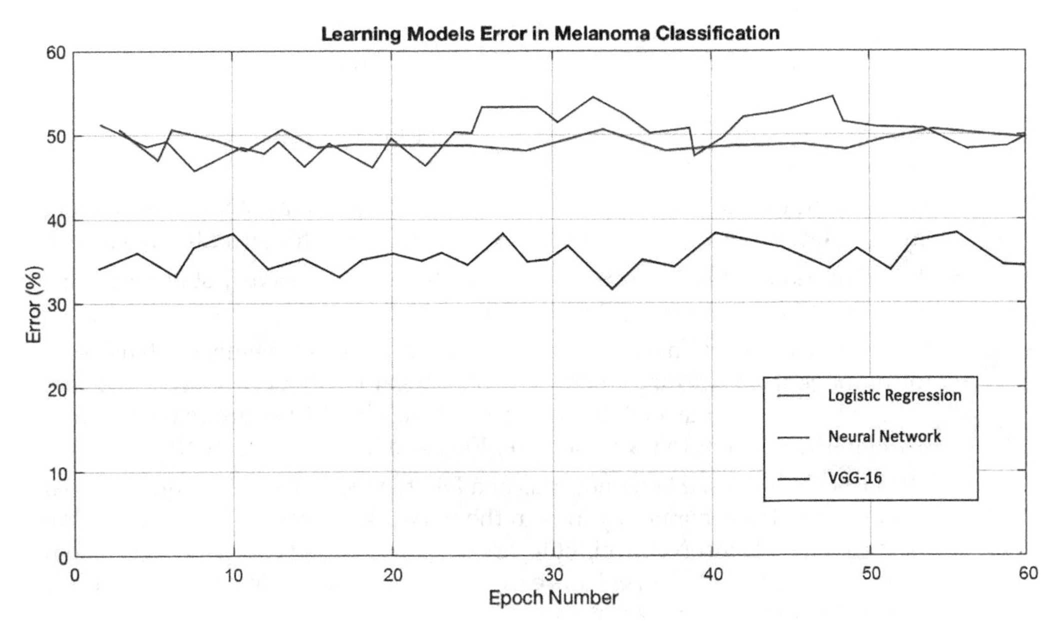

FIGURE 13.3
Comparison of test and training error for three main models.

- U-net: the baseline model
  - Because of success in biomedical segmentation tasks, we chose to implement a U-Net-based model. A standard U-Net model's architecture consists of a contraction path that downsamples an input image into high-level features, followed by an expansion path that uses this features-related information to construct a segmentation mask in which each pixel is classified into one of the possible classes. The main property of a U-Net-based model is its "copy and concatenate" links, which make the transfer of information from the initial stage (encoder) to the later stages (decoder) feasible.
  - We adopted the U-Net model for our purposes by reducing the number of downsampling layers from four to three when compared to the original model, because our input images were about half the size of those used by the authors of U-Net. To keep the image of the same shape, we have added zero paddings to our convolutions (as opposed to unpadded). We have used Keras for model implementation.
  - Dilated U-Nets: Global receptive fields.
  - Segmentation of organs requires some information on the global background, like the arrangement of organs. We observed that the deep layer neurons of the U-Net model had receptive fields that were only 68×68 pixels in dimension. When creating the segmentation mask, no part of the network could "see" the entire image, and at the same time, it added a global context. But, the initial layer neurons have receptive fields that can span the entire input image, which is called "Dilated U-Net."
  - Dilated DenseNets: Multiple scales at once.

  - To produce a global context, we used dilated convolutions only. We could use "copy and concatenate" connections between all the convolutional layers that have the same dimension.
  - This model is known as a "Dilated DenseNet," which combines two concepts: DenseNets and dilated convolutions.
  - We used the output of the first convolutional layer as an input to all subsequent layers in DenseNets, and repeated the same procedure for all additional layers.
  - At publication, DenseNets performed much better than every state-of-the-art model for CIFAR and ImageNet classification tasks.
  - However, DenseNets have a major drawback: They are memory intensive as the number of feature maps increases quadratically with network size. Instead, we used "transition layers" to decrease the number of feature maps halfway through the network to train their 40-, 100-, and 250-layer DenseNet.
  - The last convolutional layer neurons of a Dilated DenseNet can acquire global context and also generate features in the network at every previous level. We used a Dilated DenseNet of eight layers deep and varied the growth rate from 12 to 32. The Dilated DenseNet is extremely parameter efficient. There were a total of 190K training parameters.
- Training
  - We need a method to evaluate our model's performance quantitatively during training. The segmentation competition organizers have opted to use the Dice coefficient.
  - We utilized normal pixel-wise cross-entropy for the loss function, but we also evaluated the data using a "soft" dice loss.

The rounding function cannot be interpreted as a loss function because it is not differentiable. For reporting the classification performance, we use the "hard" dice coefficient. We varied the following hyperparameters in our training process:

- Batch normalization
- Learning rate
- Dropout
- Growth rate (for Dilated DenseNet)

The average pixel value of each image in the dataset was subtracted to normalize it, and then the normalized image was divided by its standard deviation. Unweighted pixel-wise cross-entropy is chosen for the loss function since it outperforms dice and weighted losses. Because the validation performance decreased, dropout was not employed. Because batch normalizing impacted performance for the U-Net and Dilated U-Net models, Dilated DenseNet employed only preactivation batch normalization. For a better balance between size and model performance, Dilated DenseNet had a rate of growth of 24. Except for Dilated DenseNet, which had a batch size of three owing to memory constraints on a 16 GB GPU, a batch size of 32 was used.

13.3.5 Lung Cancer Detection

The presence of pulmonary nodules, which are masses of tissue that grow in and around the lung, is a key feature of lung cancer. Each CT scan consists of multiple 2D slices, which are stored in a DICOM format. Figure 13.4 displays a CT scan of a patient's chest cavity cross-section (Thakur, 2018).

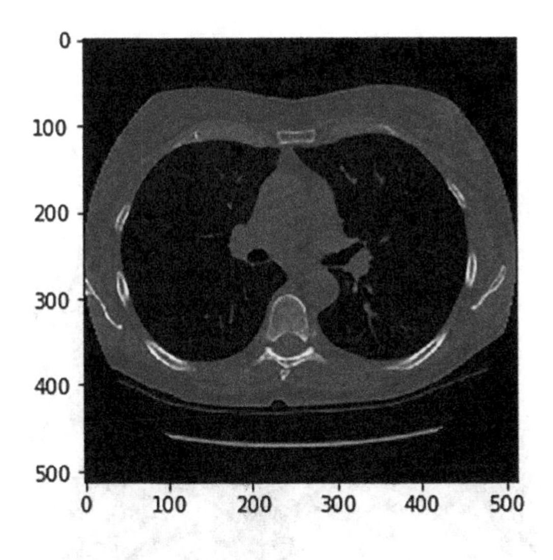

FIGURE 13.4
A chest cavity cross-section. (Kaggle Data Science Bowl 2017 dataset.)

Figure 13.5 shows a 3D visualization of CT scan images. These modules can be malignant or benign and can be visualized using a CT scan technique.

We considered the following method to design an approach for computer-aided diagnosis (CAD) for lung cancer:

- A network that learns the characteristics of cancer and predicts the risk of cancer from patient CT scans.
- A method in which we identify modules first, then classify each one's malignancy, and finally find the patient's chance of cancer.

The first approach is easy to implement but is less reliable and has not been successful in several situations, but the second approach turned out to be more effective and accurate because it relies on a multistage pipeline focusing on pulmonary nodular detection and identification.

- Modeling involves the following steps:
  - Nodule detection with LUNA16: When a grid cell interacts with a bounding box of a nodule, we treat that cell as being labeled "hasnodule"; other cells are labeled "no nodule," and then we divide the cell's bounding box into a uniform grid of each volume crop in LUNA16.
  - Malignancy detector with KDSB17: Weights of a trained nodule detector network were configured in this network, which returned the class distributions (benign, malignant, no-nodule).

Based on the patient's cancer status, cells are identified as "has-nodule" by the nodule detector and then later classified as "benign" or "malignant," and the remaining cells are categorized as no-nodule. Two separate heuristic approaches are compared for label assignment to nodules:

- patient-label strategy
- largest-nodule strategy

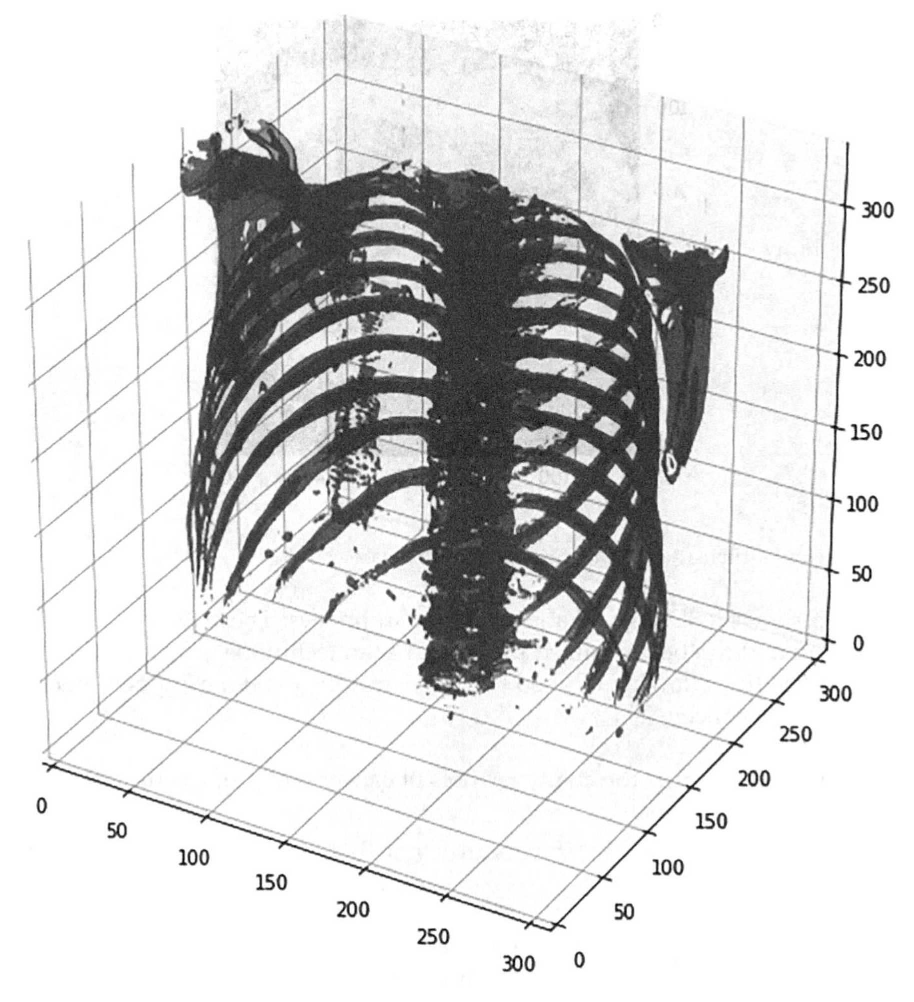

FIGURE 13.5
3D visualization for visualizing the CT scan images.

The technique of patient labeling is a simple approach, where they mark all cancer patients' nodules as malignant. The malignant sensor, on the other hand, utilizes the patient-label heuristic. It is a calculated simplification through the nodule extractor to reduce the computing overhead for backpropagation. The classifier has no such overhead since it operates on nodules that have been formed from grid cells by the extractor. Here, Figure 13.6 shows the malignancy rate in the lungs. The size of the rectangles indicates estimated malignancy.

13.4 Conclusion

EHRs are the most common data sources used for risk prediction on health issues worldwide. It also gives good opportunities and challenges. EHR dataset represents a vast, underused data resource for biomedical research. Use of awareness exploration and

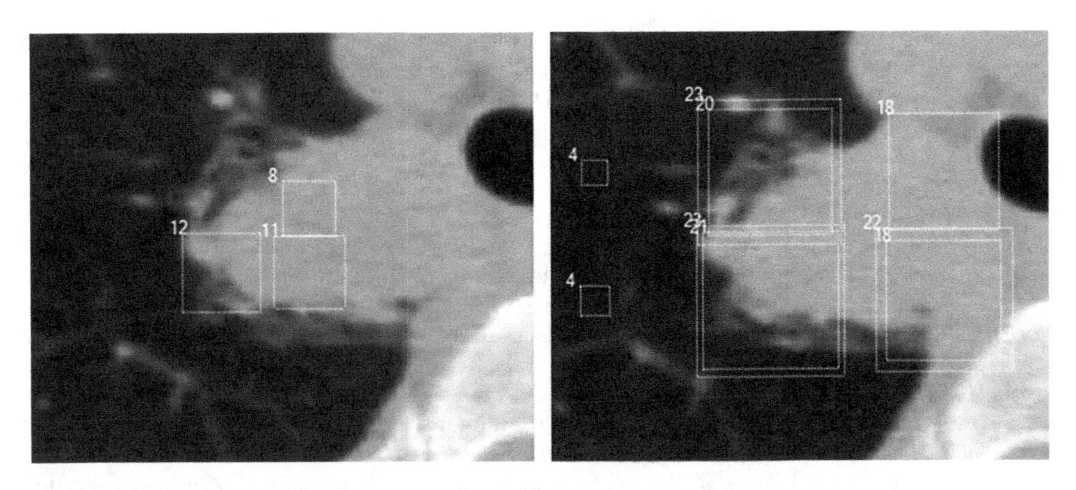

FIGURE 13.6
Malignancy is highlighted in white boxes.

machine-learning methods can be done for exploring novel arrangements in patient data and for designation and predictive purposes, such as results or risk estimation. It can extend current EHR decision support systems, which link to available patient data with clinical instructions to provide cooperation to the physician at the point of care. After evaluating four learning algorithms on skin cancer-based ISIC dataset, we can conclude that the VGG-16 model gives a 35% error rate for six epochs, while others perform poorly. For cardiac MRI segmentation, dilated dense nets with eight layers have been implemented by varying growth rate from 12 to 32 as they are extremely parameter efficient. Model services for a given structure depend heavily on the size of the prediction window and observation window size, and the dissimilarity, quantity, and frequency of the available data. DL is more precise, requires less time to treat the effects of segmentation, can segment many cell types, and can differentiate in the same image between different cell lines. This chapter highlights the working and application of DL techniques on a range of datasets. The readers may find several open challenges from related areas of automated disease detection.

References

Alwan, A. (2011). *Global Status Report on Noncommunicable Diseases 2010*. World Health Organization.

Chollet, F. (2016). Building powerful image classification models using very little data. Keras Blog, 5.

Dai, A. M., & Le, Q. V. (2015). Semi-supervised sequence learning. *Advances in Neural Information Processing Systems*, 28, 3079–3087.

Esteva, A., Kuprel, B., Novoa, R. A., Ko, J., Swetter, S. M., Blau, H. M., et al. (2017). Dermatologist-level classification of skin cancer with deep neural networks. *Nature*, 542(7639), 115–118.

Esteva, A., Kuprel, B., & Thrun, S. (2015). Deep networks for early stage skin disease and skin cancer classification. Project Report.

He, K., Zhang, X., Ren, S., & Sun, J. (2015). Delving deep into rectifiers: surpassing human-level performance on imagenet classification. In *Proceedings of the IEEE International Conference on Computer Vision* (pp. 1026–1034).

He, K., Zhang, X., Ren, S., & Sun, J. (2016a October). Identity mappings in deep residual networks. In *European Conference on Computer Vision* (pp. 630–645). Springer, Cham.

He, K., Zhang, X., Ren, S., & Sun, J. (2016b). Deep residual learning for image recognition. In *Proceedings of the IEEE Conference on Computer Vision and Pattern Recognition* (pp. 770–778).

Kuan, K., Ravaut, M., Manek, G., Chen, H., Lin, J., Nazir, B., et al. (2017). Deep learning for lung cancer detection: tackling the Kaggle data science bowl 2017 challenge. arXiv preprint arXiv:1705.09435.

Lakomkin, E., Zamani, M. A., Weber, C., Magg, S., & Wermter, S. (2019 May). Incorporating end-to-end speech recognition models for sentiment analysis. In *2019 International Conference on Robotics and Automation (ICRA)* (pp. 7976–7982). IEEE.

Li, H., Pan, Y., Zhao, J., & Zhang, L. (2020). Skin disease diagnosis with deep learning: a review. arXiv preprint arXiv:2011.05627.

Liao, H., Li, Y., & Luo, J. (2016 December). Skin disease classification versus skin lesion characterization: Achieving robust diagnosis using multi-label deep neural networks. In *2016 23rd International Conference on Pattern Recognition (ICPR)* (pp. 355–360). IEEE.

McManus, D. D., Lee, J., Maitas, O., Esa, N., Pidikiti, R., Carlucci, A., et al. (2013). A novel application for the detection of an irregular pulse using an iPhone 4S in patients with atrial fibrillation. *Heart Rhythm*, 10(3), 315–319.

Rafael-Palou, X., Aubanell, A., Ceresa, M., Ribas, V., Piella, G., & Ballester, M. A. G. (2021). Detection, growth quantification and malignancy prediction of pulmonary nodules using deep convolutional networks in follow-up CT scans. arXiv preprint arXiv:2103.14537.

Redmon, J., Divvala, S., Girshick, R., & Farhadi, A. (2016). You only look once: unified, real-time object detection. In *Proceedings of the IEEE Conference on Computer Vision and Pattern Recognition* (pp. 779–788).

Ren, S., He, K., Girshick, R., & Sun, J. (2015). Faster R-CNN: Towards real-time object detection with region proposal networks. *Advances in Neural Information Processing Systems*, 28, 91–99.

Shi, J. (2018). Lung nodule detection using convolutional neural networks. *Electrical Engineering and Computer Sciences*. University of California at Berkeley, Berkeley, CA.

Thakur, M. (2018). *Detection of Lung Cancer using Image Processing.*

Tison, G. H., Singh, A. C., Ohashi, D. A., Hsieh, J. T., Ballinger, B. M., Olgin, J. E., et al. (2017). Cardiovascular risk stratification using off-the-shelf wearables and a multi-task deep learning algorithm. *Circulation*, 136(suppl_1), A21042–A21042.

14

Time Series Analysis and Trend Exploration of Stock Market

Ratik Puri, Jagriti Bhandari, Ritik Gupta, and Adwitiya Sinha
Jaypee Institute of Information Technology

CONTENTS

14.1 Introduction and Related Study

Predicting stock prices has been an important field of research since algorithmic trading and thus stock market predictions became popular. Most of the projections use regression analysis, but the results are not so convincing. The stock market data have always been a time series data that have trend over time, which is supposed to be targeted. Stock Analysis and Trend Forecasting were done based on the study of existing methods (Menon et al., 2019). The aim was to build a high-accuracy model that is relevant to the current day and age. The conventional approaches used by some authors (Somani et al., 2014; Zhao and Wang, 2015; Sayavong et al., 2019) failed to attain real-time accuracy. Therefore, it has its limitations when it comes to sudden changes being noticed over time. This research uses the time series-based predictions as used by the authors Ryota and Tomoharu (2012) and uses the sentiments extracted from Twitter for or related posts of that company. The amalgamation of sentiment analysis with time series analysis was a step to obtain precise justifications over the stock market's volatility. Besides this, algorithms such as Moving Average, Exponential Smoothing, ARIMA, and LSTM were used to obtain the most precise predictions. In another research, the authors (Kumar and Anand, 2014) used simple ARIMA for the prediction process. Machine learning techniques are also widely used to analyze time series stock data (Sharma et al., 2017).

This chapter focuses on the building of the best model for the analysis of stocks. The use of conventional approaches such as fundamental and technological analysis does not guarantee the prediction's reliability. The project aims at predicting the stock prices based on the costs of the previous days. It uses the time series-based predictions and the sentiments extracted from Twitter for or related posts about a company. This helps when a very well-performing stock suddenly falls because of certain reasons, and this could be

DOI: 10.1201/9781003046431-14

extracted by analyzing the sentiments of the Twitter posts. The application combines the sentiments extracted from Twitter posts and the prediction from the data to provide more accurate results.

14.2 Dataset Description

In our research, we have incorporated the trends and news that might affect the stock prices of a particular company. The stock prices depend not only on the previous trends, but also on many other factors that include the announcements made by a company or the announcements made by the government for a particular company, or perhaps elections in the country. We have tried to pick the Twitter posts and analyze the sentiments of the bars and what effect it will have on the stock prices and have used both these values to predict the final pricing of the share.

The dataset was extracted by the *Yahoo Finance* API, which gave various stock attributes, such as its opening, closing price, etc., and the stock symbol and dates over which analysis was to be done. The research was done over multiple stocks, and a Web app was made for easy access and analysis of the data by any user. Stock data could be extracted for any particular company using parameters such as its stock symbol, start date, and end date of the analysis period. The data returned by default were for each day between the dates specified. A table enlisting all the stock symbols mapped by the company was also generated in the Web app to ensure a smoother user experience.

14.3 Requirement Analysis and Solution Approach

Predicting stock prices has been an important field of research since algorithmic trading and thus stock market predictions became popular. Most of the projections are done using regression, but the results are not so convincing. Stock market data have always been a time series data that have the trend over time and are supposed to be targeted. We have used the best of the algorithms to predict the stock prices accurately and have incorporated the trends and news that might affect the prices of stocks of a particular company. The stock prices depend not only on the previous trends but also on many other factors that include the announcements made by the company or the announcements made by the government of a particular company, or perhaps elections in the country. We have tried to pick the Twitter posts and analyze the sentiments of the bars and what effect it will have on the stock prices and have used both these values to predict the final pricing of the share.

In 2018, Elon Musk tweeted that he was contemplating whether to make his company private, which ultimately cost him his chairmanship. Again, a tweet was posted on May 2, 2020, stating that the stock price of Tesla was too high. This resulted in its valuation to lower by $14 Billion. Moreover, all these tweets have had a lot of impact on the share prices, which needs to be incorporated into the prediction of stock prices that need to be addressed. Therefore, we have used Twitter posts to study the sentiments about a particular company's stock prices, resulting in a much better prediction and more reliable results.

The project uses live data to predict stock prices. This has been done to make a correct and precise prediction of the expenses.

- We have extracted live data from an API that takes the stock symbol and the start and end dates of the duration that we wish to analyze.
- These data, when retrieved, consist of the opening price, closing price, volume of stocks, etc. The analysis is done based on the closing prices of the stores.
- We have drawn the graphs to study the trends and patterns the experts can use to predict if they wish to do so themselves.
- Since some algorithms need the data to be stationary, the variance and point should be constant with time. Therefore, it is mandatory to test whether the data that we have are static or not. To test its stationarity, we have used a test called the Dickey–Fuller Test (DFT).
- If the data are not stationary, we apply transformations like taking logarithms of the data or taking the difference between observations to make it fixed.
- Once the data are stationary, maybe by applying transformations or by passing the DFT, we feed the data to various algorithms. Notably, a few algorithms, such as LSTM, do not require the data to be stationary. Therefore, in those cases, the data are directly fed to them.
- Every algorithm function returns three values, i.e., the price on the nth day, the forecasted value for the $(n+1)$th day, and the error model. We have also plotted graphs to analyze the trends of original and forecasted values.
- We also retrieved a few tweets from Twitter that mentioned the stock symbols we are interested in examining.
- Based on those tweets, we perform a sentiment analysis that helps us calculate and conclude whether the polarity of that stock is positive or negative in the market.
- Finally, we suggest whether investing in stocks is a good option or not based on the calculated values.

We have used the following algorithms after DFT testing to predict future values. All these algorithms forecast values for our stock symbol, but we have considered the one with the minimum value of error. Some of the algorithms used are as follows:

- **Moving Average:** One of the simplest algorithms to analyze data, moving average, is commonly used to smooth out short-term fluctuations and highlight longer-term trends or cycles. Predictions are generated using simple averaging over a fixed number of previous values (window). It is often used in economics to analyze the gross domestic product, wages, or other time series macroeconomics. A simple moving average (SMA) in essential services is the unweighted mean of the preceding n results. In maths and science, though, the standard is typically taken from an equivalent amount of data on each side of a central attribute. This means that the mean differences are associated with the data changes rather than overtime.
- **Exponential Smoothing:** Exponential smoothing is used with the exponential window function to smooth time series data. Whereas past results are weighted similarly in the basic moving average, exponential functions allocate weights that decline steadily over time. It is a quickly understood and easily implemented

technique to render any user-dependent decision, such as seasonality, based on prior assumptions. Exponential smoothing is also used to evaluate the data from the time sequence.

- **ARIMA:** ARIMA stands for Autoregressive Integrated Moving Average. It is a powerful algorithm for forecasting. It is either used to predict future values or it is used to better understand the data. ARIMA's AR part signifies that the evolving interest variable is being regressed at its own lagged (i.e., prior) values. The MA part suggests that the regression error is in essence a continuous mixture of error factors, whose values happened in the past at various periods at the same moment. The I (for "integrated") indicates the difference (and this process of differentiation may have been carried out more than once). Each of these functions aims to make the model fit the data as well as possible.
- **LSTMs:** Long short-term memory networks—typically referred to as "LSTMs"—are a specific form of recurrent neural networks that can know long-term dependencies. Hochreiter and Schmidhuber (1997) presented them, and other people improved and popularized them in subsequent research. They perform remarkably well on a broad range of issues and are now commonly used. The LSTMs were expressly built to prevent the problem of long-term dependency. Remembering knowledge is their natural action for long periods, not something they try to understand. Both recurrent neural networks take the shape of a series of neural network repeated modules.

The following section highlights the implementation details and experimental outcomes of our case study on stock market analysis.

14.4 Modeling and Implementation Details

This section highlights the model building for our case study with stock market analysis. Figure 14.1 explains the workings of our approach. We retrieved live data from an API taking the stock symbol and the start and end dates of the period intended to be analyzed.

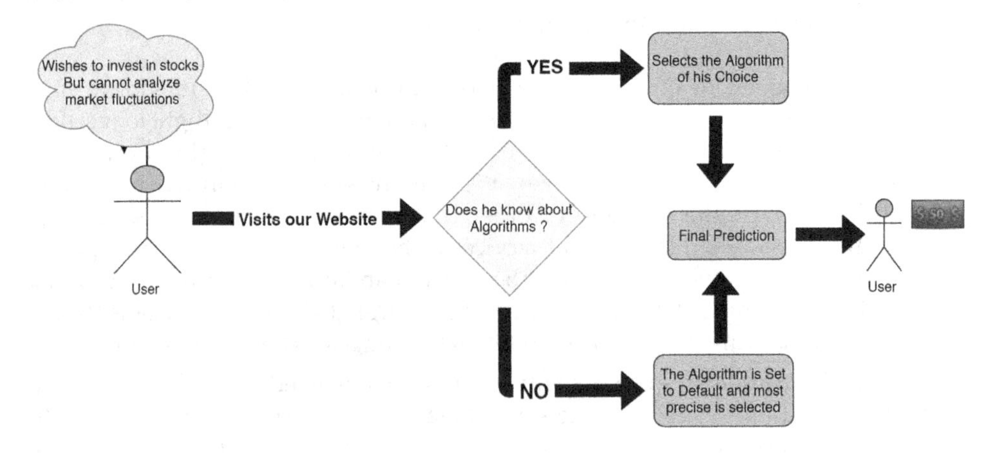

FIGURE 14.1
Case stuty with stock market data.

The data collection process involves several related stock attributes, including opening price, closing price, stock volume, etc. The data analysis is maily based on the stock closing price. Our data source is *Yahoo Finance*, and the experimental study performed is based on an analysis of Google stocks (Figures 14.2 and 14.3).

Initially, graphs were drawn to study the trends and patterns of the data that experts may use directly to determine the movement. For more concrete analysis, we analyzed further. In our research, we found out that specific algorithms require the data to be stationary, meaning the mean and variance should be constant over time, so we started checking whether the data extracted was standing or not. We explored a method known as the DFT to assess the stationarity of the data. A null hypothesis is assumed to mean that our TS is nonstationary. The result of the DFT comprises a critical value and a test statistic. If the critical value is more significant than the test statistics, the null hypothesis is rejected, and the time series is confirmed to be stationary (Figures 14.4 and 14.5).

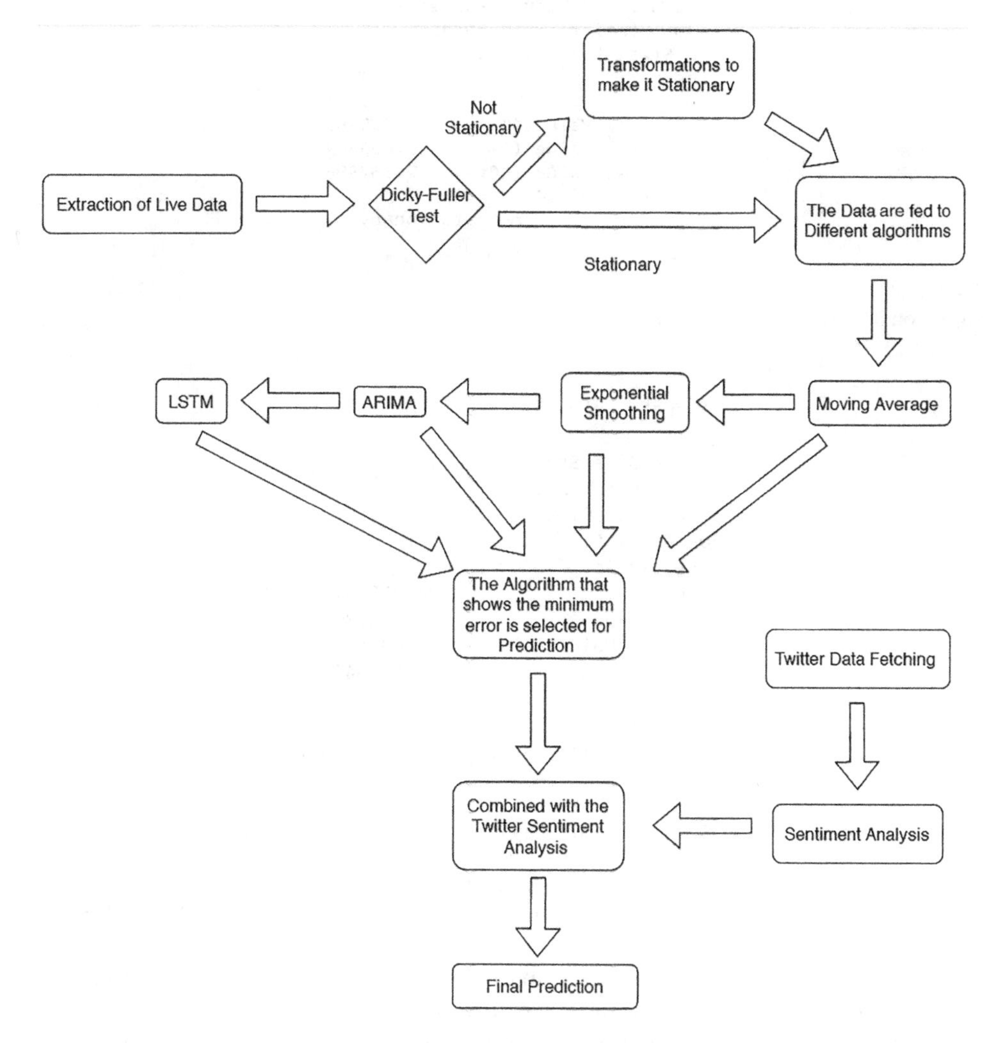

FIGURE 14.2
Control flow for time series stock market case study.

Date	High	Low	Open	Close	Volume	Adj Close
2012-01-03	332.827484	324.966949	325.250885	331.462585	7380500.0	331.462585
2012-01-04	333.873566	329.076538	331.273315	332.892242	5749400.0	332.892242
2012-01-05	330.745300	326.889740	329.828735	328.274536	6590300.0	328.274536
2012-01-06	328.767700	323.681763	328.344299	323.796326	5405900.0	323.796326
2012-01-09	322.291962	309.455078	322.042908	310.067780	11688800.0	310.067780

FIGURE 14.3
Dataframe of Google stock data.

```
test_stationarity(ts)

Results of Dickey-Fuller Test:
ADF Statistic: -0.754365
p Statistic: 0.832161
open                          -0.754365
high                           0.832161
Critical Value (1%)           -3.435563
Critical Value (5%)           -2.863842
Critical Value (10%)          -2.567996
dtype: float64
Time series is not stationary
```

FIGURE 14.4
Comparison of different values of stationary data.

```
tss=make_data_stationary(ts)

Making Data Stationary....
Log Transformations
Results of Dickey-Fuller Test:
ADF Statistic: -34.209551
p Statistic: 0.000000
open                          -34.209551
high                            0.000000
Critical Value (1%)            -3.435567
Critical Value (5%)            -2.863844
Critical Value (10%)           -2.567997
dtype: float64
Reject H0, Time series is Stationary
```

FIGURE 14.5
Comparison of different values of nonstationary data.

Figures 14.4 and 14.5 consist of two parts: in the first part, it is seen that when critical value>adf statistic, it is not stationary. If the data were not static, transformations were applied, such as taking the data logarithm or differencing between measurements to render it stationary. After some modifications, the critical value becomes less than the ADF statistic, and so we conclude the time series is stationary. Some of the algorithms, such as LSTM, do not demand that the data be stationary. Hence, the information is fed directly to them in those cases. Each algorithm method returns three values, i.e., the nth day amount, the expected $n+1$ day value, and the error in the model. We also have graphs plotted to analyze the original and predicted value trends.

Our main aim is to build a model that would suggest to the user whether to invest or not in a particular stock. We applied moving average, exponential smoothing, ARIMA, and LSTM for forecasting different values. Therefore, after using these models, we compared the root mean squared error (RMSE) of other models, and the one giving the least RMSE is chosen for further analysis. We have to forecast the value for the $n+1$th day. Therefore, we compare the predicted values with the nth day value; if the value is not less than or not equal to that, i.e., we see an increasing trend, then we take the model for further analysis, otherwise we suggest the user not to invest in that particular stock.

This is followed by sentiment analysis of the stocks in order to obtain the emotional reliability of the people toward stocks. We extracted the tweets with the search parameter as the stock symbol and then applied sentiment analysis. We used a library of NLTK named VADER for the same. VADER tells to what extent the review is negative or positive. It is a pretrained model. The sum of lexicon ratings, when normalized between –1 and 1, is called compound score, which is used as a metric here. We didn't preprocess this because Vader already considers the Capitalization and punctuation marks as parameters for sentiments. For example, if we write "The ambience is good," the compound score is less positive than when we write "the ambience is GOOD!!" How we use the compound score to predict our result is as follows:

- If the compound score is more significant than .05, then we predict the sentiment to be positive.
- If the compound score is between –.05 and .05, then it is neutral, else it is the negative sentiment.
- If the number of net positive sentiments is greater than the net negative emotions, we suggest the user invest in the particular stock; otherwise, investing isn't a great idea.

We also performed trend analysis using latent Dirichlet algorithm (LDA), which identifies trends from Twitter related to the stock market. The tweets of the Twitter handle *Yahoo Finance* were extracted and preprocessed, which involved tokenization, lemmatization, and stop-word removal. Further, the LDA was applied for topic modeling (Figure 14.6).

```
{'american',
 'attitude',
 'coronavirus',
 'covid-19',
 'driver',
 'economy',
 'fight',
 'highlight',
 'powell',
 'rates',
 'reopen',
 'sales',
 'shift',
 'stimulus',
 'strategist',
 'taking',
 'think'}
```

FIGURE 14.6
Results of topic modeling.

The results in Figure 14.6 show topics such as coronavirus, COVID-19, economy, etc., as the most trending topics. Further, a website was designed using Flask, Javascript, CSS3, and HTML5. Figures 14.7 and 14.8 show the snapshots of our designed website using herokuapp. The website allows any user to select a stock symbol and duration that they wish to analyze (Figure 14.9). Hence, the final output becomes a suggestion

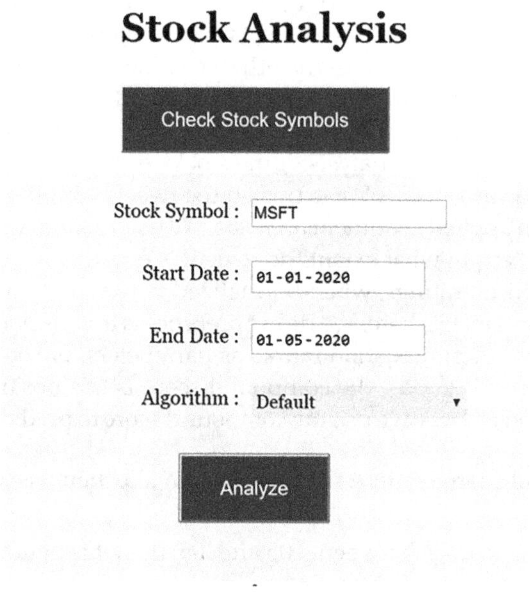

FIGURE 14.7
Homepage of stock analysis website.

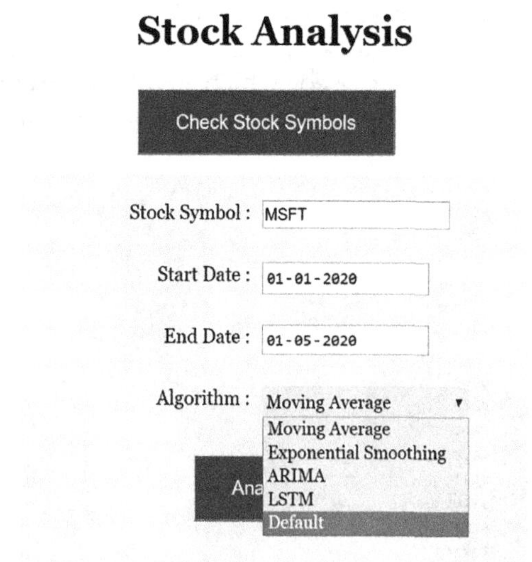

FIGURE 14.8
Different options for selecting model for stock analysis.

of whether or not to invest on a specific stock. Figure 14.10 shows an overview of the complete dataset with stock attributes.

Figure 14.11 shows the error analysis received from the prediction obtained during our experimentation. It is apparent from the graph that the actual and predicted data considerably overlap with an RMSE score of 3.6218. From Figure 14.12, we can see an

	symbol	stocks
0	A	Agilent Technologies, Inc. Common Stock
1	AA	Alcoa Corporation Common Stock
2	AAAU	Perth Mint Physical Gold ETF
3	AACG	ATA Creativity Global - American Depositary Shares, each representing two common shares
4	AADR	AdvisorShares Dorsey Wright ADR ETF
5	AAL	American Airlines Group, Inc. - Common Stock
6	AAMC	Altisource Asset Management Corp Com
7	AAME	Atlantic American Corporation - Common Stock
8	AAN	Aaron's, Inc. Common Stock
9	AAOI	Applied Optoelectronics, Inc. - Common Stock
10	AAON	AAON, Inc. - Common Stock
11	AAP	Advance Auto Parts Inc Advance Auto Parts Inc W/I
12	AAPL	Apple Inc. - Common Stock
13	AAT	American Assets Trust, Inc. Common Stock

FIGURE 14.9
Different symbols of distinct company stocks.

	High	Low	Open	Close	Volume	Adj Close
Date						
2019-12-31	157.770004	156.449997	156.770004	157.699997	18369400.0	157.270432
2020-01-02	160.729996	158.330002	158.779999	160.619995	22622100.0	160.182480
2020-01-03	159.949997	158.059998	158.320007	158.619995	21116200.0	158.187927
2020-01-06	159.100006	156.509995	157.080002	159.029999	20813700.0	158.596817
2020-01-07	159.669998	157.320007	159.320007	157.580002	21634100.0	157.150772
2020-01-08	160.800003	157.949997	158.929993	160.089996	27746500.0	159.653915
2020-01-09	162.220001	161.029999	161.839996	162.089996	21385000.0	161.648468
2020-01-10	163.220001	161.179993	162.820007	161.339996	20725900.0	160.900513
2020-01-13	163.309998	161.259995	161.759995	163.279999	21626500.0	162.835236
2020-01-14	163.600006	161.720001	163.389999	162.130005	23477400.0	161.688370
2020-01-15	163.940002	162.570007	162.619995	163.179993	21417900.0	162.735504
2020-01-16	166.240005	164.029999	164.350006	166.169998	23865400.0	165.717361
2020-01-17	167.470001	165.429993	167.419998	167.100006	34371700.0	166.644836
2020-01-21	168.190002	166.429993	166.679993	166.500000	29517200.0	166.046463
2020-01-22	167.490005	165.679993	167.399994	165.699997	24138800.0	165.248642
2020-01-23	166.800003	165.270004	166.190002	166.720001	19680800.0	166.265869
2020-01-24	167.529999	164.449997	167.509995	165.039993	24918100.0	164.590439
2020-01-27	163.380005	160.199997	161.149994	162.279999	32078100.0	161.837952
2020-01-28	165.759995	163.070007	163.779999	165.460007	24899900.0	165.009308
2020-01-29	168.750000	165.690002	167.839996	168.039993	34754500.0	167.582260
2020-01-30	174.050003	170.789993	174.050003	172.779999	51597500.0	172.309357
2020-01-31	172.399994	169.580002	172.210007	170.229996	36142700.0	169.766296
2020-02-03	174.500000	170.399994	170.429993	174.380005	30149100.0	173.904999
2020-02-04	180.639999	176.309998	177.139999	180.119995	36433300.0	179.629364
2020-02-05	184.199997	178.410004	184.029999	179.899994	39186300.0	179.409958
2020-02-06	183.820007	180.059998	180.970001	183.630005	27751400.0	183.129807
2020-02-07	185.630005	182.479996	182.850006	183.889999	33529100.0	183.389099
2020-02-10	188.839996	183.250000	183.580002	188.699997	35844300.0	188.185989

FIGURE 14.10
Aggregated stock market dataset.

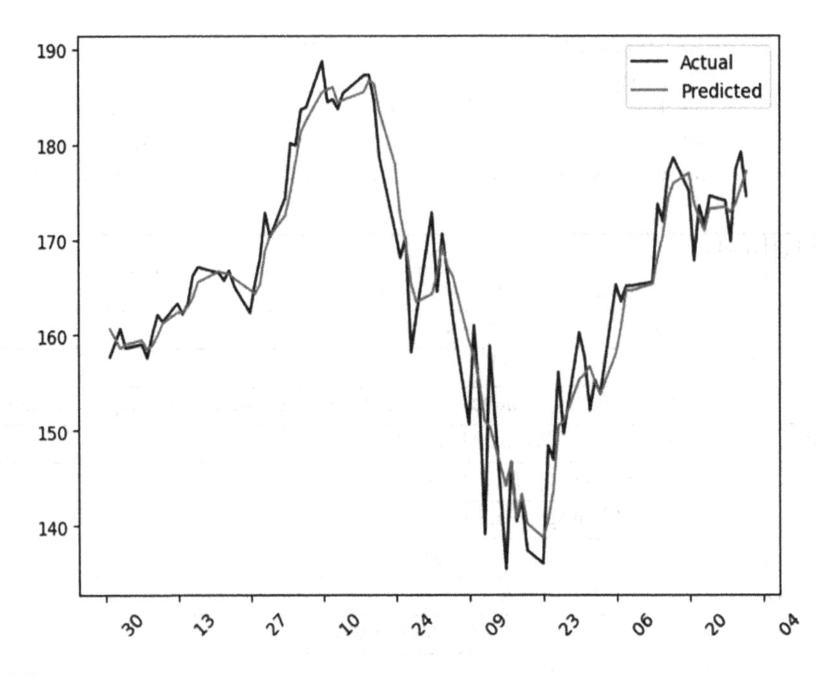

FIGURE 14.11
Error analysis and RMSE outcome.

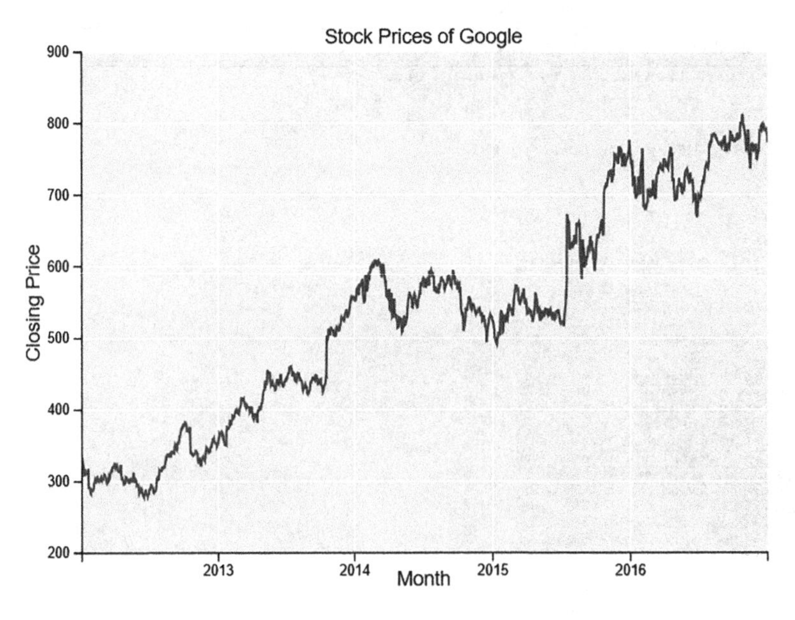

FIGURE 14.12
Closing price of stock data.

increasing trend of time series data of the stocks. After applying DFT, it is apparent that the data are nonstationary, because the mean does not seem to be constant with time. Figure 14.13 clearly shows that the mean is not constant with time and is highly varying while the standard deviation is almost constant. There we need to station the data. Figure 14.14 shows the graph of stationed data. Since some of the time series algorithms

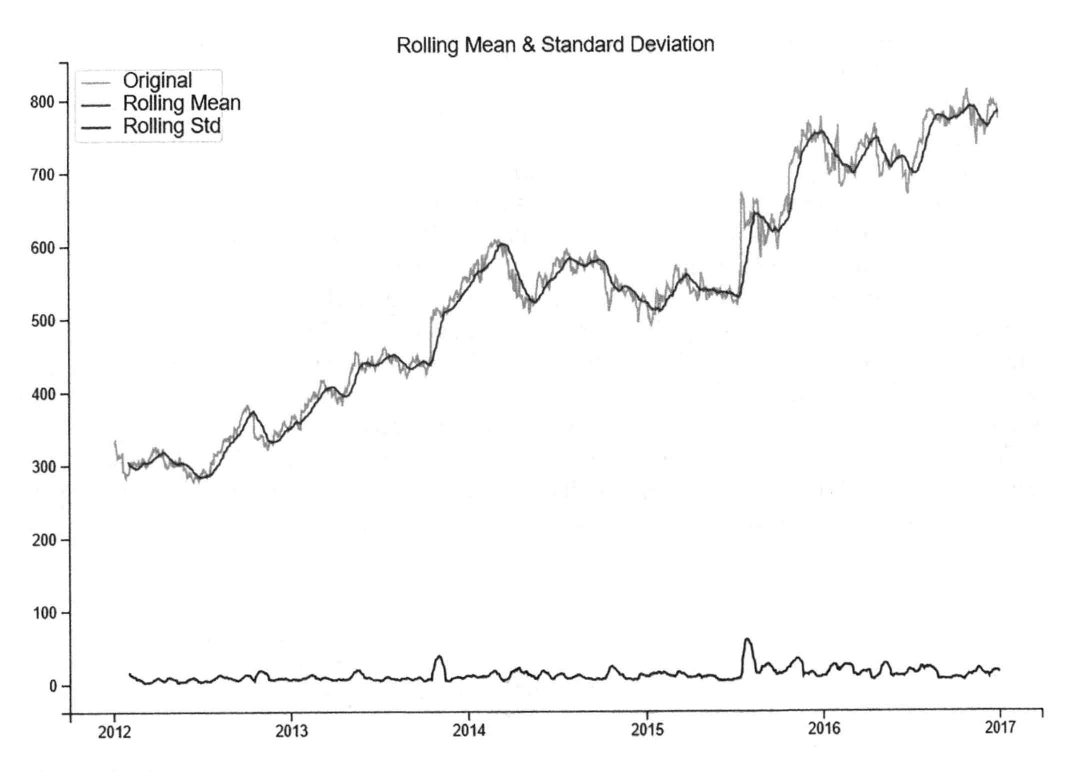

FIGURE 14.13
Rolling mean, rolling standard deviation, and closing price.

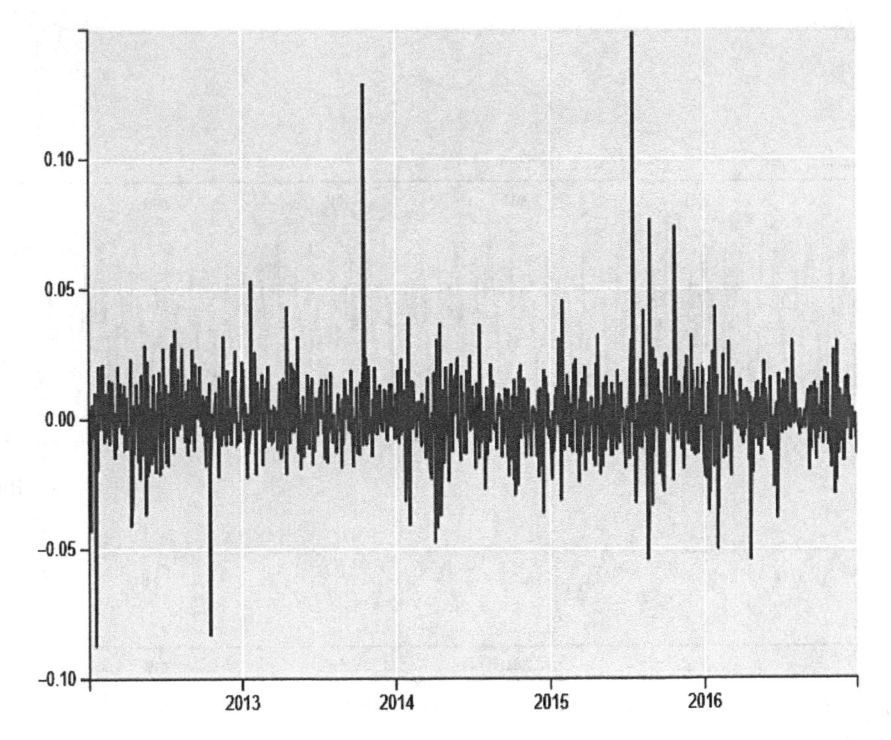

FIGURE 14.14
Stock prices of Google after data transformation.

we are applying require the data to be stationary, the data become stationary after using various logarithmic transformations.

Figure 14.15 shows the seasonality, trend, and residuals of the time series separately from the original data. Figure 14.16 shows the seasonality, residuals, and direction of time series after making it stationary. After making the data stationary, we applied various models such as moving average, exponential smoothing, and ARIMA to predict and compare with the original values. Our results included RMSE and mean absolute deviation. We have compared different algorithms based on RMSE. The models that require data to be stationary are moving average, exponential smoothing, and ARIMA.

Figure 14.17 shows that the RMSE for the moving average is 13.6546. The moving average takes a window size and calculates the price of the nth value with the average previous $n-1$ values. One of the significant drawbacks of moving average is that we have to provide window size, which may depend on different types of data.

From Figure 14.18, we find that the RMSE for exponential smoothing is 12.9451. This error is less than the moving average, so it's better. From Figure 14.19, we see that our predicted values almost coincide with actual values, so the RMSE is only 6.6447. Moreover,

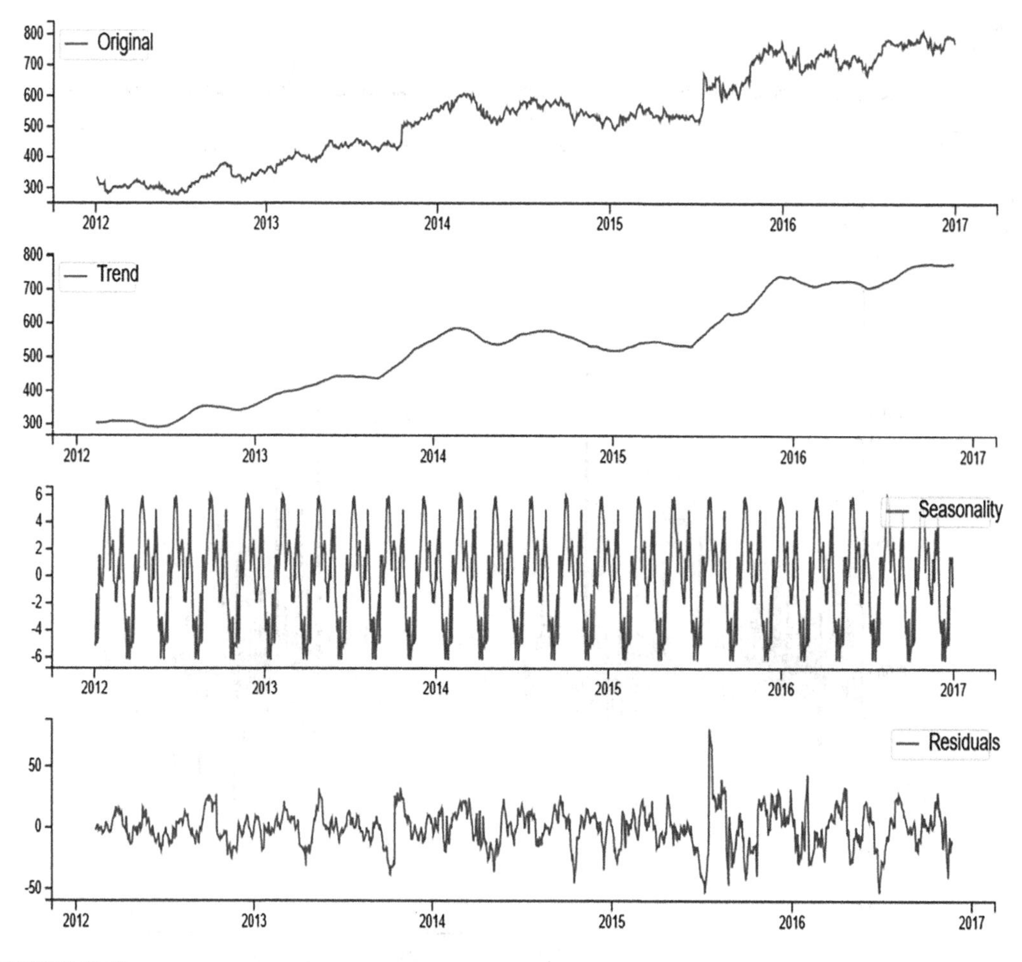

FIGURE 14.15
Components of original time series.

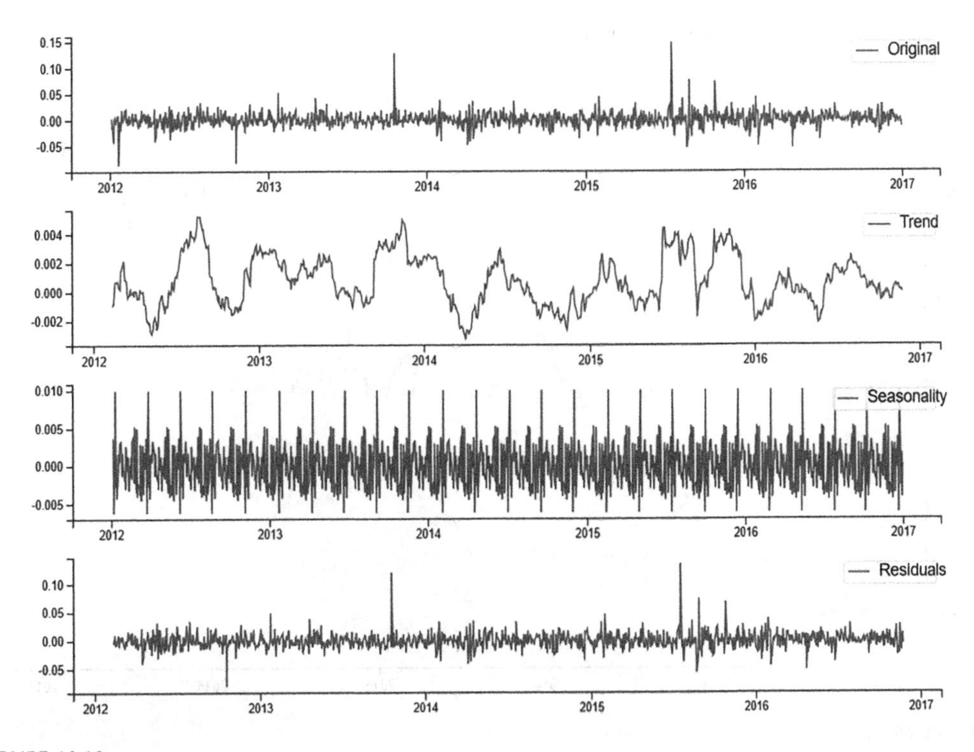

FIGURE 14.16
Components of stationary time series.

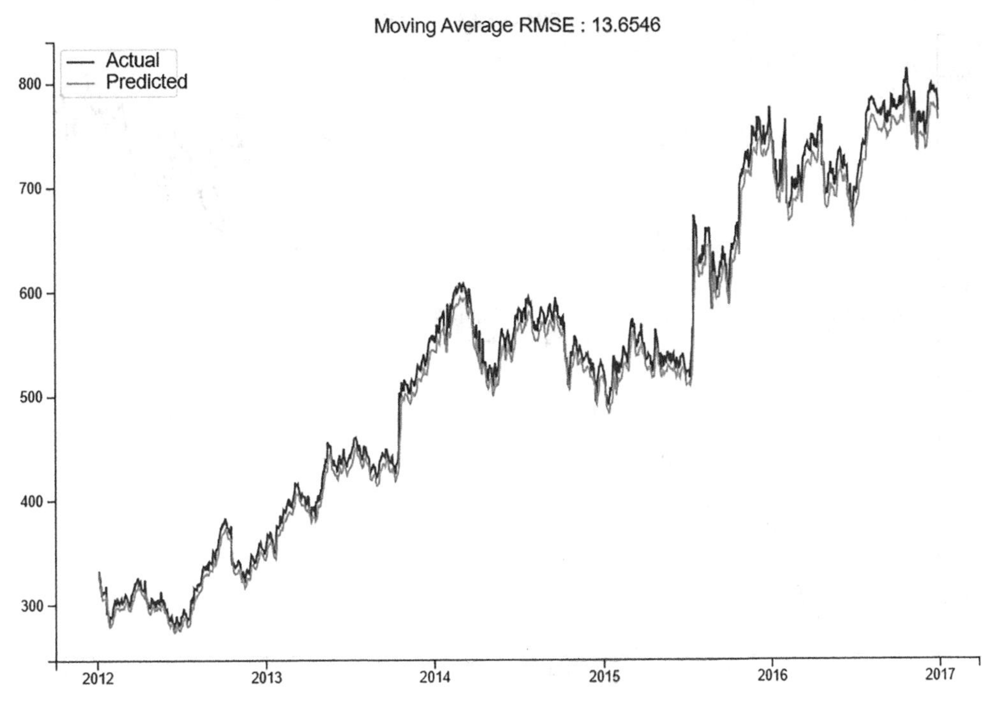

FIGURE 14.17
Graph of actual versus predicted values for moving average.

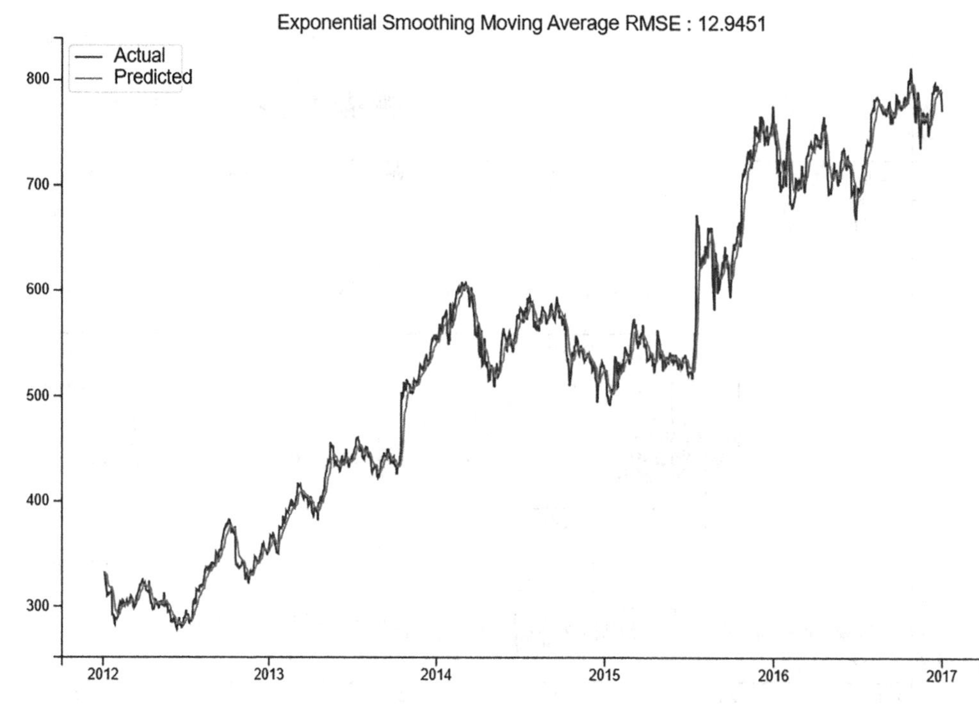

FIGURE 14.18
Graph of actual versus predicted values for exponential smoothing.

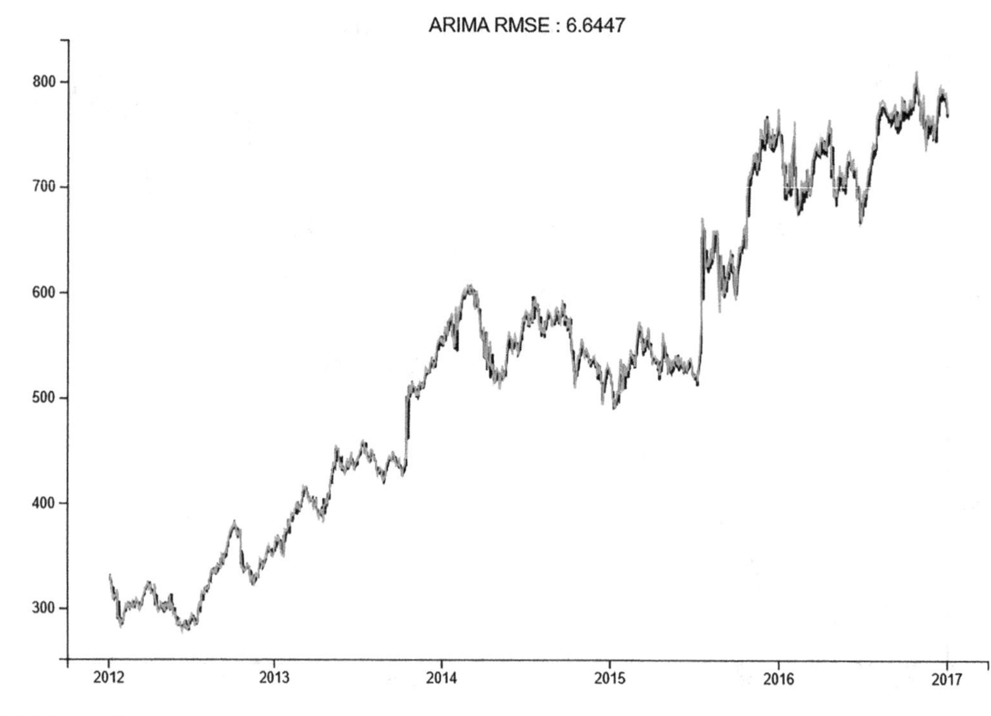

FIGURE 14.19
Graph of actual versus predicted values for ARIMA model.

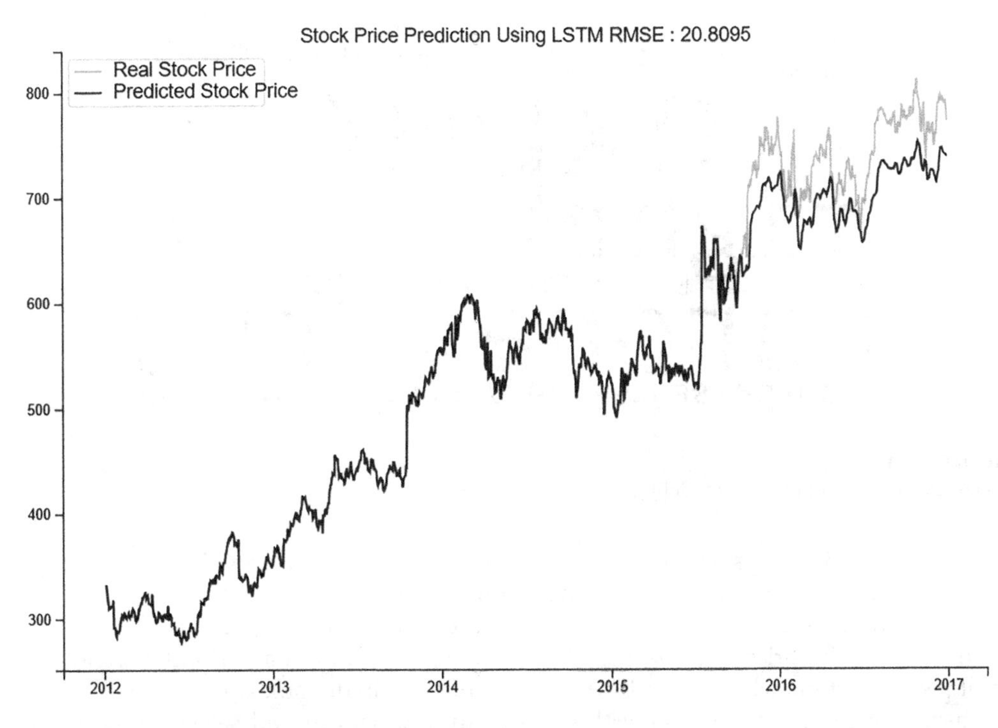

FIGURE 14.20
Graph of actual stock price versus predicted after applying LSTM.

TABLE 14.1
Root Mean Squared Errors of Various Algorithms

Algorithm	RMSE
Moving average	13.6546
Exponential smoothing	12.9451
ARIMA	6.6447
LSTM	20.8095

Figure 14.20 shows that LSTM's performance is the worst with the highest RMSE values, i.e., 20.8095. Further, based on the RMSE errors summarized in Table 14.1, we conclude that ARIMA performs the best as it has the least RMSE score.

Figure 14.21 shows the trends we identified through LDA on tweets fetched from *Yahoo Finance*'s Twitter handle. Some of the most common words include coronavirus, economy, etc. These may assist in identifying the main trends in the market and the potential reasons for change in the current economy.

14.5 Conclusion

The prediction of stock prices is considered significant for making investment decisions. Our research has provided a case study on stock predictions. With the use of sentiments

FIGURE 14.21
Trends extracted through topic modeling.

from social media posts and tweets about particular stocks, it provided better insights. Therefore, our initiative made it easier to track the stock market trend and analyze. The sentiments were extracted from Twitter that provided information regarding expected trends in the future. The price of a share of the company also follows a particular trend, hence multiple algorithms when ensembled together provide better stock prediction. Therefore, we have considered various algorithms to study the pattern and select the best one in terms of minimum prediction error. Also, a website was designed to provide a front-end for the end-users to conduct comparative analysis of the stocks, thereby assisting them in better decision making. The inclusion of tweets or the news can be a future step toward having precise predictions, and hence could add value to the stock market analysis.

References

Hochreiter, S, and Schmidhuber, J. "LSTM can solve hard long time lag problems." In *Advances in neural information processing systems* 9, MIT Press, Cambridge MA (1997).

Kumar, M, and M. Anand. "An application of time series ARIMA forecasting model for predicting sugarcane production in India." *Studies in Business and Economics* 9 1 (2014): 81–94.

Menon, A, S Singh, and H Parekh. "A review of stock market prediction using neural networks." *2019 IEEE International Conference on System, Computation, Automation and Networking (ICSCAN).* IEEE, 2019.

Ryota, K, and N Tomoharu. "Stock market prediction based on interrelated time series data." *2012 IEEE Symposium on Computers & Informatics (ISCI).* IEEE, 2012.

Sayavong, L, Z Wu, and S Chalita. "Research on stock price prediction method based on convolutional neural network." *2019 International Conference on Virtual Reality and Intelligent Systems (ICVRIS).* IEEE, 2019.

Sharma, A, D Bhuriya, and U Singh. "Survey of stock market prediction using machine learning approach." *2017 International Conference on Electronics, Communication, and Aerospace Technology (ICECA).* Vol. 2. IEEE, 2017.

Somani, P, S Talele, and S Sawant. "Stock market prediction using hidden Markov model." *2014 IEEE 7th Joint International Information Technology and Artificial Intelligence Conference.* IEEE, 2014.

Zhao, L, and L Wang. "Price trend prediction of the stock market using outlier data mining algorithm." *2015 IEEE Fifth International Conference on Big Data and Cloud Computing.* IEEE, 2015.

15

Medical Search Engine

Palak Arora, Shresth Singh, and Megha Rathi
Jaypee Institute of Information Technology

CONTENTS

15.1 Introduction

Data Analytics, the way toward investigating datasets to obtain shrouded data and draw conclusions with its help, has the potential to resolve various problems, including health. Having health issues has become more common in today's world. So, for every individual, it is important to take a precautionary measure to check whether a person has any chance of getting any disease by knowing the symptoms that they may be having. For this purpose, we use deep learning techniques to predict whether a person has a disease or not, while additively instructing them on the precautions and treatments. It is attractive as the results are obtained through an Android application installed on a mobile device (Krishna Priya et al., 2015).

Data analytics tools help in finding out the future trends, helping businesses to take decisions accordingly. They search throughout the data to find meaningful trends and behaviors that experts may miss because of out-of-scope expectations.

Current technology has been used to improve the quality of human life in different ways. One of the most benefited areas is medicine (Ramesh et al., 2016). However, despite the great progress that the field has made, problems, such as the full coverage of the general population, which is present, particularly in developing countries, remain unresolved. Although technology has contributed substantially to the way health services are

DOI: 10.1201/9781003046431-15

provided, they have not been able to eliminate certain problems, such as the barriers that a patient has to overcome to access the medical service, the wait times, or the management of all cases presented in the emergency department (Rosenberg and Andersson, 2000). One of the solutions that technology proposes is the use of automatic learning methods, that is, applications of computing that are trained with data and are capable of performing the same activities as a doctor. These types of systems have been proposed since the eighties using different methods, such as the use of artificial intelligence (Clancey and Shortliffe, 1984). Our study is focused on reducing the time it takes to find the ideal specialist in organizations that provide medical services, to focus the valuable time of the specialists in determining and solving the affliction, leaving the first phase of identification or diagnosis to a computer system. The system developed by us consists of an Android application that asks the user to enter the symptoms they are experiencing and then sends that data to our Web API, which predicts the disease based on the symptoms and also returns possible treatments and care tips for the user to employ against the disease.

15.2 Literature Survey

Shankar et al. (2015) discussed a way to predict diseases and their healing times based on the symptoms shown and the severity of such symptoms. First, they assign different coefficients to every symptom and then they sort the dataset by the severity score of each symptom, which is given by the user. The values calculated above become the basis of the identification of the diseases. Reinforcement learning was then used to predict the cure time of the diseases. The weights are defined based on the similarity of the conditions of the user and other people who had a similar conditions. They also define the conditions of the user by comparing with people who have suffered the same disease. Inputs from the students in their college were used to create the dataset. They contend that this approach performs much better than the previously suggested algorithms as their way mimics the actual interaction between a patient and a doctor.

Pinango and Dorado's (2014) investigation tends to the issue of the forecast of maladies based on the revealed side effects to give better clinical consideration to patients. To display an indication relationship with populace-based examinations, they suggest using a Bayesian framework. Correlation between specific symptoms is described by using a Bayesian probabilistic model. The effectiveness of the model is checked using the training data and a similarity measure. A dataset of 500,000 observations of 5,000 people taken from the openMRS study was used. They claim to achieve high accuracy with this dataset.

Shofi et al. (2017) proposed to build an Android-based application that asks the user some questions and then predicts their diseases based on their answers. They used a forward chaining inference trained on a dataset extracted from the book "Doctor in Your Home" by Dr. Tony Smith. They claim that the predictions of their model were highly accurate as validated by experts.

Palaniappan and Awang's (2008) study discussed the efficiency of machine learning techniques, namely the naive Bayes classifier, Decision Trees, and Neural Networks in predicting the occurrence of heart disease in a patient using categorical data from the patient's medical history. The dataset was extracted from the Cleveland Heart Disease database and all numerical data were converted to categorical data. The results show that the neural network and the naive Bayes classifier were able to predict the occurrence of disease related to the heart more accurately than the decision tree.

A study by Ramesh et al. (2016) summarized the part played by big data analysis in healthcare and the many shortcomings of contemporary machine learning algorithms. They contend that the ideal approach to extricating helpful data from big data is to use a fusion of two or more data-mining techniques, as they give the best possible results in the least amount of time.

Chen et al.'s (2017) study suggested an ingenious convolutional neural network-based multimodal disease risk prediction (CNN-MDRP) algorithm, which uses both organized just as unstructured information gathered from emergency clinics. Their model yielded an accuracy of 94.8% when tested on a local disease of cerebral infarction.

Prabakaran and Kannadasan (2018) contributed a robust algorithm by combining decision trees and clustering techniques to create a system to predict cardiac arrest risk.

Gavhane et al. (2018) utilized key attributes such as age, sex, pulse rate, etc., to develop a heart disease-predicting application. They used the Cleveland database from the UCI library to collect their dataset and trained a multilayered perceptron on it, which yielded high accuracy in predicting heart disease.

Weston et al. (2018) put forth a method to find factors that influence symptom onset age and course of disease in Autosomal Dominant Alzheimer's Disease (ADAD) and make a proof-based strategy to foresee side effects beginning in ADAD. Family ancestry along with transformation type can clarify huge segments of fluctuation seen with age at side effects beginning in ADAD, which offers exact help to the utilization of such information to foresee beginning in clinical exploration.

In yet another novel work, Tsanas et al. (2011) proposed to quantify average Parkinson's disease symptom severity in a clinically useful way by using nonlinear speech algorithms that were mapped to a standard metric.

Khan et al. (2015) employed quality articulation marks to group malignant growths to explicit analytic classifications utilizing an artificial neural network (ANN). Little, round blue cell tumors (SRBCTs) were used to prepare the ANN. The diseases studied originated from four particular analytic classifications and regularly caused demonstrative difficulties in clinical practice. All examples were effectively characterized using the ANN and the most applicable qualities were additionally found using the ANN.

Tu (2017) compared two known data-mining algorithms, ANNs and logistic regression, on their applicability for predicting medical outcomes. The points of interest and burdens of these procedures are talked about for this reason.

Krishna Priya et al. (2015) emphasized information mining as a systematic procedure intended to investigate information looking for designs that were not identified with formal connections. Information mining is frequently depicted as a procedure of discovering designs, blends, patterns, or connections by scanning for a lot of information put away in vaults, databases, and information stores. In this venture, a classification calculation such as C4.5 was utilized to characterize the Pima diabetes dataset. Results were acquired using Android, among other factors, and afterward approved utilizing the outcomes of the examples obtained in the new informational collection. Obtaining data from databases is fundamental to a viable conclusion. The reason for information mining is to separate the data put away in a dataset and produce clear and justifiable examples. This examination targets finding a tree of choice in the forecast of diabetes. Readiness is utilized to improve information quality. During preprocessing, higher information characteristics are considered to foresee sugar. This is a significant factor to consider. The choice tree calculation utilized for grouping likewise delivers higher precision results compared with other arrangement calculations. Finally, the program results are accessible in the most valuable Android application for the current age.

Pattekari et al. (2014) built an intelligent system utilizing an information-mining process called Naive Bayes. Utilized as an online application for this client, it responds to pre-characterized questions. Returns concealed data from a database and think about client esteem with a prepared informational collection. It can respond to complex inquiries for diagnosing cardiovascular ailments and in this way assist wellbeing with caring experts who take sound clinical and solid decisions sincerely based on strong systems. By giving compelling treatment, it additionally assists with decreasing the expense of treatment. The framework can distinguish and extricate secret data related to illnesses (heartbeat, malignant growth, and diabetes) from a database of coronary illness history. Backing for the choice of the cardiovascular disease prediction program is created utilizing the Naive Bayesian classification process. The program extricates the concealed data from the verifiable coronary illness database. This is the best model for predicting patients with coronary illness. This model can respond to complex inquiries, each in its ability to demonstrate translation, access to nitty-gritty data, and precision. It can likewise incorporate different strategies for information mining. They may likewise incorporate other information pressure strategies, e.g., series, clustering, and association rules. Relentless data can in like manner be used as opposed to part data.

Hussain et al. (2018) notified that versatile health (mHealth) applications are promptly available to the normal individual on cell phones, and although mHealth applications can improve access, buy, and proficiency of medicinal services conveyance, they handle delicate wellbeing data and, in this way, can convey critical dangers to security and the protection of their clients. Candidates are regularly mysterious, and clients don't have the foggiest idea of how their information is overseen and utilized. This is combined with the rise of new dangers because of a need for the advancement of versatile applications or the ambiguity of the current portable working framework. Numerous sorts of portable applications are accessible in the market, yet the Android stage has increased a great deal of prevalence. In any case, the Android security model is short for totally guaranteeing the insurance and security of customers, information, including mHealth application information. Notwithstanding the security features provided by Android, for instance, assents and sandboxing, mHealth applications are up, till now stressed over critical assurance and security issues. These security issues ought to be tended to improve the affirmation of mHealth applications among customers and the reasonability of mHealth applications in human administration structures. In this way, this chapter provides a calculated structure in improving the security of clinical data identified with *m-bazaar* Android applications, just as ensuring the protection of their clients. In light of a writing survey that recommended the requirement for a focus on the security system, three separate segments were presented and sequenced, each depicted in different segments. Initially, in a stage-1 advancement process, they talked to build up a security structure for mHealth applications to guarantee the security and protection of delicate wellbeing data. The subsequent stage was a conversation of who can accomplish the usage of the prototypic confirmation of-idea system. Finally, the third area examined the assessment procedure for the proficiency and viability of the proposed system. In a recent study (Sinha and Rathi, 2021), analysis of COVID-19 data was done to uncover the factors that led to death due to COVID-19. In another significant work (Gupta et al., 2021), AI-based techniques applied for analyzing COVID-19 outbreak. In yet another novel work by Sinha (2021), a novel learning model was developed for the analysis of lockdowns in India during the COVID-19days. Another research (Saxena et al., 2021) on coronavirus predicted infection and segmentation using advanced deep learning techniques.

15.3 Methodology

Awareness is still not spread among people about the things that may be affecting their physical health and the diseases causing it. Then, usually, they end up underestimating the symptoms they are experiencing. Most people believe that minor illnesses, such as a cold or diarrhea, do not require proper examination or good treatment and people end up ignoring them, which may cause severe illness in the future. Therefore, one should know whether or not they show the symptoms of a harmful disease and what exactly they should do about it. To make it simpler for somebody to analyze the symptoms or indications they are experiencing, in this process, we will build an Android-based application that uses a neural network trained on a disease symptom database to predict the disease afflicting the user, along with providing relevant treatment solutions to the user, which they can use to judge whether they can do their treatment by themselves or consult a doctor immediately; all of this is done quickly and efficiently using an Android device.

15.3.1 Dataset Description

Dataset is a mixed dataset composed of 1,867 instances with 404 symptoms such as abdominal pain, cough, fast_heart_rate, etc., containing binary data and 132 diseases, such as diabetes, malaria, etc., containing categorical data. The symptoms that are the cause of a disease are marked as "1" and those that are not the cause of a disease are marked as "0". Appropriate data cleaning and preprocessing techniques are used to convert this dataset into 4,962 instances with 132 symptoms and 44 diseases, for less redundancy and better prediction by the deep learning model.

15.3.2 Approach

Characterization comprises anticipating a specific result dependent on a piece of given information. To foresee the result, the algorithm needs the dataset that has the training data and the output called the goals or prediction attribute. All of these are done by finding relevant and direct relations between the attributes. The prediction algorithm that we used is a deep learning-based ANN.

15.3.3 Artificial Neural Network

An ANN is a computational model having the same functionality as the human brain. It is the computer vision of the thinking and computational characteristics of natural neural systems. ANN learns based on the inputs and outputs provided to it, therefore acting as an arbitrary capacity guess apparatus. ANNs are viewed as nonlinear factual information displaying instruments where the unpredictable connections among data sources and yields are demonstrated or designs are found. These instruments help in arriving at the best and financially savvy arrangements by taking information tests instead of a whole dataset that sets aside both time and cash.

ANNs can have numerous layers that are interconnected. The primary layer comprises info neurons. These neurons send information to the shrouded layers, which thus send information to the yield neurons of the last layer.

15.3.4 System Architecture

The framework of engineering describes the flow of the study work. The first step in the process is the collection of the data required for the work. Here, the dataset used is the information of patients at New York-Presbyterian Hospital conceded during 2004, which is collected in the first step. The next step in the process is preprocessing of the data. Here, we convert the raw data into an understandable format. Now the preprocessed data are classified into an ANN model to predict the person's disease. The user enters the details to know his or her results for the test into an Android app installed on his or her mobile device. The attributes entered by the user are compared with the pretrained ANN model and the results are generated, thus further generating the treatment and prevention of the disease (Figure 15.1).

15.3.5 Proposed Solution

The dataset has been trained with a four-layer dense ANN using Keras classifier with the input layer having 132 input symptoms. Three hidden layers contain 65, 33, and 33 neurons from the first layer, respectively. The activation functions used in all the layers starting from the first layer are relu, tanh, relu, and softmax, respectively. The output class, i.e., a prognosis containing 44 classes of categorical type, was converted into the binary format by using OneHotEncoder, for better prediction of results to remove redundancy. The ANN training is done with a batch size of 25 and 500 epochs. The accuracy obtained is 0.8769, i.e., approximately 88%.

15.3.6 Overfitting

Keeping redundant and irrelevant datasets may result in overfitting, which means that the dataset is trained on a particular set of data, which may or may not fit any other dataset, furthermore leading to unobservable efficiency of the model and hence should be removed. In our study, we have tackled this problem by using dropout in all the layers. The trained Keras model is then converted to a "model.h5" file, which is further converted to the TensorFlow Lite file "model.tflite," which is easily compatible with Android applications. The "model.tflite" is then added in the Assets directory of the Android Studio; the

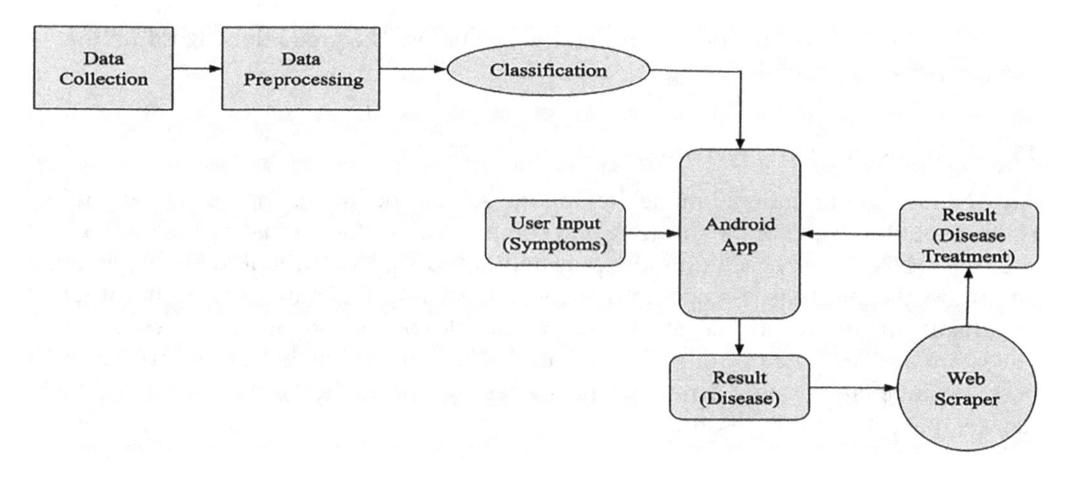

FIGURE 15.1
System Architecture of the proposed solution.

Android application is made with basic spinners taking symptoms information from the user and then passing it into the model and then the predicted disease is displayed on the app along with the probability of the actual occurrence of the disease. This predicted disease is then passed to the Web Scraper, which is made with Beautiful Soup in a Python file hosted on pythonanywhere. The data scraped from it are the treatment, management, or prevention measured and are displayed on the app as well.

15.3.7 Validating Model's Effectiveness

The measure used for validating the prediction results from the model is the "Classification Confusion Matrix." It tells the recurrence of right and wrong forecasts. It thinks about the genuine qualities in the test dataset with the anticipated qualities in the prepared model. The corner-to-corner esteems show the right forecasts. In this example, accuracy_score, a function of the sklearn library, is used to find out the value of the confusion matrix parameter. The accuracy_score function computes the accuracy in the form of the fraction, which is the count of the number of right predictions divided by the total number of predictions. The accuracy_score turns out to be 0.96, i.e., 96% of the predictions are correct.

15.4 Heart Rate Calculator

Heart Rate Calculator is an extra feature added in the Android app that uses the camera and flashes light of the user's phone to calculate heart rate by calculating the amount of red in preview frames and calculating an average.

Working:

- Gets the amount of red in each preview frame.
- Decodes and gets the average amount of red components in the image.
- Calculates the rolling average.
- Takes data in chunks of 10 seconds.
- Currently using texture view instead of surface view for removing preview of the camera from user's eyes. The user can also see the condition of this heart, i.e., where the heart is working Good, Average, Poor, etc., based on the coming heart rate by looking through the reference guide provided in the app (Figure 15.2).

15.5 Results and Analysis

Developed framework will confer a forecast result of the sickness, which the individual may deal with. The framework gives an insight concerning the heart status of the user by estimating their heart rate in real time. If the user has any chance of having the disease, then the application can be used to showcase the treatment and prevention measures to prevent or cure the disease. The results obtained after the creation of the classifier are stated in Figures 15.3–15.5.

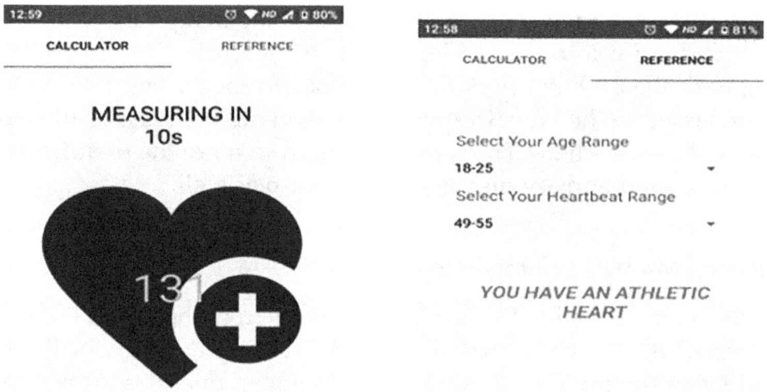

12:59 80%
CALCULATOR REFERENCE
MEASURING IN
10s
131

12:58 81%
CALCULATOR REFERENCE
Select Your Age Range
18-25
Select Your Heartbeat Range
49-55
YOU HAVE AN ATHLETIC
HEART

FIGURE 15.2
Functioning of real-time heart rate calculator.

Layer (type)	Output Shape	Param #
dense_1 (Dense)	(None, 65)	8645
dropout_1 (Dropout)	(None, 65)	0
dense_2 (Dense)	(None, 33)	2178
dropout_2 (Dropout)	(None, 33)	0
dense_3 (Dense)	(None, 33)	1122
dropout_3 (Dropout)	(None, 33)	0
dense_4 (Dense)	(None, 44)	1496

Total params: 13,441
Trainable params: 13,441
Non-trainable params: 0

FIGURE 15.3
Summary of classifier after model training.

```
In [19]: data = pd.DataFrame(l,columns=['Disease','Treatment'])

In [20]: data
Out[20]:
```

	Disease	Treatment
0	adenocarcinoma	<h2>Brain Cancer Treatment Overview</h2><p>Tre...
1	adhesion	<h2>Adhesions Treatment - Self-Care at Home</h...
2	bronchitis	<h2>What Are the Treatments?</h2><p>Most of th...
3	carcinoma	<h2>Treatments</h2><p>Squamous cell carcinoma ...
4	confusion	<h2>Treatment for Sudden Confusion</h2><p>Doct...
5	degenerative polyarthritis	<h2>How Is Arthritis Treated?</h2><p>Treatment...
6	endocarditis	<h2>How Is Hypertrophic Cardiomyopathy Treated...
7	exanthema	
8	hyperbilirubinemia	<h2>How Is It Treated?</h2><p>In adults, jaund...
9	adenocarcinoma	<h2>Brain Cancer Treatment Overview</h2><p>Tre...
10	adhesion	<h2>Adhesions Treatment - Self-Care at Home</h...

FIGURE 15.4
Result of Web scraper for showing treatment and prevention.

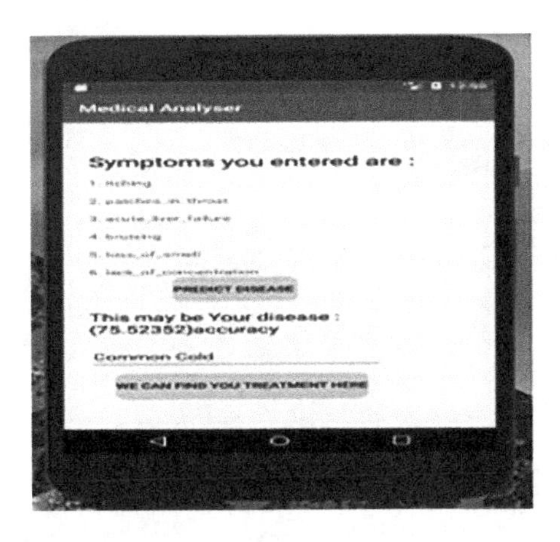

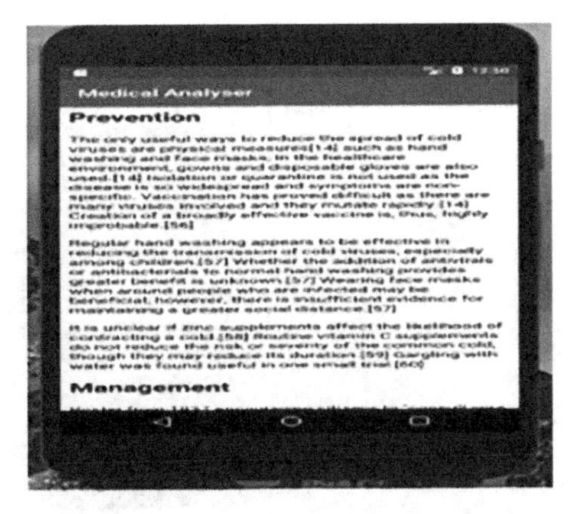

FIGURE 15.5
Working of Android application.

15.5.1 Data Visualization

The most prominent symptom that is the cause of the maximum number of diseases in our dataset is found out by using Tableau. By plotting the sum of occurrences of "1" for each symptom on the X-axis and all the symptoms on the Y-axis. Fatigue is found out to be the most prominent symptom (Figure 15.6).

15.6 Conclusion

The Medical Search engine using a deep learning algorithm through ANN gives its clients a forecast aftereffect of the condition of the client prompting a sickness. Because of rigorous work in the field of artificial intelligence and great advancements in machine learning algorithms, they are quite efficient and accurate in providing realistic results. Hence, we use ANN, which is a kind of Multi-Layer Perceptron (MLP) in our proposed solution. The application hence made is easy to use and if utilized by a large number of people it will help in acknowledging the heart status of the user as well as the probability of having a disease. This can eventually help in reducing the number of people dying due to the same.

15.7 Future Work

Another similar forecast system can be trained, which includes more amount of diseases, with the assistance of refreshed innovations like fuzzy logic, image processing, and so on. Accuracy and effectiveness can be increased. Big data technologies like Hadoop, Cloud computing, etc. can be put to good use for storing the data of the users. Along with this, the additional heart rate measurement feature that provides heart rate in real time can be added up in the dataset to calculate results based on the heart rate as well.

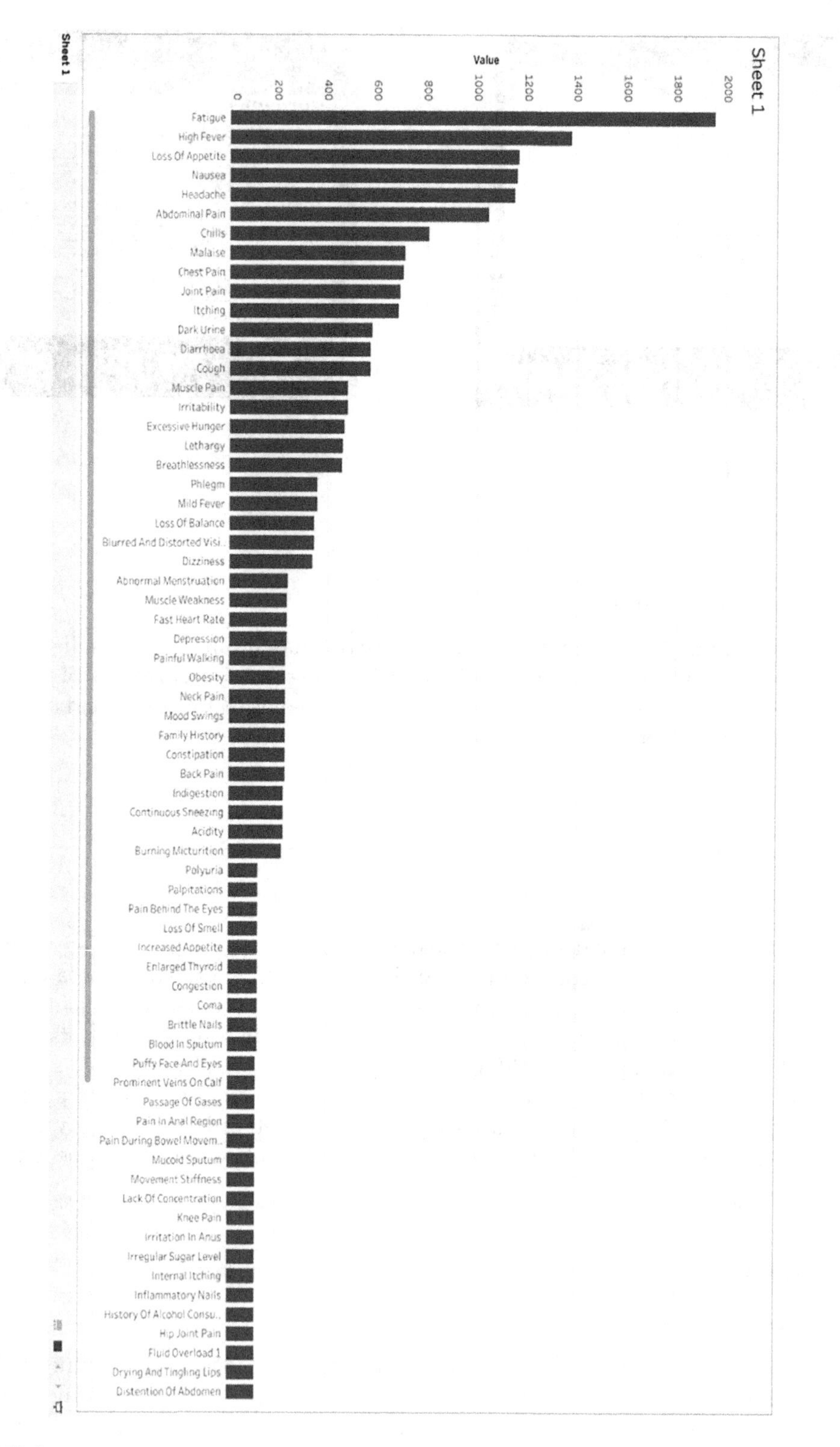

FIGURE 15.6
Histogram, to show the most prominent symptoms in the dataset (fatigue).

References

Breiman, L., Friedman, J., Olsen, J., Stone, C. Classification and Regression Trees, *Engineering & Technology*, 1(6), 122–134, 1984.

Chen, M., Hao, Y., Hwang, K., Wang, L., Wang, L. *Disease Prediction by Machine Learning Over Big Data from Healthcare Communities* (pp. 8869–8879). IEEE Access, 5, 2017.

Chen, M., Wang, L., Disease Prediction by Machine Learning over Big Data from Healthcare Community, *Digital Object Identifier*, accepted April 5, 2017.

Clancey, W. J., Shortliffe, E. H. *Readings in Medical Artificial Intelligence: The First Decade*. Addison-Wesley Longman Publishing Co., Inc., 1984.

Gavhane, A., Kokkula, G., Pandya, I., Devadkar, K. Prediction of Heart Disease Using Machine Learning. *2018 Second International Conference on Electronics, Communication and Aerospace Technology (ICECA)* (pp. 1275–1278). IEEE, 2018.

Gupta, M., Singhal, Y., Sinha, A. Assessing Spatiotemporal Transmission Dynamics of COVID-19 Outbreak using AI Analytics, *Springer 2nd Doctoral Symposium on Computational Intelligence (DoSCI)*, (pp. 1–8), 2021.

Hussain, M., Zaidan, A. A., Iqbal, S., Ahmed, M., Albahri, O. Conceptual Framework for the Security of Mobile Health Applications on Android Platform, *Telematics and Informatics*, 35(5), 1335–1354, 2018.

Khan, J., Wei, J. S., Ringner, M., Saal, L. H., Ladanyi, M., Westermann, F., et al. Classification and Diagnostic Prediction of Cancers Using Gene Expression Profiling and Artificial Neural Networks. *Nature Medicine*, 7(6), 673, 2015.

Krishna Priya, V., Monika, A., Kavitha, P. *Android Application to Predict and Suggest Measures for Diabetes, DM Techniques*. Rajalakshmi Engineering College, Chennai, 2015.

Palaniappan, S., Awang, R. Intelligent Heart Disease Prediction System Using Data Mining Techniques. *2008 IEEE/ACS International Conference on Computer Systems and Applications* (pp. 108–115). IEEE, 2008.

Pattekari, S. A. Praveen, A. Prediction System for Heart Disease Using Naïve Bayes, *International Journal of Ethics in Engineering & Management Education*, 1(1), 2348–4748, 2014.

Pinango, A., Dorado, R. A Bayesian Model for Disease Prediction Using Symptomatic Information. *2014 IEEE Central America and Panama Convention (CONCAPAN XXXIV)* (pp. 1–4). IEEE. 2014.

Prabakaran, N., Kannadasan, R. Prediction of Cardiac Disease Based on Patient's Symptoms. *2018 Second International Conference on Inventive Communication and Computational Technologies (ICICCT)* (pp. 794–799). IEEE. 2018.

Ramesh, D., Suraj, P., Saini, L. Big Data Analytics in Healthcare: A Survey Approach. *2016 International Conference on Microelectronics, Computing, and Communications (MicroCom)* (pp. 1–6). IEEE. 2016.

Rosenberg, H., Andersson, B. Repensar la protección social ensaluden América Latina y el Caribe. *Revistapanamericana de saludpública*, 8, 118–125, 2000.

Saxena, N., Singh Chahal, E., Sinha, A., Chand, S. Coronavirus Infection Segmentation & Detection Using UNET Deep Learning Architecture. *IEEE International Conference INDICON, IIT Guwahati, 19–21 December 2021* (pp. 1–6), 2021.

Shankar, M., Pahadia, M., Srivastava, D., Ashwin, T. S., Reddy, G. R. M. A Novel Method for Disease Recognition and Cure Time Prediction Based on Symptoms. *2015 Second International Conference on Advances in Computing and Communication Engineering* (pp. 679–682). IEEE. May 2015.

Sinha, A. PSIR: A Novel Phase-wise Diffusion Model for Lockdown Analysis of COVID-19 Pandemic in India. *System Assurance Engineering & Management* (pp. 1–17). Springer. October 2021, doi: 10.1007/s13198-021-01477-1.

Sinha, A., Rathi, M., COVID-19 Prediction Using AI Analytics for South Korea. *Applied Intelligence* (pp. 1–19). Springer, 2021, doi: 10.1007/s10489-021-02352-z.

Tsanas, A., Little, M. A., McSharry, P. E., Ramig, L. O. Nonlinear Speech Analysis Algorithms Mapped to a Standard Metric Achieve Clinically Useful Quantification of Average Parkinson's Disease Symptom Severity, *Journal of the Royal Society Interface*, 8(59), 842–855, 2011.

Tu, J. V. Advantages and Disadvantages of Using Artificial Neural Networks Versus Logistic Regression for Predicting Medical Outcomes. *Journal of Clinical Epidemiology*, 49(11), 1225–1231, 2015.

Weston, P. S., Nicholas, J. M., Henley, S. M., Liang, Y., Macpherson, K., Donnachie, E., ... Fox, N. C. Accelerated Long-Term Forgetting in Presymptomatic Autosomal Dominant Alzheimer's Disease: A Cross-Sectional Study, *The Lancet Neurology*, 17(2), 123–132, 2018.

16

Assessing Impact of Global Terrorism Using Time Series Analysis

Satyam Saini, Vidushi Tripathi, Kartik Tyagi, and Adwitiya Sinha
Jaypee Institute of Information Technology

CONTENTS

16.1 Introduction

Terrorism can be characterized as the felonious utilization of brutality and terrorizing elements, especially against ordinary people, to achieve political gains (Terrorism, 2020). This definition in itself is ambiguous and subjective. In this case, the question of subjectivity means that there is no subtle definition prevalent globally as far as terrorism is considered. Any action that is violent or threatens violence with the intent for political, social, religious, and economic gains should be regarded as an act of terrorism. These attacks usually target residential buildings, airports, public transportation, commercial sites, land border crossings, large crowds, and religious settings/ceremonies, often leading to significant casualties. When carried out on a broad scale, for example, the coordinated attacks on September 11 or 12 and the coordinated shootings across Mumbai on September 26, these actions not only affect the immediate victim but also have far-reaching damages,

DOI: 10.1201/9781003046431-16

both physically and psychologically (North, 2002; Klitzman and Freudenberg, 2003). Little to no action to estimate the human reactions of such attacks, beyond counts of death and injuries, further worsens the situation (Arce, 2018).

Global Terrorism Index states that the global consequences of terrorism are diminishing after peaking in 2014; attacks against civilians hiked by 17% between 2015 and 2016, which primarily targeted private citizens and property (Jost, 2017). Countries across the globe are pouring in increasing amounts of money to provide a terrorism-free life for their citizens. In 2016, an estimated 2.2% of the global GDP was pushed into world military expenditure, around $1,686 billion (Perlo-Freeman et al., 2016). However, the inability of worldwide governments to tackle this seemingly ever-growing challenge by themselves is visible. Across the globe, nations believe that genuinely successful counter-terrorism systems benefit from including nearby locals, the private segment, the media, and other public institutions. They likewise empower the trading of intelligence, data, and expertise between national offices and across borders. Recent advancements in technology can significantly leverage the fight against terrorism. The developments in data mining permit the preparation and recognizable proof of crucial data, which can counter fear-based oppressor activities with effectiveness (Shacheng, 2012). Detecting patterns in such activities and neutralizing the threat early on so that the perpetrators of the incidents could be brought to justice has become the priority in more and more countries.

The chapter aims to push the boundaries of existing work and analyze the various factors that can help predict terrorist activities. We propose a framework for characterizing the patterns of attacks for understanding practices, analyzing association in terrorist activities, and predicting potential terrorist attacks by terrorists to prevent loss of life and property. The rest of the chapter is organized as follows: Section 16.2 overviews this region's relevant research or literature. Section 16.3 covers discussions regarding the datasets and insights from them. The proposed methodology is discussed in Section 16.4. Additionally, the machine learning algorithms used are also discussed in this section. The resultant outcomes are discussed in Section 16.5 and concluded with future work in Section 16.6.

16.2 Related Work

There are several researches conducted toward the potential identification of possible terrorist attacks. Talreja et al. (2017), in their paper, showed how machine learning algorithms could correctly predict four out of five offenders using tree-based algorithms and support vector machines. In their research, Gao et al. (2019) used more than 30,000 data values on terrorist attacks from the GTD database for 2015 and 2016 to set up several standard machine learning algorithms such as Logical Regression, Random Forest, Gauss Bayesian Network, Decision Tree, and AdaBoost to predict one or more terrorist actors. Gundabathula and Vaidhehi (2018), in their work, presented the best machine learning models such as Decision Tree trained using the C4.5 algorithm, Instance-based learning, etc. These data can be used to identify the most successful terrorist group responsible for a historical evidence-based attack. Ozgul et al. (2009), in their work, proposed a Group Detection Model (GDM) that can predict as well as identify perpetrators of unsolved terrorist activities.

The research conducted by LaFree and Dugan (2007) has created the GTD data set that provides a comprehensive collection of terrorist attacks starting from 1970, constituting domestic and international incidents. Ma et al. (2012) examined quantitative responses of stable nighttime lights from time series DMSP/OLS imagery to shifts in urbanization variables such as population, electricity consumption, and GDP in China. Wucherpfennig et al. (2011), in their paper, presented the GeoEPR dataset, which lists ethnic groups across space and time that are politically important. This research helped in analyzing the role that ethnic groups of a region play in a terrorist event. A real-time terrorist activity database was also developed from reliable sources and used these data to train a risk model capable of calculating the risk level of various locations by Toure and Gangopadhyay (2016). Nordhaus (2006), in his work, tried to experimentally establish a link between economic activity and geography.

Our study differs from previous works because we analyze various factors that might influence terrorism, such as social, economic, and geographic factors. We then compare popular and reliable machine learning results with ensemble classifiers and deep learning approaches in their ability to predict the threat of global terrorist attacks based on multiple tools and globally distributed datasets. The historical Global Terrorism Database (GTD) is adopted as the primary data source and six other standard databases for training and testing machine learning models. Additionally, time series analysis is performed by training models on data from 1970 to 2017 to identify trends and other data characteristics.

16.3 Dataset Description and Analysis

The following section describes the dataset description used for conducting our study on global terrorism and its impact.

16.3.1 Global Terrorism Database (GTD)

The GTD published by the University of Maryland includes incidents and respective details about terrorist attacks from 1970 to 2017 (LaFree and Dugan, 2007). The data are provided in a tabular form with a column each for the various aspects of the observed terrorist attack. Latitude, longitude, incident location, date, target information, claims of responsibility, perpetrator information, etc. are some of the important features out of 170 other attributes. Each row corresponds to the data sample, which is 1,700,000.

16.3.2 GeoEPR Dataset

The GeoEPR (Ethnic Power Relations) dataset, published by International Conflict Research, tries to identify all the ethically relevant ethnic groups worldwide (Wucherpfennig et al., 2011). It contains data in the form of a polygon that follows the WGS84 coordinate system, describing the location of the ethnic groups on a digital map, which can be visualized using QGIS software (QGIS, 2020).

16.3.3 G-Econ

The G-Econ dataset contains various geographical factors at a one-by-one degree latitude and longitude cell (Nordhaus et al., 2006) Distance to major navigable lakes, rivers, the ice-free oceans along with temperature, and mean precipitation (mm/year) are a few features that have been sampled starting from 1980 to 2008. The distance values reported in the dataset use the kilometer scale. The dataset used covers approximately 27,500 terrestrial grid cells.

16.3.4 Nighttime Lights

This dataset is published by the Earth Observation Group in a grid format (Lights, 2011). The files were constructed using all the available archived DMSP-OLS smooth resolution data for calendar years. Fleeting events, like fires, have been discarded. Regions with no cloud observations are depicted by the value 255, whereas the rest of the values range between 1 and 63 (both inclusive). Night lights provide essential information on the economic growth of different countries and for forecasting poverty.

16.3.5 Population Density

The population of a particular region can be a significant factor in predicting terrorist attacks. The gridded population density dataset, published by NASA's Earth Observatory, is a converted global grid of quadrilaterals with a resolution of 2.5 arc minutes depicting the density at those particular coordinates instead of actual population data (CIESIN, 2005). Figure 16.1 shows population density plotted over the world map, with black color indicating high density and white indicating low density.

16.3.6 Topography

Topographic maps depict the height, location, and shape of land features such as mountains and valleys, rivers, and even the craters of volcanoes. The gridded topography dataset, published by NASA's Earth Observatory, is a converted global grid of quadrilaterals at a resolution of 2.5 arc minutes (Rodriguez et al., 2005).

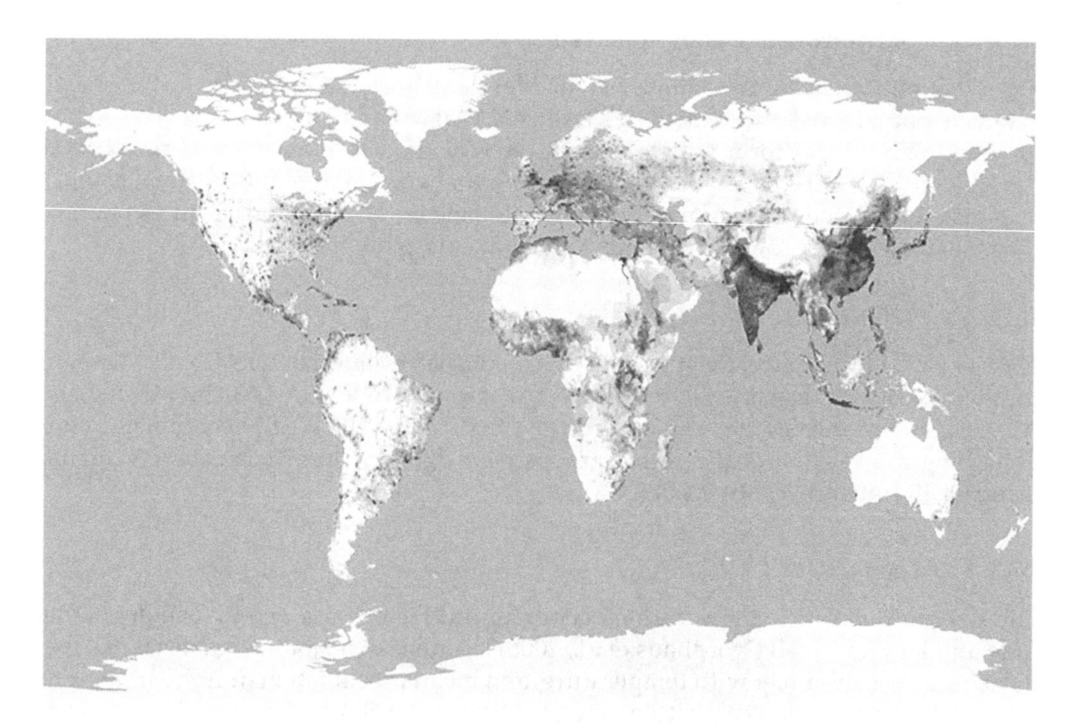

FIGURE 16.1
Population density plot.

16.3.7 Happy Index

United Nations Sustainable Development Solutions Network ranks countries by the happiness levels of the residing citizens (Helliwell et al., 2016). As many governments and organizations have started using happiness indicators to develop strategies, the study is globally recognized. The index is primarily based on life expectancy, economic productivity, social support, freedom, absence of corruption, and generosity. The values in the other columns are relative to the values of a hypothetical country dystopia, with the lowest national averages of all six factors.

16.3.8 Overall Analysis

GTD dataset, in its proper form, covers terrorist activities around the globe. Here, we analyze the dataset concerning South Asia, which itself includes several incidents since 1970. Figure 16.2a–c shows how terrorism spread in the years 1990–2017 in Southern Asia specifically. In 1990, we can see the dark red spots over Sri Lanka predominantly because of the Liberation Tigers of Tamil Eelam (LTTE). LTTE was one of the numerous organizations that came into existence to fight for Tamil rights in Sri Lanka. In 2006, we can see dark spots in Mumbai because of the massive Mumbai train bombing attacks in November 2006 that shook India.

Additionally, the figure depicts the dark red highlights over Pakistan (because of Jaish-e-Mohammed) and Afghanistan (because of the Taliban), implying the widespread terrorism in both countries. Figure 16.3 shows a detailed analysis about the attacking methods in the South Asian countries shows that Bombing Explosions has been the most often used attacking method, which is considerably high compared to the Armed Assaults method. Hijacking seems to be the lowest cause of attacks, owing to the increased security at airports and strict protocols at such places.

Among South Asian countries, Pakistan tends to be the most affected country, with the most attacks since 1978, closely followed by the two countries that share its border, Afghanistan and India. This trend proves that geographic neighbors usually tend to suffer from the increasing terrorism-related activities inside a nation. Figure 16.4 shows South Asian countries affected by different terrorist groups.

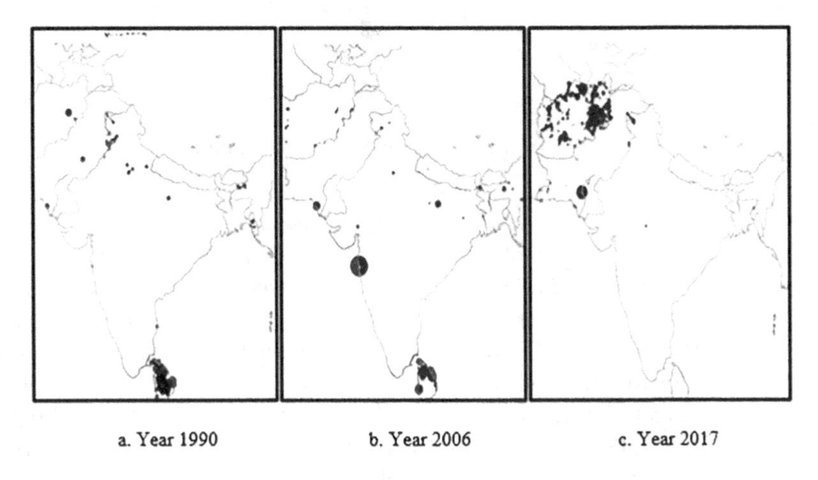

FIGURE 16.2
Terrorist activities in South Asia for year (a) 1990, (b) 2006, and (c) 2017.

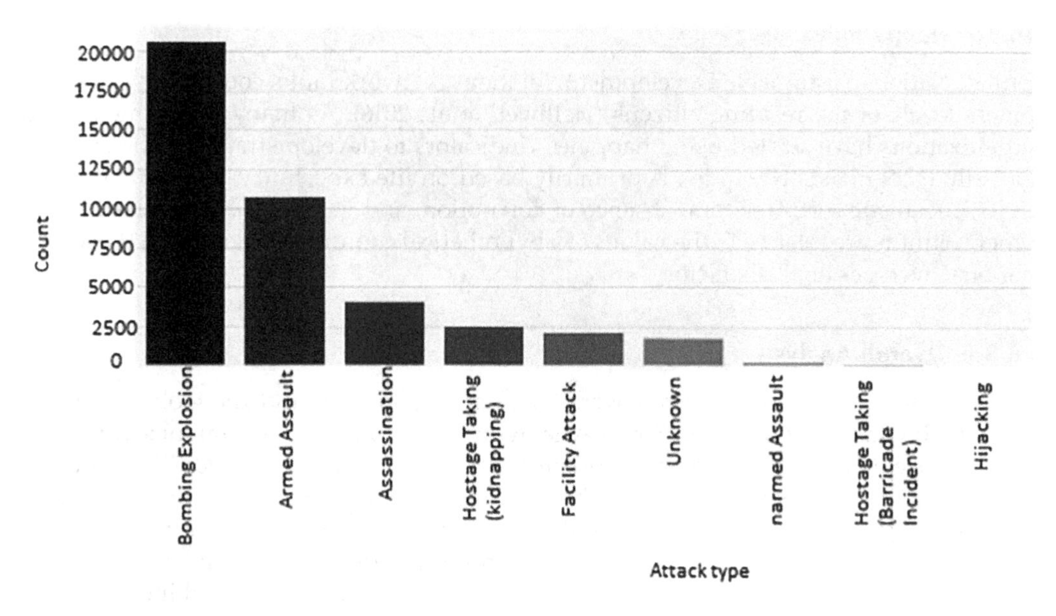

FIGURE 16.3
Weapons used by terrorists for organized crime in South Asia.

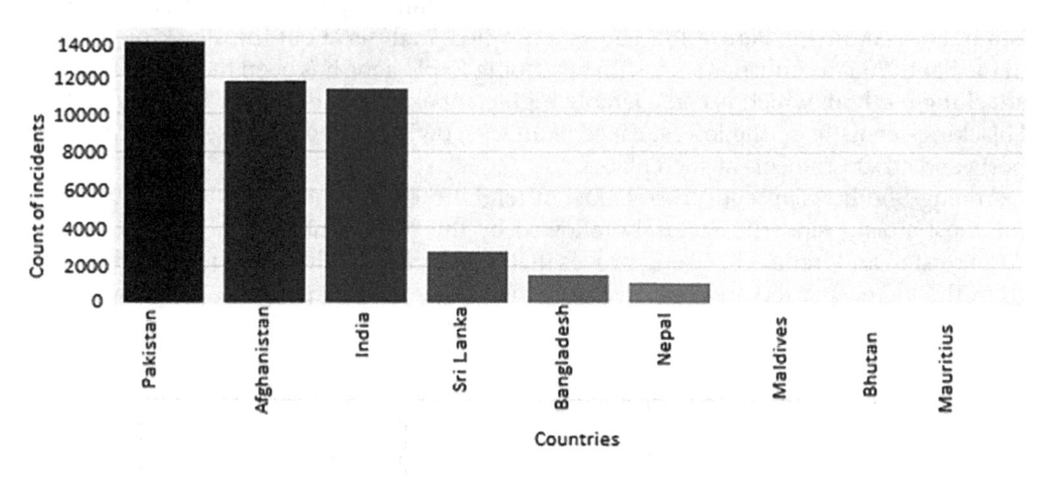

FIGURE 16.4
Plot depicting the count of successful terrorist activities in South Asian countries.

From the last analysis, Pakistan, Afghanistan, and India tend to be most affected by terrorism. Hence, a terrorist group centered on their borders must be dominantly responsible for the most attacks as per the analysis. It is closely followed by the CPI-Maoist and Maoist, which affect India along its Eastern border (which it shares with China). Southern India and Sri Lanka are affected by LTTE, which secures the fourth spot as shown in Figure 16.5.

The time series plot for the rise of terrorism in South Asia can be seen in Figure 16.6. Afghanistan continues to be the most affected state, with 3,388 attacks in the given year. Afghanistan is closely followed by Pakistan, with 2,568 attacks. On the other hand, India has put counter-terrorism efforts in the right place, with just 262 attacks in 2012. Sri Lanka continues to be a peaceful country as it was the year before, in 2011.

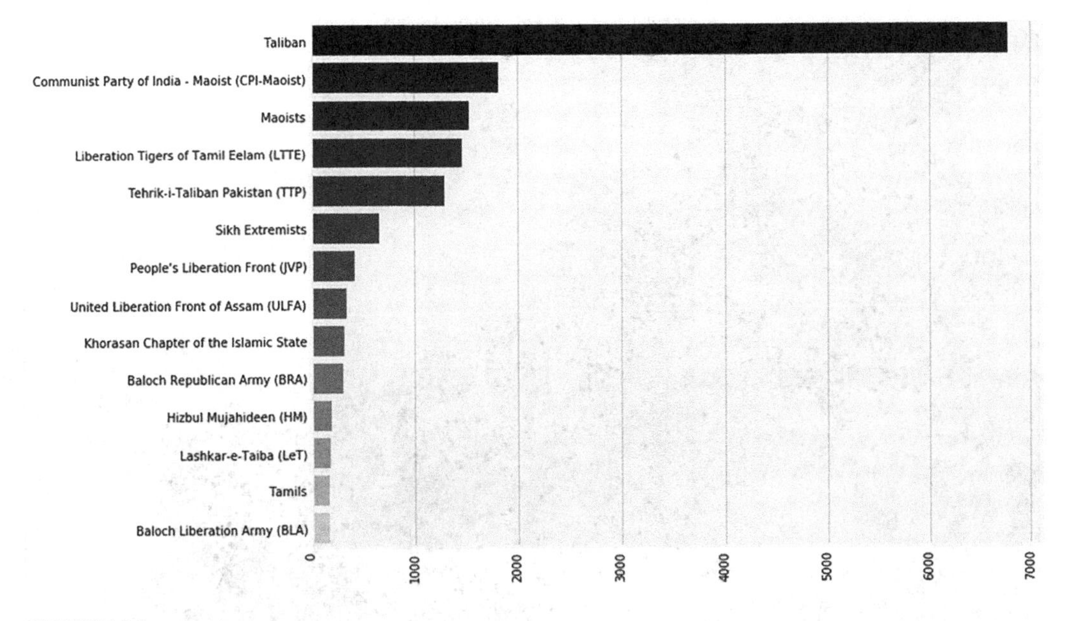

FIGURE 16.5
Terrorist groups based on the number in South Asia.

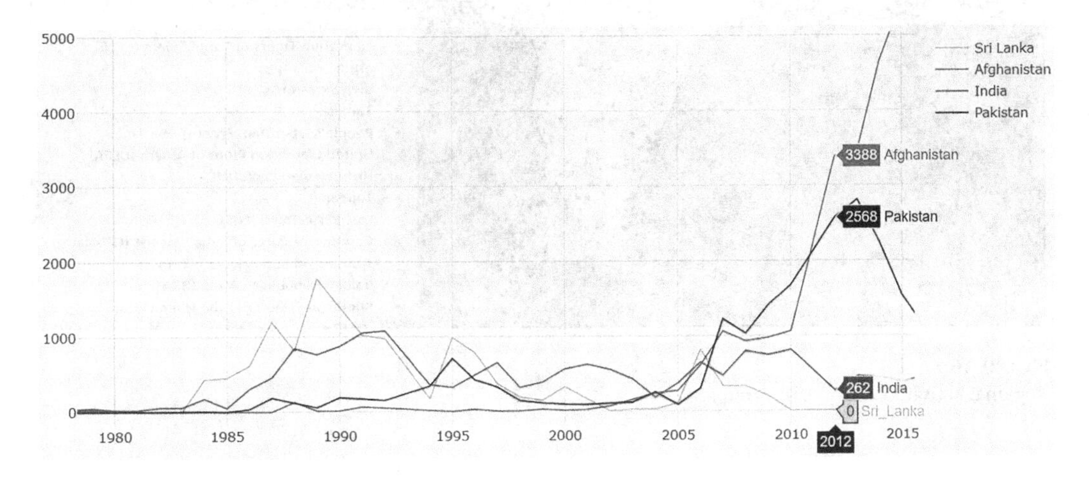

FIGURE 16.6
Time series plot for terrorist activities in South Asia.

The activities of prominent groups active in South Asian nations can be seen in Figure 16.7. The heavy density in the North and East depicts the activeness of the Taliban, CPI-Maoists, and other upcoming ethnic group-based terrorist camps.

A timeline plot with a count of terrorist attacks from 1970 to 2017 is shown in Figure 16.8. The figure shows attacks to be on the rise since the database came into being in 1970. However, recent anti-terrorist solid measures have been effective as the attacks have seen a dip since 2014.

The terrorist activities inflict casualties and destruction inflicted on the population residing in these areas. The damage can be seen from Figure 16.9, which underlines the

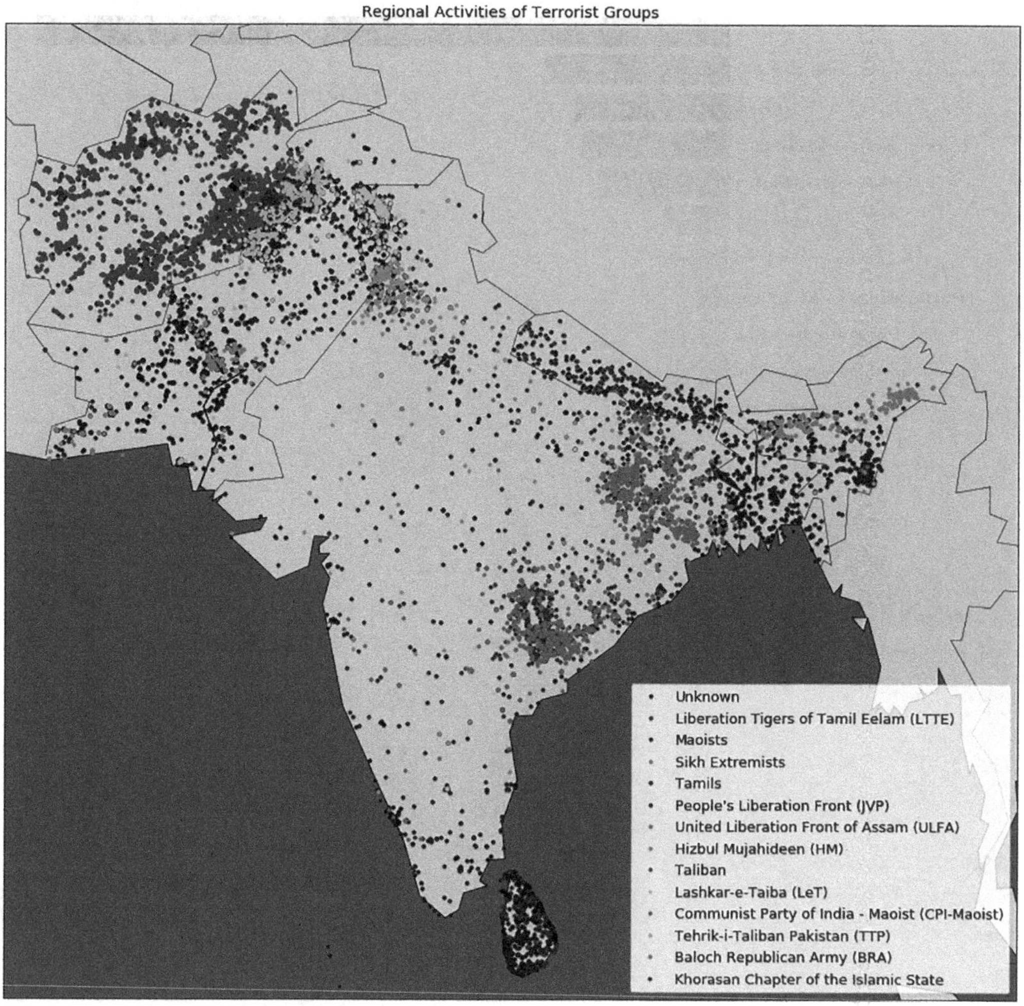

FIGURE 16.7
Regional activities of terrorist groups.

facts stated in Figures 16.7 and 16.8. Blue dots depict attacks with less than a hundred but not zero casualties, whereas red dots show samples with high casualties.

Analysis of South Asian countries shows that India continues to be the most tolerant and ethnically welcoming nation, with more than 20 ethnic groups living peacefully. The same can be seen in Figure 16.10.

Figure 16.10, along with Figure 16.5, helps in understanding that among all the South Asian countries, India, Afghanistan, and Pakistan are the most ethnically diverse. At the same time, Pakistan, Afghanistan, and India are the most affected countries by the terrorist attacks giving slight hints about their positive correlation. This statement is supported by the correlation coefficient between the number of ethnic groups, and the number of attacks in a country came out to be 0.65759.

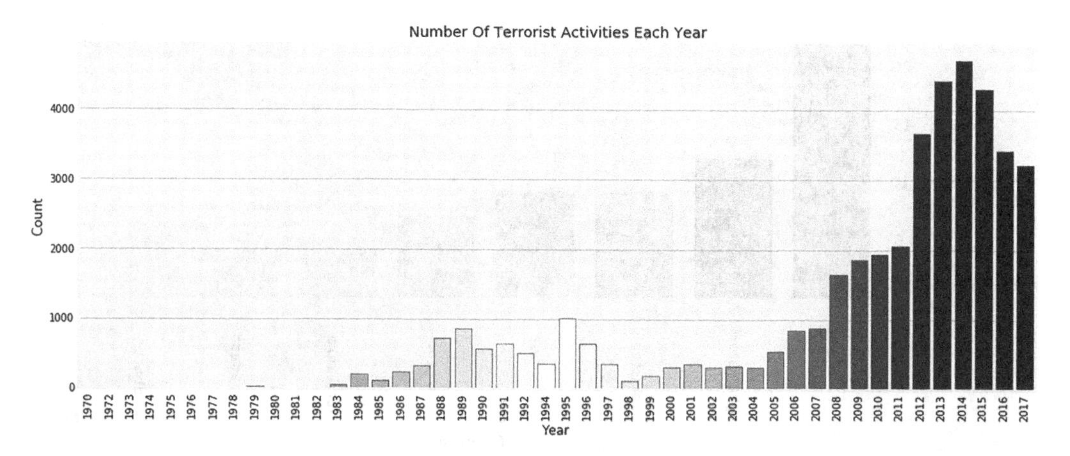

FIGURE 16.8
Total number of terrorist attacks from 1970 to 2017.

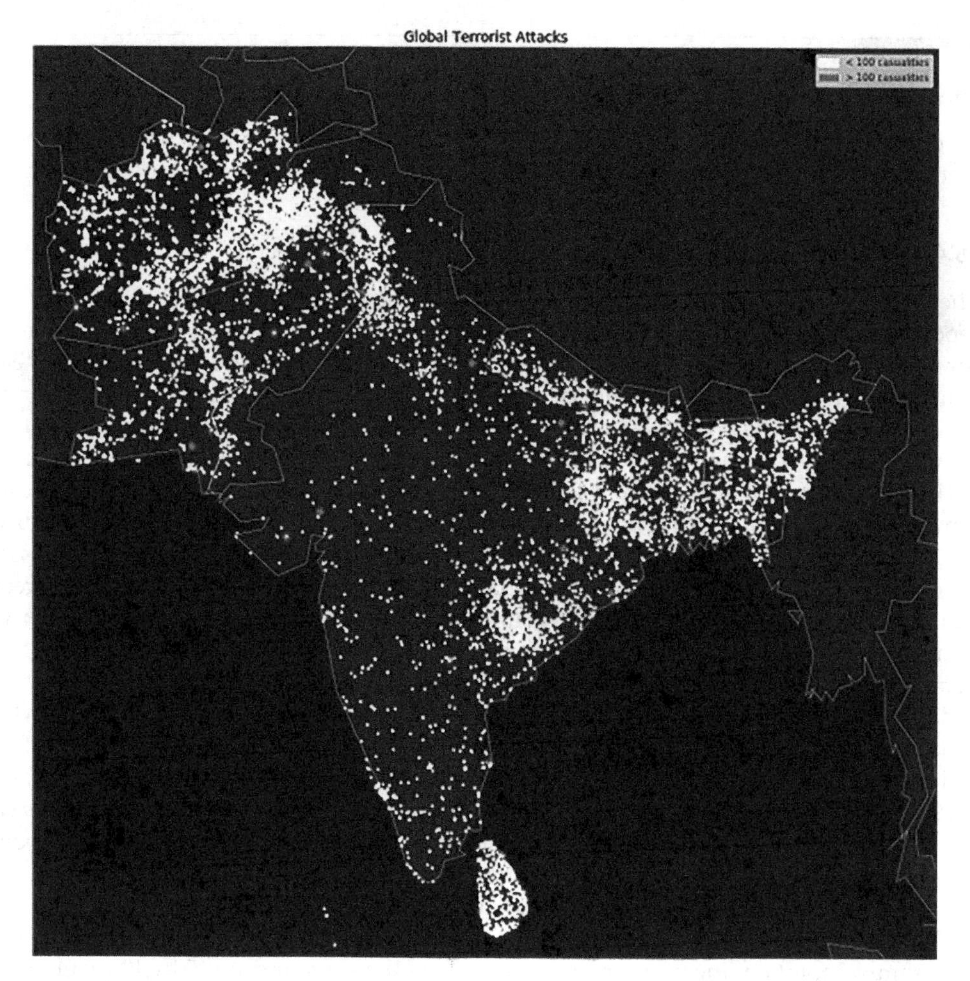

FIGURE 16.9
Casualties inflicted by terrorist attacks.

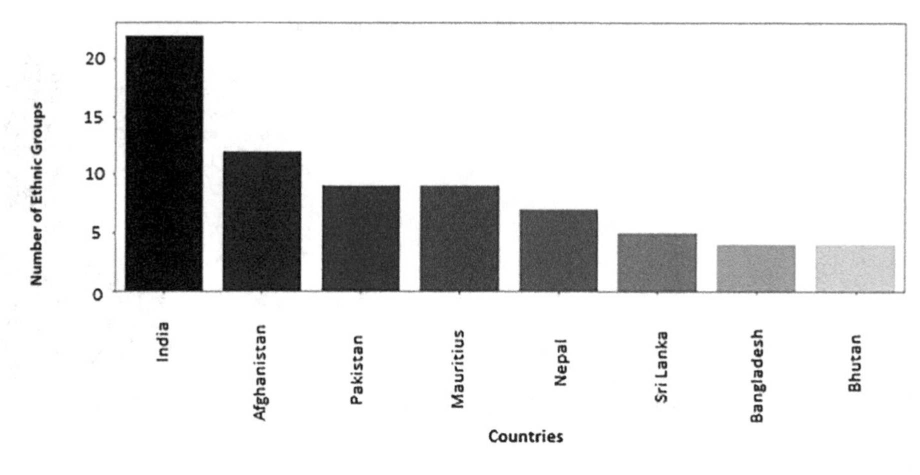

FIGURE 16.10
Top ethnically diverse countries in South Asia.

16.4 Proposed Methodology

Section 16.4.1 highlights data sampling with normalization techniques and the proposed methodology.

16.4.1 Data Sampling

The sources and format of data presented in the datasets used in our chapter have been reported in Section 16.3.1. Each dataset presents an enormous number of informative features. However, for the scope of this chapter, selected features are extracted from each dataset. Additional preprocessing is done for sample points with no value under the required feature in a particular dataset. Features across each dataset are resampled at a uniform pixel resolution of 0.1×0.1 degrees and merged, resulting in the actual data.

The location of each incident from 1970 to 2017 data is extracted from the GTD database and converted to raster form. Any sample point is considered a point of high attack probability if casualties occur and is given class 1, otherwise 0. We obtain five features based on the G-Econ dataset: distance to a major navigable river, major navigable lake, ice-free ocean, average precipitation, and average temperature. At each sample point, the count of ethnic groups was derived from GeoEPR at a 0.01 by 0.01-degree latitude resolution using the QGIS join attributes function. From NASA's Earth Observatory system, information with regard to population density and topography were derived. Specific rows in the merged training dataset, specifically where attacks occurred along the coasts, were already allotted distinguishing values because of limited spatial resolution. Plots verifying the findings are shown in Figure 16.11 for both datasets. For such attribute values in the population dataset, the training unit was not dropped. Instead, the average values of geographically adjacent samples were filled in place. Since the points were coastal units for the topological data, their Digital ElevationModEM value was adopted to be 0 (as their height from sea level is 0 km), as shown in Figure 16.12.

The nighttime light data were resampled at the mentioned standard resolution, with each sample point having an average luminosity of the surrounding 10 by 10-pixel region. Happiness indices of 155 countries were mapped with the location of the attacks using the K-nearest neighbor algorithm for locations with no happiness values.

It is uniformly spread worldwide before splitting the complete dataset into training and validation, as shown in Figure 16.13. Each point corresponds to 12 attributes.

To perform time series analysis the data were further preprocessed to include other relevant factors from the GTD dataset, namely "eventid," "Year," "Month," "city," "Day," "extended," "Country," "Region," "State," "latitude," "longitude," "specificity," "vicinity," "crit1," "crit2," "crit3," "doubtterr," "success," "multiple," "suicide," "AttackType," "Target_type," "Target," "Group," "Weapon_type," "Killed," "Wounded," "property," and "ishostkid." A new feature was added depicting the relative time passed for every incident since the first attack. Only those groups were considered whose attack count was more significant than 200 over the years. On top of such groups was the count of "Unknown" attacks, for which no one took responsibility. We removed these columns from our dataset for time series analysis.

16.4.2 Normalizing

The mentioned potential features have different units. Hence, they were normalized so that each feature contributed approximately proportionately toward the final results.

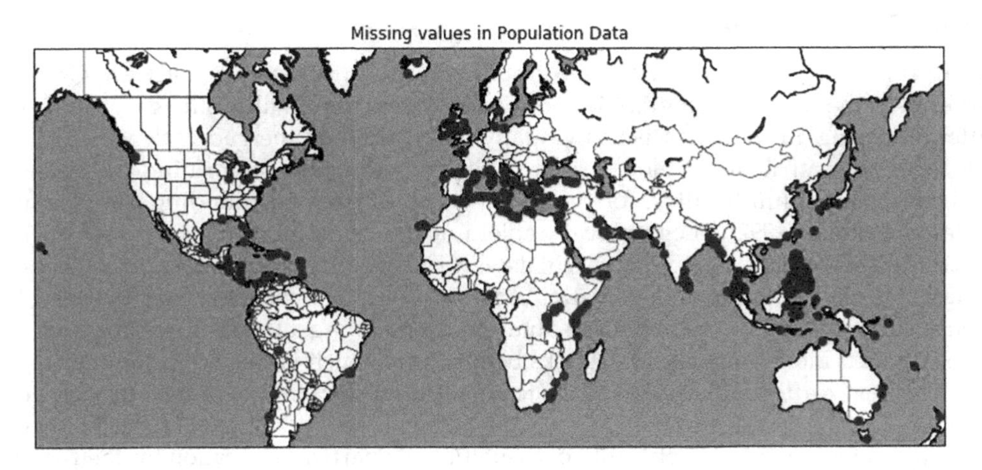

FIGURE 16.11
Missing values in population dataset.

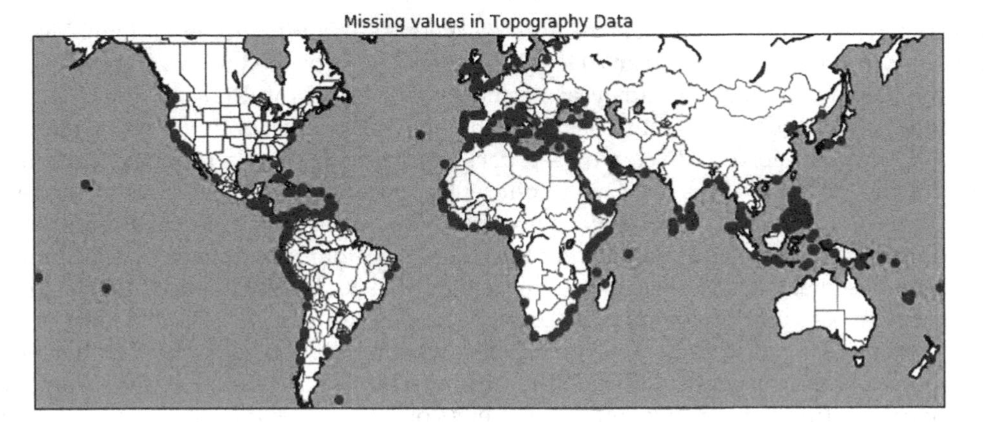

FIGURE 16.12
Missing values in topological dataset.

FIGURE 16.13
Sample points plot over world map.

16.4.3 Dataset Design

Geospatial merging of data was performed using the open-source software QGIS [17]. WGS-84 is the uniform geographic coordinate system across all of the geographic data, after preparing all the datasets individually. They are then merged with the GTD dataset using join attributes by location in QGIS and explicitly performing the K-Nearest Neighbors algorithm to fill in for the missing values from surrounding samples in the datasets (Altman, 1992). Geographically adjoining attack locations form a part of the same statistical unit because of the changes in dataset resolution. A sample row in the processed dataset (used for training and validation) contains attributes collected from the exact location at each data layer. For time series analysis, after adding a few features, the final dataset is divided into two parts, i.e., attacks before 2018, which is treated as the training set, and the attacks after 2018, treated as the testing set. The aim for the trained models here is to predict, based on features of location and learning from past trends of the training dataset, either a high or a low probability of casualties at a particular location for the year 2018.

16.4.4 Machine Learning Algorithms

Five relatively popular and robust machine learners (XGBOOST, CatBoost, LightGBM, SVM classifier, and Random Forest) and DeepNeuralNets were trained to tune their associated parameters and perform accuracy assessments. The "sklearn" Keras (using Tensorflow backend) and a few other built-in packages in the 64-bit version of Python 3.3 were used.

For training DeepNeuralNet, we used backward propagation, a robust algorithm that tweaks the "hidden" units' weights to efficiently represent salient features of the resulting domain (Rumelhart et al., 1986). This efficiency makes it feasible for training multilayer networks. The training is further backed using Adam Optimizer, which improves gradient-based stochastic objectives (Kingma and Ba, 2014). While model training, backpropagation was used to identify the count of hidden units in each hidden layer and the decay rates were tuned. Relu and sigmoidal activation functions were used in the hidden and output layers, respectively. For SVM training, a radial basis function (RBF) kernel was considered in this chapter. Two hyperparameters of RBF, cost, and sigma, impact classification tasks (Awad and Khanna, 2015). *Cost* is the price paid by the model for misclassification.

A high cost means low bias and high variance because we penalize the cost of misclassification a lot (Ali et al., 2012). A higher value of gamma ends up in more accuracy but biased results and vice versa. For Random Forest, 50 trees were used, as increasing the number of trees did not affect the performance (Breiman, 2001). Recently, gradient tree-boosting techniques have outshined other machine learning methods in many applications such as weather forecasting, Web search, etc. (Friedman, 2001; Wu et al., 2010; Zhang and Haghani, 2015). xgboost, a scalable machine learning system for tree boosting, was trained with a tree-based model (gbtree) (Chen and Guestrin, 2016). After utilizing grid search, a learning rate of 0.1 was adopted with 100 estimators and a max_depth of 3. Because of imbalanced data (as most attacks had casualties and hence labeled 1), the scale_pos_weight parameter was set to balance the positive and negative weights. For the rest of the parameters, default values gave satisfactory results. While most popular gradient-boosting algorithms use depth-wise tree growth, LightGBM, introduced by Microsoft research, uses the leaf-wise tree growth algorithm, which can converge much faster (Ke et al., 2017). Grid searching for learning rate gave 0.1 as the best value along with no min_width defined.

Another popular gradient-boosting algorithm, CatBoost, or Categorical Boosting, shows promising results on benchmark datasets (Prokhorenkova et al., 2018). Python CatBoostClassifier was trained with Tree depth parameter value experimentally set to 8 and the learning rate set to 0.0374. The model took 1 minute 15 seconds for training. For the time series analysis, a voting classifier was trained. This classifier is a meta-classifier that pools similar or different machine learning classifiers for classification using different voting methods (Raschka, 2015). This classifier allows for soft and hard voting. In hard voting, the output label is the one that has the highest probability of being predicted by each classifier. While in soft voting, the output label is predicted based on each classifier's average probability given to that class. Four classifiers (Logistic Regression, Gaussian Naive Bias, Random Forest, SVM classifier) were selected for our ensemble process with the "n_estimator" parameter set to 50.

16.5 Result Analysis

We try to analyze the invidious power of each factor in our dataset. "feature_importance" function in Python libraries was used for this analysis. The maximal number of essential source runs was set to 1. Figure 16.14 shows the Feature importance of different attributes for the prediction of terrorist activity. The figure depicts the importance of social, geographical, and economic factors while predicting terrorist activities. Latitude and longitude have the highest importance, with values of 0.130 and 0.132, respectively, followed by topology elevation and nighttime light intensity.

A correlation matrix depicts correlation coefficients between sets of variables in a tabular form. Each feature is treated as a random variable (Xi) in the table, which is correlated with every other feature of the table, allowing us to visualize the pairs with the highest correlation. Figure 16.15 shows the pairwise dependency of each factor on the other.

A Receiver Operating Characteristic (ROC) depicts the performance of a classification model at all classification thresholds using two parameters, namely True-Positive rate (TPR) and False-Positive rate (FPR).

TPR or sensitivity or recall is the ratio of the total count of relevant terrorist activity instances predicted correctly and given by Equation 16.1.

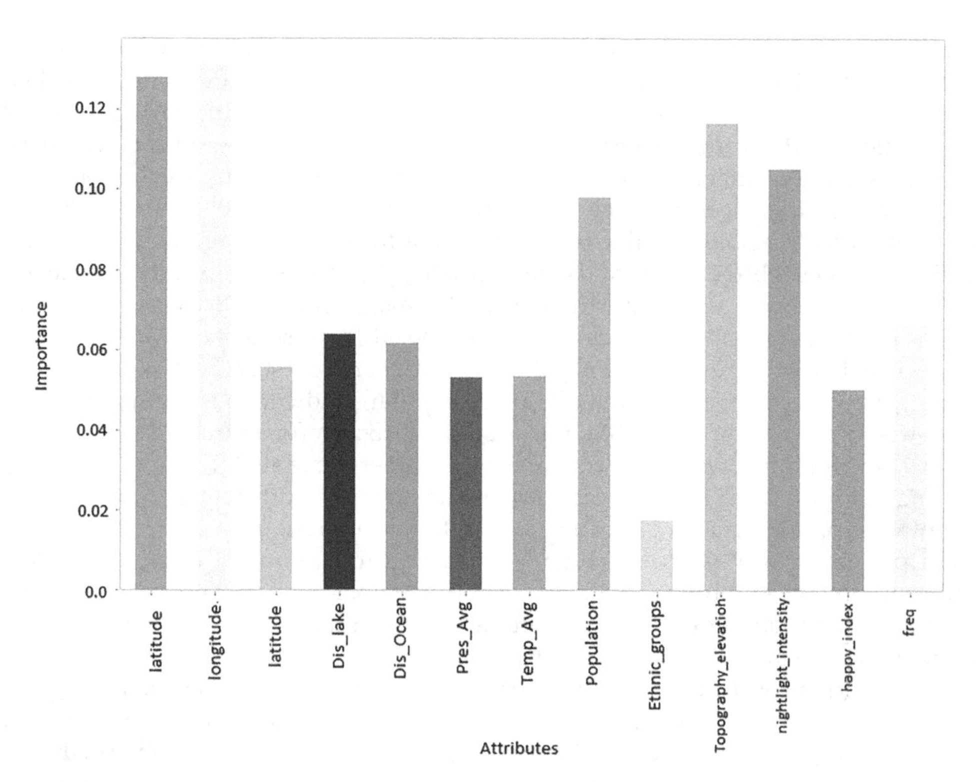

FIGURE 16.14
Plot depicting the importance of each attribute toward the prediction of terrorist attack.

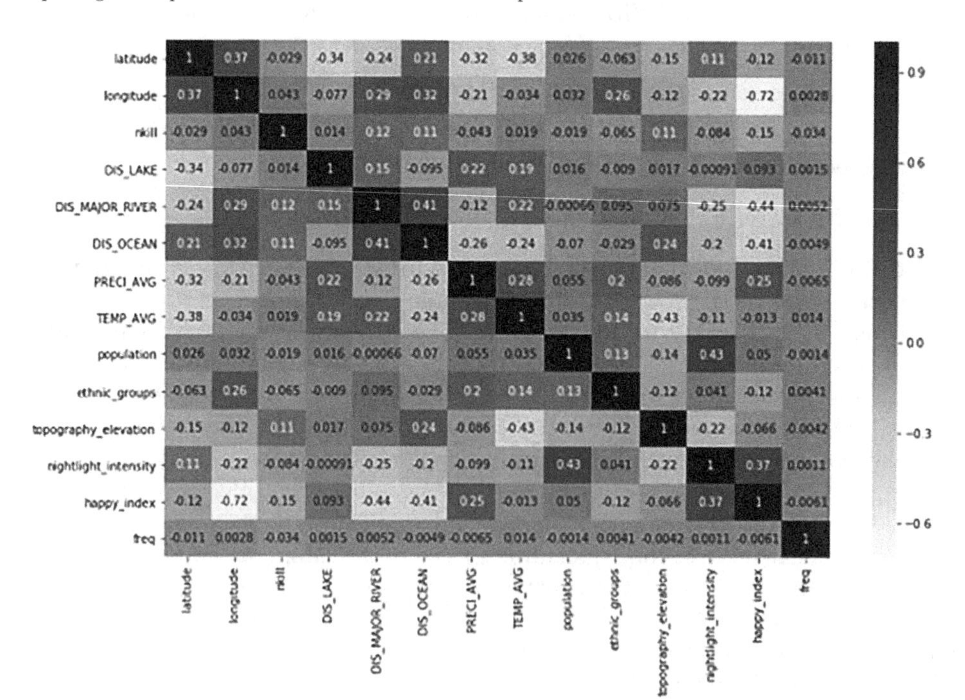

FIGURE 16.15
Correlation matrix of the dataset.

$$\text{TPR}\left(\text{sensitivity}\right)=\frac{\text{True positive}}{\text{True positive}+\text{False negative}} \tag{16.1}$$

FPR is the probability of falsely rejecting the null hypothesis for a particular test. It is given by Equation 16.2.

$$\text{FPR}\left(1-\text{specificity}\right)=\frac{\text{False positive}}{\text{True negative}+\text{False positive}} \tag{16.2}$$

The Area Under the Curve (AUC) is a measure for calculating the performance of models with increasing AUC values suggesting greater accuracy. Given this measure, we chose models with optimized tuning boundary parameters. The ROC curve (Figure 16.16) shows that the LightGBM (0.83) model was able to capture the intricacies of the attributes in the best way, followed by CatBoost (0.78) and XGBoost (0.75). The AUC values of each model have been shown in the bottom right of the figure.

The classification report helps us measure the quality of predictions made by our trained models.

The precision will be the fraction of relevant activities among the retrieved instances, given by Equation 16.3:

$$\text{TPR}\left(\text{sensitivity}\right)=\frac{\text{True positive}}{\text{True positive}+\text{False positive}} \tag{16.3}$$

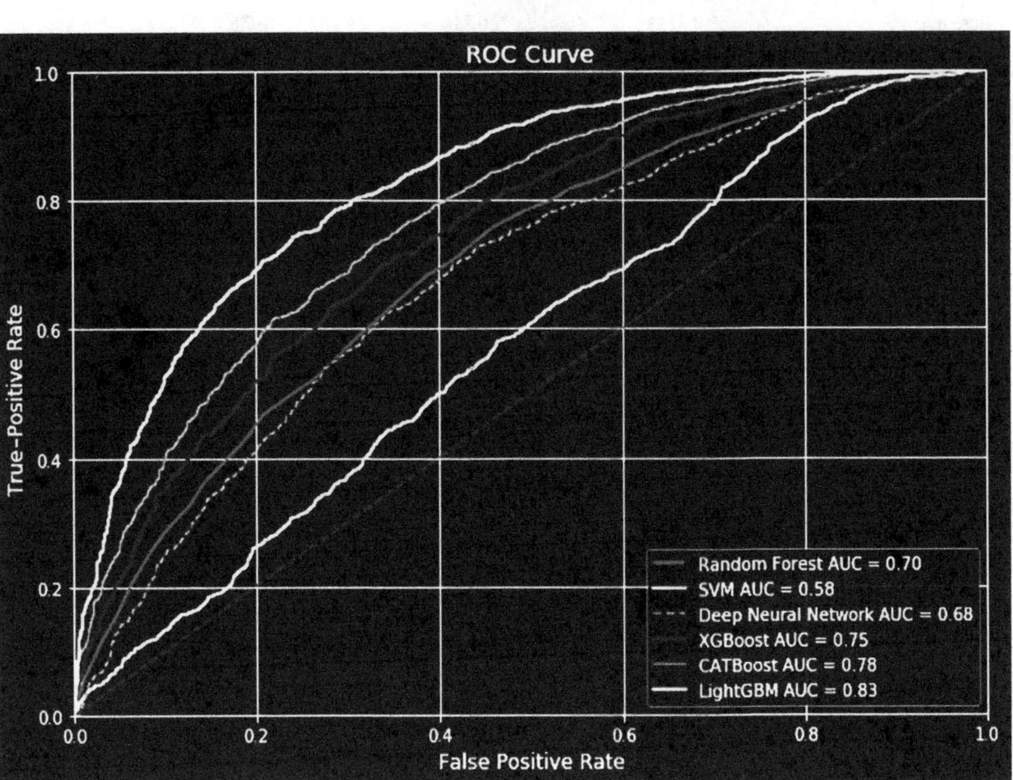

FIGURE 16.16
ROC curve of six models applied over the curated dataset.

The $F1$ score is a weighted harmonic mean of precision and lies between 1.0 and 0.0, higher being better. It is given by Equation 16.4:

$$F1 - \text{score} = \frac{\text{Precision} \times \text{Recall}}{\text{Precision} + \text{Recall}} \tag{16.4}$$

The support is the number of occurrences of the given class in the validation dataset. Tables 16.1—16.6 compare the validation results of the six models defined in the section trained on 13,198 samples and validated on 3,300 samples using a classification report. SVM classifier gave the highest precision (0.75) for the no casualty (low probability of attacks) class closely followed by LightGBM (0.74). Random Forest has the highest recall (0.42) and an F1-score (0.48) value for zero casualty labels. For high-probability incidents (class labeled as 1), SVM has the highest recall value (0.99), whereas Random Forest has the highest precision value (0.75), which is slightly more than the value for LightGBM (0.74). For the same class, CatBoost, XgBoost, and LightGBM gave the same F1-score value (0.82).

TABLE 16.1

Evaluation Metric for Deep Neural Network

	Precision	Recall	F1-Score	Support
0	0.64	0.22	0.33	1,074
1	0.71	0.94	0.81	2,226
Micro avg	0.71	0.71	0.71	3,300
Macro avg	0.68	0.58	0.57	3,300
Weighted avg	0.69	0.71	0.65	3,300

TABLE 16.2

Evaluation Metric for CatBoost

	Precision	Recall	F1-Score	Support
0	0.66	0.31	0.42	1,068
1	0.74	0.92	0.82	2,232
Micro avg	0.72	0.72	0.72	3,300
Macro avg	0.70	0.62	0.62	3,300
Weighted avg	0.71	0.72	0.69	3,300

TABLE 16.3

Evaluation Metric for XGBoost

	Precision	Recall	F1-Score	Support
0	0.66	0.26	0.37	1,068
1	0.73	0.94	0.82	2,232
Micro avg	0.72	0.72	0.72	3,300
Macro avg	0.69	0.60	0.60	3,300
Weighted avg	0.71	0.72	0.67	3,300

TABLE 16.4

Evaluation Metric for LightGBM

	Precision	Recall	F1-Score	Support
0	0.64	0.33	0.43	1,068
1	0.74	0.91	0.82	2,232
Micro avg	0.72	0.72	0.72	3,300
Macro avg	0.69	0.62	0.62	3,300
Weighted avg	0.71	0.72	0.69	3,300

TABLE 16.5

Evaluation Metric for SVM Classifier

	Precision	Recall	F1-Score	Support
0	0.75	0.09	0.16	1,074
1	0.69	0.99	0.81	2,226
Micro avg	0.69	0.69	0.69	3,300
Macro avg	0.72	0.54	0.48	3,300
Weighted avg	0.71	0.69	0.60	3,300

TABLE 16.6

Evaluation Metric for Random Forest

	Precision	Recall	F1-Score	Support
0	0.55	0.42	0.48	1,074
1	0.75	0.83	0.79	2,226
Micro avg	0.70	0.70	0.70	3,300
Macro avg	0.65	0.63	0.63	3,300
Weighted avg	0.69	0.70	0.69	3,300

For the time series analysis, we then try to analyze the invidious power of each factor in our dataset. The "feature_importance" function in Python libraries was used for this analysis. The maximal number of essential source runs was set to 1. Figure 16.17 shows the importance of the selected features toward time series analysis. Happiness Index got a maximum score with a value of 0.144, followed by latitude (0.135), longitude (0.12), and days passed (0.114), respectively.

Given the substantial importance of these additional features and the initial ones, we performed a time series analysis. Random Forest, Decision Tree, Gradient Boosting, and Voting technique are the four algorithms that are trained and used for prediction. The model's output (high or low probability of casualties) is compared with the ground-truth values obtained (after preprocessing) from the GTD dataset. As summarized in Table 16.7, the results show the trained models' ability to predict the probability of high or low casualties for the year 2017 based on selected features. The comparative results of the algorithms show that almost every algorithm can give state-of-the-art results (close to 95%) for the data.

TABLE 16.7
Accuracies of Various Algorithms

Algorithm	Accuracy (%)
Random Forest	94
Decision Tree	95
Gradient Boost	95
Voting Classifier	95

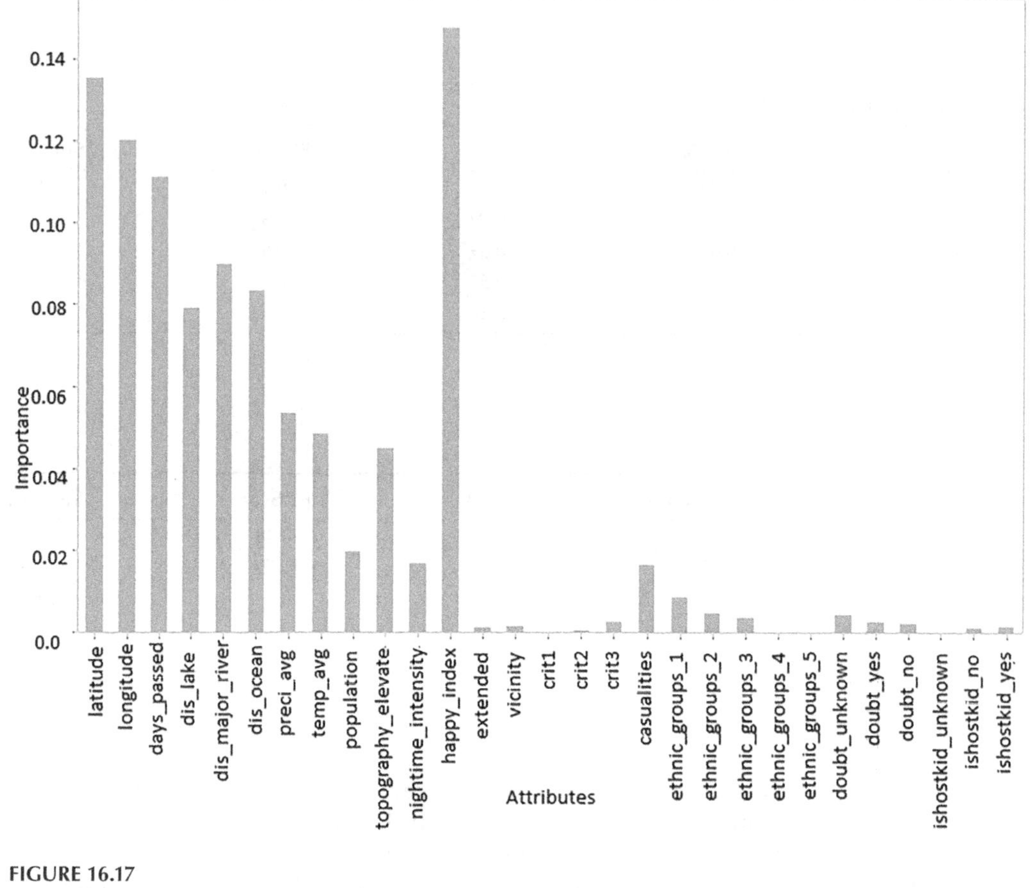

FIGURE 16.17
Feature importance of time series analysis.

16.6 Conclusion

Our research advocates a promising future for data mining and machine learning in the fight against terrorism. Machine learning techniques convey immense potential in forecasting terrorism activities, thereby saving numerous human lives. The time series analysis reinstates the fact that the selected factors can tremendously predict the chances of a region being at a high or low risk of terrorist activity. All the machine learning models were very much able to learn from the multidimensionality of the dataset without overfitting the data. Three (Gradient Boost, Decision Tree, and Voting Classifier) of the four models exhibit an accuracy level of 95%. The remaining Random Forest model falls short marginally with 1% as compared to others. Although in our chapter, we try to broaden the horizon of factors involved in predicting the dynamics of terrorism, we certainly do fall short on specific parameters. One such is the limited social, economic, and physical factors we have included in our chapter. Other factors under each category, when included, may result in better results. Another factor is the unavailability of any high-precision geographic dataset, because of which each dataset had to be scaled up, resulting in loss of valuable information.

References

Ali, J., Khan, R., Ahmad, N., et al. (2012). Random forests and decision trees. *International Journal of Computer Science Issues (IJCSI)*, 9(5), 272.

Altman, N. S. (1992). An introduction to kernel and nearest-neighbor nonparametric regression. *The American Statistician*, 46(3), 175–185.

Arce, D. (2018). On the human consequences of terrorism. *Public Choice*, 178(3–4), 371–396. doi:10.1007/s11127-018-0590-9.

Awad, M., & Khanna, R. (2015). Support vector machines for classification. In *Efficient Learning Machines* (pp. 39–66). Apress, Berkeley, CA.

Breiman, L. (2001). Random forests. *Machine Learning*, 45(1), 5–32.

Chen, T., & Guestrin, C. (2016). Xgboost: A Scalable Tree Boosting System. In *Proceedings of the 22nd Acm Sigkdd International Conference on Knowledge Discovery and Data Mining* (pp. 785–794).

CIESIN, F. (2005). *Gridded Population of the World, Version 3 (GPWv3): Population Count Grid*. Center for International Earth Science Information Network (CIESIN), Columbia University.

Friedman, J. H. (2001). Greedy function approximation: a gradient boosting machine. *Annals of Statistics*, 1189–1232.

Gao, Y., Wang, X., Chen, Q., et al. (2019). Suspects prediction towards terrorist attacks based on machine learning. In *2019 5th IEEE International Conference on Big Data and Information Analytics (BigDIA)* (pp. 126–131).

Gundabathula, V. T., & Vaidhehi, V. (2018). An efficient modelling of terrorist groups in India using machine learning algorithms. *Indian Journal of Science and Technology*, 11(15), 1–10.

Helliwell, J. F., Huang, H., & Wang, S. (2016). The distribution of world happiness. *World Happiness*, 8.

Jost, J. (2017). Institute for economics & peace: global terrorism index 2015. *SIRIUS Zeitschrift für Strategische Analysen*, 1(1), 91–92.

Ke, G., Meng, Q., Finley, T. et al. (2017). Lightgbm: a highly efficient gradient boosting decision tree. In *Advances in Neural Information Processing Systems* (pp. 3146–3154).

Kingma, D. P., & Ba, J. (2014). Adam: a method for stochastic optimization. arXiv preprint arXiv:1412.6980.

Klitzman, S., & Freudenberg, N. (2003). Implications of the world trade center attack for the public health and health care infrastructures. *American Journal of Public Health*, 93(3), 400–406. doi:10.2105/ajph.93.3.400.

LaFree, G., & Dugan, L. (2007). Introducing the global terrorism database. *Terrorism and Political Violence*, 19(2), 181–204.

Lights, D. O. N. (2011). *Image and Data Processing by NOAA's National Geophysical Data Center.* DMSP Data Collected by the US Air Force Weather Agency.

Ma, T., Zhou, C., Pei, T., et al. (2012). Quantitative estimation of urbanization dynamics using time series of DMSP/OLS nighttime light data: a comparative case study from China's cities. *Remote Sensing of Environment*, 124, 99–107.

Nordhaus, W., Azam, Q., Corderi, D., et al. (2006). *The G-Econ Database on Gridded Output: Methods and Data*. Yale University, New Haven, 6.

Nordhaus, W. D. (2006). Geography and macroeconomics: new data and new findings. *Proceedings of the National Academy of Sciences*, 103(10), 3510–3517.

North, C. (2002). Research on the mental health effects of terrorism. *JAMA*, 288(5), 633. doi:10.1001/jama.288.5.633.

Ozgul, F., Erdem, Z., & Bowerman, C. (2009). Prediction of unsolved terrorist attacks using group detection algorithms. In *Pacific-Asia Workshop on Intelligence and Security Informatics* (pp. 25–30). Springer, Berlin, Heidelberg.

Perlo-Freeman, S., Wezemen, P., & Wezeman, S. (2016). *Trends in World Military Expenditure, 2015*. SIPRI.

Prokhorenkova, L., Gusev, G., Vorobev, A., et al. (2018). *CatBoost: Unbiased Boosting with Categorical Features*.

QGIS Development Team (2020). QGIS Geographic Information System. Open Source Geospatial Foundation Project. http://qgis.osgeo.org.

Raschka, S. (2015). *Python Machine Learning*. Packt Publishing Ltd.

Rodriguez, E., Morris, C. S., Belz, J. E., et al. (2005). *An Assessment of the SRTM Topographic Products*.

Rumelhart, D. E., Hinton, G. E., & Williams, R. J. (1986). Learning representations by back-propagating errors. *Nature*, 323(6088), 533–536.

Shacheng, W. (2012). Data mining: study on intelligence-led counterterrorism. In *Proceedings of the 2011 2nd International Congress on Computer Applications and Computational Science* (pp. 87–93). Springer, Berlin, Heidelberg.

Talreja, D., Nagaraj, J., Varsha, N. J., et al. (2017). Terrorism analytics: learning to predict the perpetrator. In *2017 International Conference on Advances in Computing, Communications and Informatics (ICACCI)* (pp. 1723–1726). IEEE.

Terrorism Dataset. (2020). In Lexico Dictionaries. https://www.lexico.com.

Toure, I., & Gangopadhyay, A. (2016). Real time big data analytics for predicting terrorist incidents. In *2016 IEEE Symposium on Technologies for Homeland Security (HST)* (pp. 1–6). IEEE.

Wu, Q., Burges, C. J., Svore, K. M., et al. (2010). Adapting boosting for information retrieval measures. *Information Retrieval*, 13(3), 254–270.

Wucherpfennig, J., Weidmann, N. B., Girardin, L., et al. (2011). Politically relevant ethnic groups across space and time: introducing the GeoEPR dataset. *Conflict Management and Peace Science*, 28(5), 423–437.

Zhang, Y., & Haghani, A. (2015). A gradient boosting method to improve travel time prediction. *Transportation Research Part C: Emerging Technologies*, 58, 308–324.

17

Sustainable Statistics for Death Cognizance Analysis

Megha Rathi, Dhananjay Jindal, Manan Thakral, Rishabh, and Palak Arora
Jaypee Institute of Information Technology

CONTENTS

17.1 Introduction

According to the United Nations (UN) world population prospects report, approximately 7,452 people die in the United States every day (Mokdad et al., 2004) approximating to a death every 12 seconds in the US alone. When we talk about all around the globe, 151,600 people die each day (Danaei et al., 2009), making the numbers even worse with approximately a couple of deaths every second.

These numbers are very real, and as the world population increases, the number of deaths also increases along with it. We need to at least stagnate this ratio, and self-preservation is the first rule we are taught and this helps to implement it.

In this chapter, we are tracing the cause of death as the output using the combination of attributes as input variables, namely age, sex, race, and month of death. These factors can hugely affect the cause of death in their individual and adverse nature (Xu et al., 2010). The dataset we have gathered from this analysis is a large dataset with over 2 million recorded deaths every year (CDC Dataset, 2019) along with reasons of death, age at the time of death, race, education, past injuries, and over 70 such attributes. We are using the data from the years 2005, 2010, and 2015 only. Almost all of these attributes impact the nature of one's death and the symptoms they present. For example, for an older person, heart disease may be a significantly higher possibility than a younger person. On the other hand, when we talk about cancer, it follows no such rule of age and impacts the young adults the same way it impacts the elderly.

We use a Decision Tree Algorithm to trace this behavior of the cause of death and predict it based on the queries that come in. This will not only help us in predicting a single probable cause of death but will allow us to give the list of probable causes of death based on the priority the person needs to take precautions against for that person at that moment.

DOI: 10.1201/9781003046431-17

17.2 Literature Review

A system is developed for finding out the death ratio because of cardiac arrest (Rowlandson, 2007). Patent work is presented in this study by acquiring medical data of patients for prediction of death because of cardiac arrest. The developed system is a combination of various small interconnected modules for conducting the cardiac test for death prediction.

In another significant work by the authors (Deo et al., 2016), the death prediction model was developed for predicting and validating deaths because of an unexpected heart attack. Unexpected death because of cardiac attacks occurs in young people as well as the elderly. The proposed model aims to predict sudden cardiac death in US adults. The proposed study confirmed a generalizable likelihood score for finding out the probability of cardiac death in US adults.

Patients with asymptomatic heart disease account foremost for demise because of heart attack. There is an urgent requirement for the development of a cost-effective automated tool for the prediction of death in asymptotic individuals. An overall electric diverse risk score is computed in the research work for the prediction of unexpected cardiac deaths (Waks et al., 2016). Deviant electrophysiological substrate scaled by overall electric non uniform parameters is separately linked with sudden heart death in the general population. Unexpected cardiac death prediction accuracy upgrades by the addition of global diverse electric parameters. In another significant study (Sinha and Rathi, 2021), analysis on COVID-19 data was done for uncovering the factors that lead to the death because of COVID-19.

In a recent study (Alonso et al., 2019), machine learning-based system was developed for finding out the risk associated with heart disease death grounded on adenosine myocardial perfusion SPECT (MPS) and correlated medical data, and results were compared and validated using logistic regression. Experimental results showed that LASSO exceeds in terms of precision and complexity to logistic regression, whereas Support Vector Machine provides the best area under the curve (AUC) for death due to cardiac arrest.

In another latest study (Bihorac et al., 2019), machine learning-based approach is utilized for computing the probability of death risk or complications after surgery. Grounded on machine learning analytics, we developed an automated model with high preferential potential for estimating the risk of post-surgical problems and demise by utilizing the available EHR data. Model compute risk score for eight divergent post-surgery complications with AUC ranging between 0.82 and 0.94. Also, predictive models forecast death risk with AUC ranging between 0.77 and 0.83.

Another examination (Ayman et al., 2013) was done to find a relationship between the markers of cardiomyocyte injury in theoretical investigations and abrupt heart demise: sudden cardiac death (SCD). The pathophysiology of SCD is unpredictable yet is accepted to be related to an unusual heartbeat much of the time. The connection between biomarkers of cardiomyocyte injury in wiped-out subjects and SCD has not been examined. All the passings, including the SCD, were chosen by a focal undertakings council. In the middle of 13.1 years, 246 members had SCD. Early hsTnT levels are unequivocally connected with SCD, essential bookkeeping hazard factors, and frequency of cardiovascular breakdown and cardiovascular breakdown. Residents were isolated into three gatherings dependent on starting hsTnT levels and SCD chance. A change in hsTnT levels was related to the danger of SCD (totally balanced HR to +1 pg/mL for every year from pattern). The discoveries recommend a connection between cardiomyocyte injury in freak contemplates and a higher danger of SCD than those related to the conventional hazard.

In another study, Buxton et al. (2007) decided the changeability of various factors to anticipate disastrous passing and the danger of complete demise in patients with coronary supply route malady and left ventricular septal inadequacy. They at that point built up a calculation to foresee the danger of abrupt demise and passing. Numerous elements notwithstanding the discharge part (EF) impact the predominance of patients with coronary supply route illness. In any case, there are a couple of apparatuses to utilize these data to control clinical choices. They at that point created chance-taking calculations to evaluate the effect of the changeability of every one of the arranged passing and the danger of mortality. The factors that contributed most to the excellent examination were significant level movement, history of cardiovascular breakdown, ceaseless precarious tachycardia disconnected to sidestep medical procedure, EF, age, restraint from different firmness, and inappropriate enlistment. The model shows that patients with end-of-life EF≤30% have a fixed danger of death of 2%. Numerous species add to freak mortality and the danger of complete passing. Patients with EF≤30% yet no other hazard factor for unsurprising deadly injury. Patients with EF>30% and other hazard variables may have higher mortality and a higher danger of unexpected demise than other patients with EF≤30%. In this manner, the danger of unexpected passing of patients with coronary supply route ailment relies upon numerous extra changes in EF.

Another study (Danaei et al., 2011) discovered the consequences of the rise of infection explicit mortality, throughout the years, from orderly audits and meta-breaks down of symptomatic examinations that have been tended to (i) of significant expected entanglements, and (ii) where conceivable because of the cold counter. They gauge the quantity of explicit sickness passing's brought about by all the uncalled for paces of introduction to each hazard, age, and sex. In 2005, smoking and hypertension represented an expected 467,000 passings, which is the loss of life of, in any event five or six grown-ups in the United States. Weight and physical dormancy each slaughtered around 1 of every 10. Over-the-top salt admission, omega-3 unsaturated fats, and high-fat eating regimens were the dangers of an eating regimen with the best passing outcomes. Albeit 26,000 individuals kick the bucket from ischemic coronary illness, ischemic stroke and diabetes are as of now disturbed by liquor use; they are disheartened by the passing of 90,000 individuals of other heart sicknesses, malignant growth, cirrhosis of the liver, pancreatitis, liquor misuse issues, traffic stream, and different wounds, and brutality. Smoking and hypertension, with both dynamic intercessions, have the most noteworthy number of passings in the United States. Other dietary ways of life and metabolic dangers of interminable maladies likewise cause the most elevated number of passings in the United States.

The variation from the norm of transient cardiovascular estimation is viewed as identified with the hazard factor that puts your wellbeing in danger for patients with cutting-edge, incessant interminable coronary illness.

Speculating the expectation of heart rate briefly modified territories of estimation for a great many people utilizing the properties underneath is obscure. An arbitrary example of 325 investigation subjects, matured 65 or more seasoned, who had broad involvement with gambling clinical preliminaries, lab tests, and 24-hour recording was followed for a long time. Pulse, including standard and moderate proportions of pulse fluctuation, was examined. This investigation is great, yet in the two examinations beneath, we have a similar arrangement of the issue. They have a predetermined number of guinea pigs and do character tests. Every individual's body has various issues and each burrows and reacts to the next individual unexpectedly. In this manner, leading an individual report is viewed as a prerequisite, yet there is no uncertainty that exploration depends exclusively on prior conditions (Mäkikallio et al., 2001). Our examination varies from the above in that it takes a gander at all the reasons for death referenced by the model, i.e., 39 rejections by the Centers for Disease Control and Prevention (CDC) and we bolster that for individuals with low status, way of life, and way of life your sexual qualities, age, and race ought to be included. Another issue with the subsequent study is that it just takes grown-ups more than 65 years old. Even though it might be a generally excellent exercise for grown-ups, we are attempting to make it dependent on age. Grown-ups are only one finish of the range.

Another significant study (Mitushi et al., 2021) provides insight on the usage of advanced machine learning techniques for the detection of Suspicious Activity in Surveillance Applications. A latest study (Paras et al., 2019) utilized Sustainable Technique for the prediction of forest fire. Authors in the research work (Sinha et al., 2020) employ Sustainable Time Series Modeling Technique for Traffic trend prediction in metro cities.

17.3 Problem Formulation

We are utilizing a dataset that CDC released from the year 2005 through 2015 every year for the United States under the National Vital Statistics Systems. Putting the domain of prediction aside, analyzing mortality data and the reasons for it is essential to understanding the complex circumstances of death in the country. The dataset can also be used to measure life expectancy, which the US government is already doing.

1. Assumptions:
    - The dataset is enough to predict the nature of the death of a person given the necessary attributes.
    - The data are valid for everyone if they live even outside the United States if they take input as to their country's data.
    - Part of the dataset taken for training/testing is large enough and not biased.
    - Errors of over-fitting and under-fitting have not existed.
2. Formulated Problem: Columns of the dataset Used for Decision Tree Creation: education_1989_revision, education_2003_revision, education_reporting_flag, month_of_death, sex, detail_age, place_of_death_and_decedents_status, day_of_week_of_death, injury_at_work, manner_of_death, activity_code, X39_cause_recode, race.

The basic algorithm used was the "Decision Tree Algorithm." We didn't just use the Decision Tree Algorithm; we also did trend analysis.

We have various machine learning algorithms at our disposal, but we used the Decision Tree classification algorithm, as for some reason or the other, this algorithm made more sense. When it comes to the Decision Tree, it does consider the likelihood of attributes but doesn't discard any and goes till the leaf node matching the result. If it reaches the leaf node, it predicts the value even if a certain attribute's probability is zero. Data used in the model have all types of data such as nominal data, binary data, numerical data, and ordinal data.

In a Decision Tree, all types of data can be treated easily and give more accurate output fast because it builds the fastest Decision Tree. In Decision Tree, data may be over-fitted or over-classified but only if a small sample is tested. It may be expensive to produce a tree. Seeing the versatility in our dataset, decision tree is the most suitable algorithm. Table 17.1 represents the advantages and disadvantages of various other existing algorithms.

TABLE 17.1

Comparative Analysis of Decision Tree over Other Algorithms

Model	Description	Advantages	Disadvantages
Linear regression (LR)	Linear regression, a statistical tool, is a linear method to model the relationship between a response/output and multiple predictor variables/input variables. The case of one explanatory variable is called simple linear regression.	The good thing about this is that the values can be ordinal, interval, and ratio-level too.	A linear regression model, even while the assumptions are met, shows the relationship between a few variables over the range of values tested within the dataset. Going outside the range of the training values is not advisable in a linear regression equation. This makes the dataset limited to a certain extent as if it is not trained again for new input, it cannot handle the out-of-range values.
Logistic regression (Log. R.)	Logistic regression is a modeling algorithm used to classify data and to explain the relationship between one predicted/output binary variable and one or more predictor variables.	When it comes to logistic regression, it is a very useful tool unless we have already identified all-important input/independent variables.	Using the Decision Tree Algorithm, the model can be applied to the whole dataset and it will find out the important columns itself by pruning. We do not need to identify the association rules before applying the model.
Naive Bayes (NB)	Naive Bayes is a machine learning classifier based on the concept of conditional probability by Bayes' theorem. It considers each attribute of the dataset to be an independent one and all attributes contribute to the output at the same time. Since each attribute is independent, the probability is the product of the probabilities of each attribute.	When the inputs are independent of each other, a Naive Bayes classifier performs better as compared to other models. Also, we need less training data. It performs well in the case of categorical input.	If the class has a probability as 0, the whole dependence on other attributes plays no effect anymore as the conditional probability is 0. This is bad because, in the case of a strong relationship to an attribute, the whole entry is ignored due to one entry and is not studied while that specific entry may not play a role in prediction. Naive Bayes cannot handle that.

17.4 Observations from the Dataset

The dataset without machine learning can give us a lot of information and, when plotted with very specific information, can show trends. One such example is shown in Figure 17.1, which shows the number of deaths in 2005, 2010, and 2015 across the 12 months of a year.

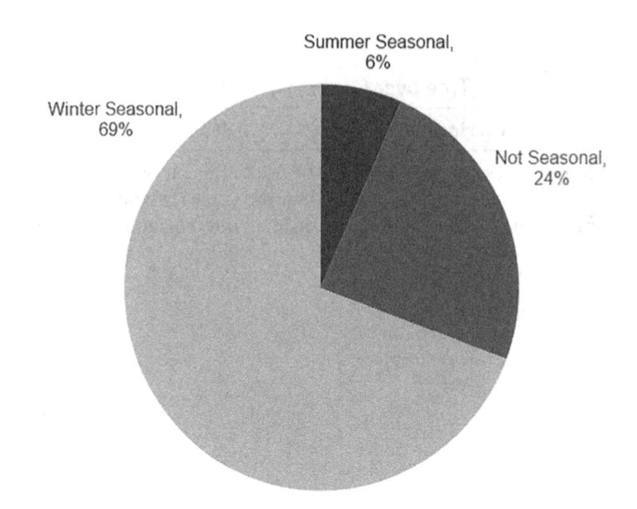

FIGURE 17.1
Seasonal death percentage (in the USA).

The below graph clearly shows the trend of deaths and it is pretty much consistent across the years considered (2005, 2010, 2015). From January to February, there is a rapid decrease in the number of deaths and the reason can be deduced as the presence of very heavy winters causing a lot of deaths as mentioned by legacy.com.

Another trend that shows up is the increase in deaths from February to March and constant decrease till June, starts rising for a month, stagnates, and falls reaching its bottom in September. Thereon, the mortality rate constantly starts rising again till January of next year, when it is at its peak. This fact is accentuated by the fact that the winters are generally more deadly than any other season. Figure 17.2 (top and bottom) shows this fact. Some causes of death do not show any relation to the season like cancer, and are defined as not seasonal, while some other diseases like flu and pneumonia, strike in winter months, the Winter Seasonal group.

Figure 17.3a depicts the most significant causes of death. Heart disease is the most common cause by far followed by lung cancer and other respiratory ailments. Heart failure might be a symptom of America's major obesity problem but respiratory highlight another major killer, which are pollution and smoking.

Figure 17.3b considers the same causes of death, but we now view them across the male and female genders. An interesting trend is that the cause of death remains mostly the same for both genders, but women are more likely to die of heart diseases and Alzheimer's. This can be justified by the fact that women seem to live longer and therefore are more likely to die from diseases that commonly affect older people.

Another trend that can be seen from the below graph is that as time passes by, the mortality rate keeps on increasing. As mentioned before, this is the reason for the increasing world population. As more people exist, the chances of one of them dying virtually remaining the same, the number of deaths is rising at an increasing rate. When we talk about 39 Cause recode, it has about 42 causes of death dominated by mainly disease-based deaths, leaving the remaining niche as all other reasons of deaths. As we can see from the decision tree shown later in Figure 17.5, if the X39 cause recodes is below 37 but at the same time below 40, the cause of death is undoubtedly accident. If the condition is not satisfied, the person most certainly died of natural causes. Now, if the person has the 39 cause recode below 39 and he or she had an injury at work, he or she most probably died from an external cause. Else if

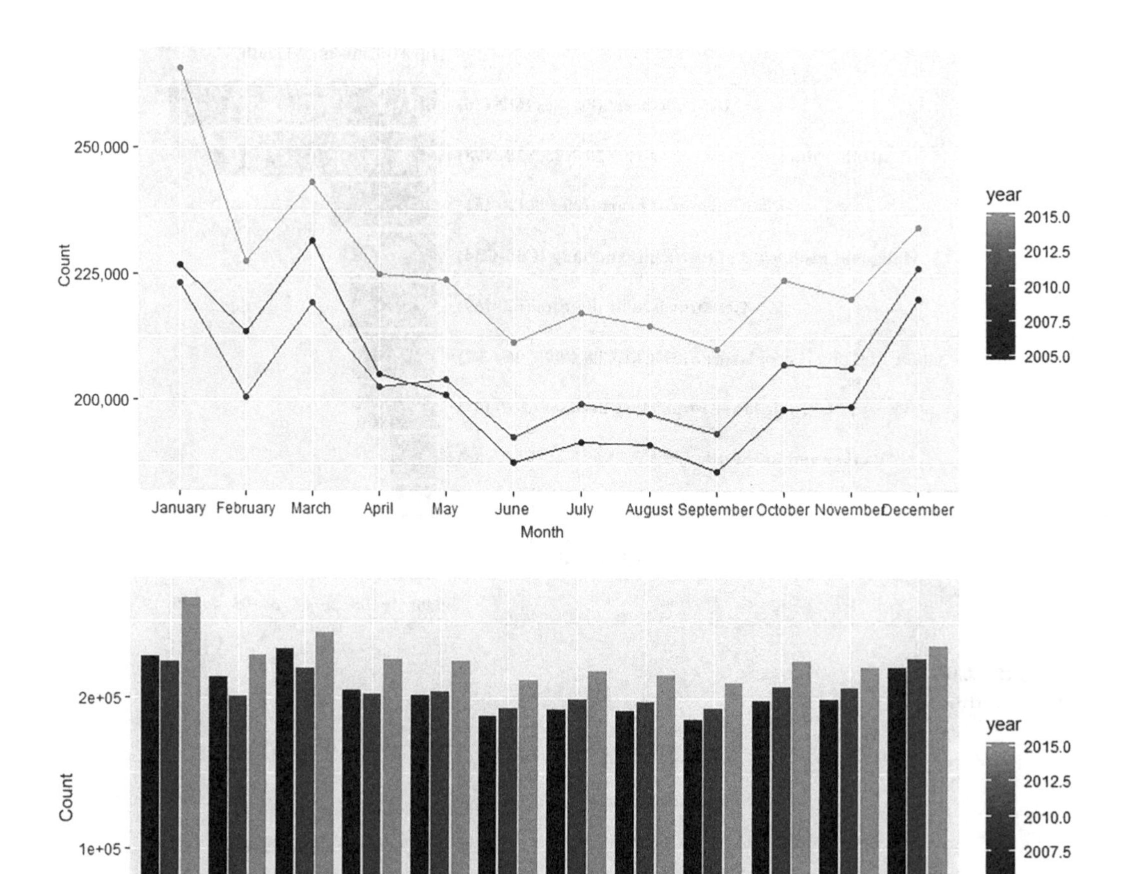

FIGURE 17.2
(Top) Death counts rising over the years. (Bottom) Number of deaths per month (in the USA) for 2005, 2010, and 2015.

the person has a 39 cause recode equal to 40, it is death due to suicide/self-harm, but when recode is 41, the reason of death is assault while the reason of death when recode is 42 is the unknown reason for death. When it comes to being older than 56, the person may have died of natural causes. On the other hand, if not, the cause of death is self-harm (Figure 17.4).

17.5 Experimental Results

The Decision Tree Algorithm is used as a prediction model and achieves an accuracy of 98.9%. Prediction accuracy is high because of the number of entries in the training set. Figure 17.5 presents the decision tree for the predictive model used in the study.

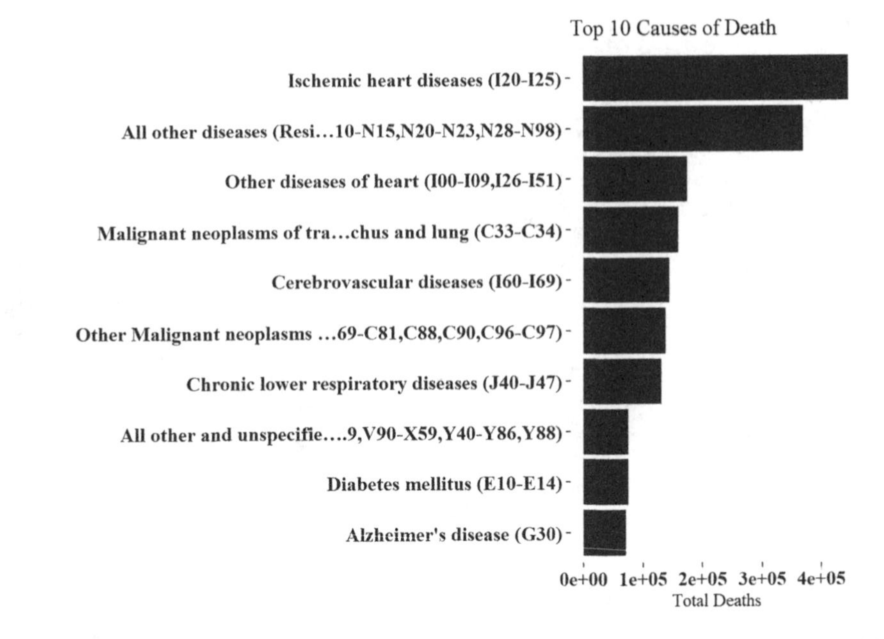

FIGURE 17.3a
Total deaths.

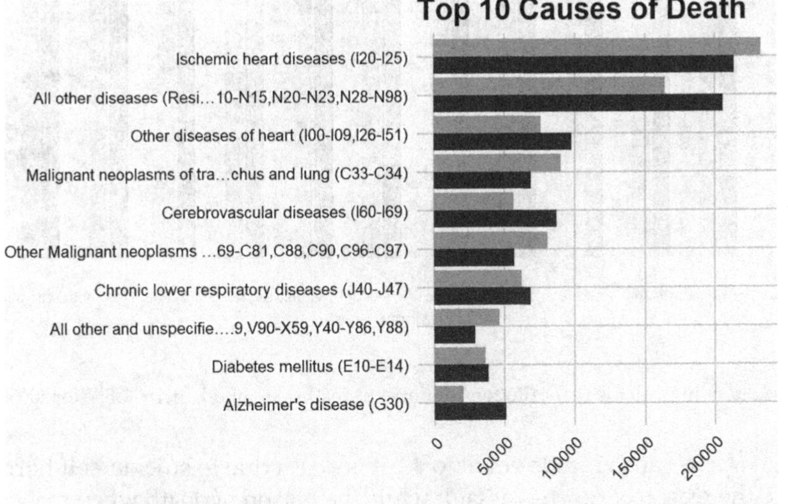

FIGURE 17.3b
Number of deaths per month (in the USA) for 2005, 2010, and 2015.

As seen in the above scenarios, the rate of deaths of people in the United States every year had the same pattern for the years we saw. Accidents account for 41% of deaths among people in the 15–24 age group. When we talk about the dataset, it was for the years 2005, 2010, and 2015 and monthly mortality rate patterns were maximum in winters and minimum in summers. Just as the world population is increasing, the total number of deaths is increasing

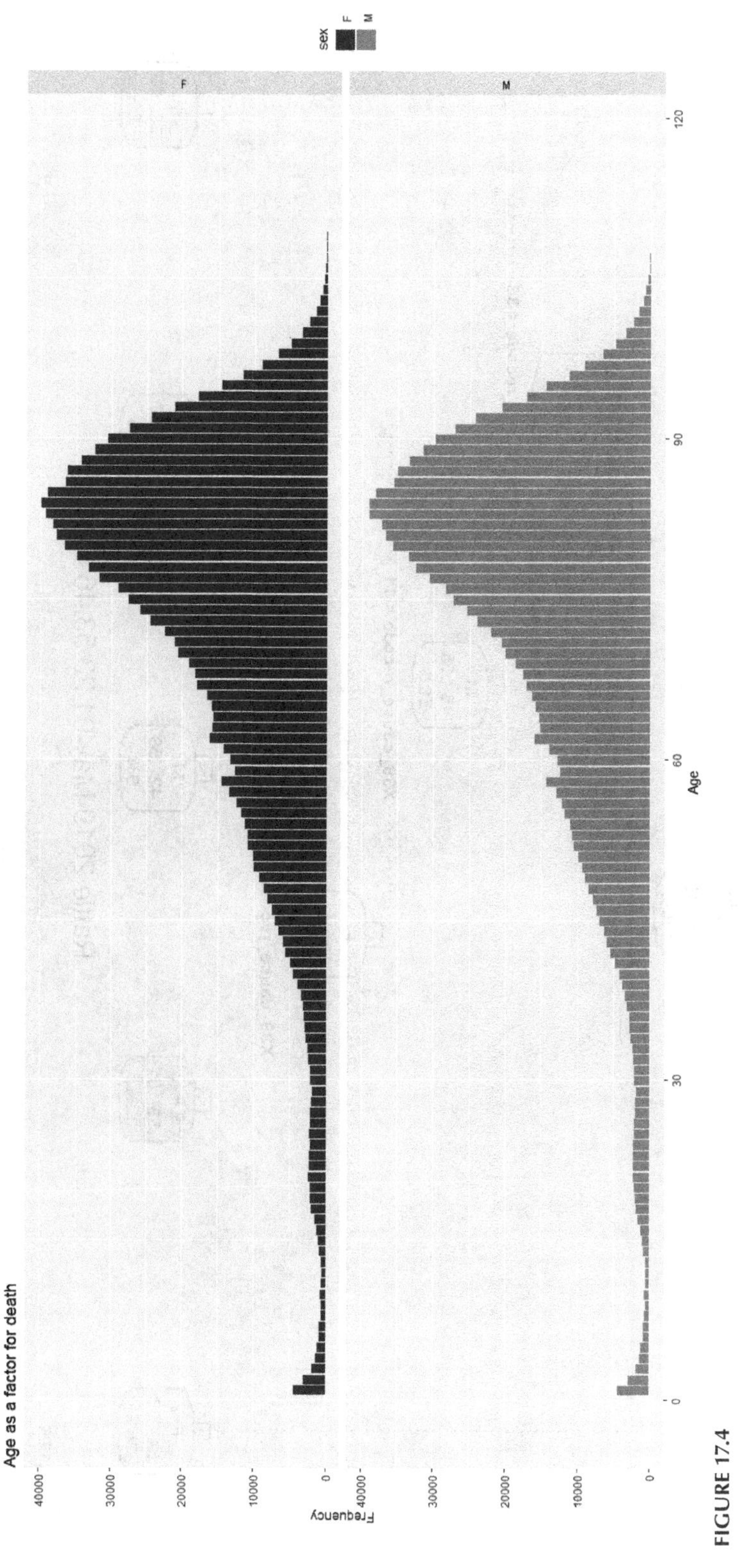

FIGURE 17.4
Death by age for males and females.

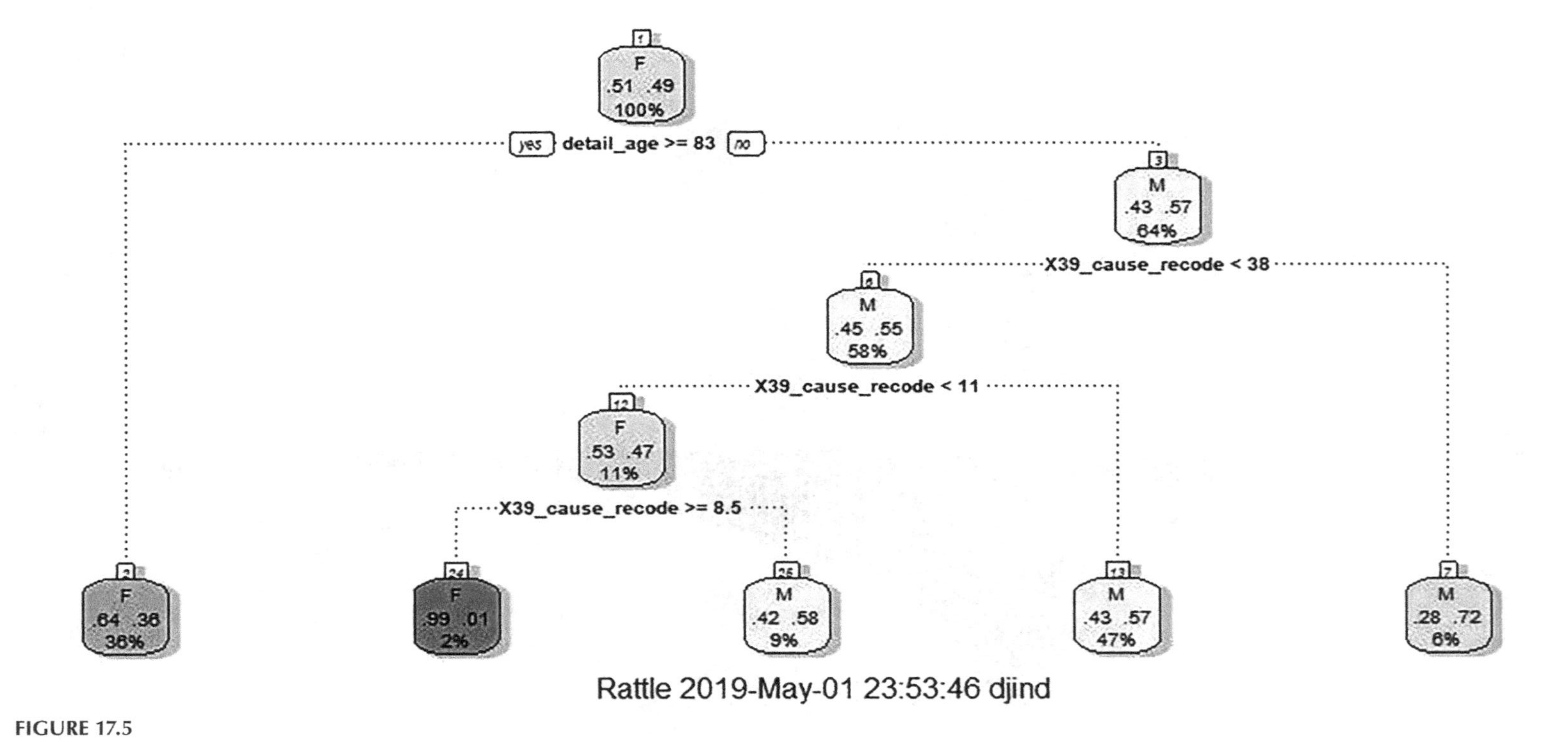

FIGURE 17.5

Male or female? Classifying gender by death.

TABLE 17.2

Property Effect on Race
100.00% → injury_at_work
10.86% → X39_cause_recode
10.18% → education_1989_revision
9.94% →place_of _death_and_decedents_status
9.57% →day_of _week_of _death
9.55% → month_of _death
0.78% → manner_of _death
0.53% → detail_age
0.52% → sex
0.42% → education_2003_revision
0.22% → education_reporting_flag

causing the whole death count to go up in 2015 as compared to 2005 in a similar pattern. We also came to know that when a person dies below the age of 56 years and he or she didn't die of natural causes, he or she most likely dies due to self-harm. Many other rules were generated above. For instance, when we see the results of Rule-3, the race is a huge factor when it comes to workplace injuries. Table 17.2 presents the effect of property on race.

Manner of death of a person can tell us many things about the person. The kind of job he or she used to do, how healthy he or she was physically a while before their death, and many more. If we use machine learning on the manner of death of a person, we might be able to tell the characteristics of a person and vice versa. Tabular results derived from the model are shown in Table 17.3.

17.6 Conclusion

The body of a human is very unpredictable and difficult to understand. Therefore, various factors like sex, age, race, and others can be coined to be very helpful. How a person dies can tell us many things about a person. The type of work he or she used to do, how healthy he or she was before the time of their death, and many more. If we use machine learning in the form of human death, we may be able to tell the personality traits and argue against them.

As per the methodologies applied, a very effective accuracy of 98.9% was achieved. Various rules and traits identified are places summarized in Tables 17.2 and 17.3.

We also know that if a person dies under the age of 56 and does not die of natural causes, he or she may die of self-injury. Many other rules were made above. This research can be very effective in saving lives in no time.

TABLE 17.3

Results Derived from the Model

Injury at Work	Cause	Education	Place of Death	Day of Death	Month of Death	Output: Race
-	Not external causes	3+ years of high school	Home	Weekday	Not in late winters	White
No	-	-	-	-	-	White
Yes	-	-	-	-	-	White
-	TB or syphilis	High school graduate	Home	-	-	Black
No	-	Uneducated	Hospital	-	-	Black
No	Assault	Elementary school graduate	-	-	January	Black
Unknown	Cancer	Uneducated	Hospital	-	-	Black
-	HIV	Elementary school graduate	-	-	-	Black
-	TB or syphilis	Uneducated	-	-	-	Black
-	Syphilis or HIV	<3 years of high school	Home	-	-	Black
No	Assault	-	-	Sunday	Second half	Black
No	Assault	Elementary school graduate	-	-	-	Black
Unknown	Assault	Elementary school graduate	-	-	Not January	Black
-	Assault	1–2 years of elementary school	Home or hospital	Not Sunday	January	Black
-	Not TB or syphilis	Uneducated	Hospital	Weekday	Mid-year	Black
-	HIV	Elementary school	Hospital	-	-	Black
-	HIV	Elementary school	Home	-	-	Black
No	Assault	High school	-	-	-	Black
No	Assault	-	Hospital	-	-	Black
Unknown	TB, syphilis, HIV or Cancer	Uneducated	Hospital	-	-	Black
Unknown	Not assault	Uneducated	Hospital	Second half	-	Black
-	Disease	Uneducated	Hospital	Sunday or Monday	-	Black
-	TB, syphilis or HIV	-	-	-	-	Black
-	Disease	Uneducated	Hospital	-	-	Black
-	Assault	-	-	-	-	Black
-	-	-	-	-	-	White

17.7 Future Scope

In a huge group of one careful medical procedure understanding, we created and approved a machine-meaningful calculation that utilizes existing clinical information in electronic wellbeing records progressively to anticipate the danger of significant inconveniences and demise after a medical procedure with high affectability and particularity. Given the connection between an enormous number of postoperative intricacies and symptoms are expenses, there is a basic need to separate the danger of post-employable hazard from the board. Further exploration is expected to affirm this methodology without indicating the chance of utilizing this calculation in a genuine clinical setting to evaluate whether the utilization of the calculation could prompt more mind and results contrasted with the current practice.

References

Alonso, D. H., Wernick, M. N., Yang, Y., Germano, G., Berman, D. S., Slomka, P. (2019). Prediction of cardiac death after adenosine myocardial perfusion SPECT based on machine learning. *Journal of Nuclear Cardiology*, 26(5), 1746–1754.

Ayman, A. H., Gottdiener, J. S., Bartz, Sotoodehnia, N., Defilippi, C., Dickfeld, T., et al. (2013). Cardiomyocyte injury assessed by a highly sensitive troponin assay and sudden cardiac death in the community: the cardiovascular health study. *Journal of the American College of Cardiology*, 62(22), 2112–2120.

Bihorac, A., Ozrazgat-Baslanti, T., Ebadi, A., Motaei, A., Madkour, M., Pardalos, P. M., et al. (2019). MySurgeryRisk: development and validation of a machine-learning risk algorithm for major complications and death after surgery. *Annals of Surgery*, 269(4), 652.

Buxton, A. E., Lee, K. L., Hafley, G. E., Pires, L. A., Fisher, J. D., Gold, M. R., et al. (2007). Limitations of ejection fraction for prediction of sudden death risk in patients with coronary artery disease: lessons from the MUST study. *Journal of the American College of Cardiology*, 50(12), 1150–1157.

Centers for Disease Control and Prevention (CDC), Death in the United States [Dataset], 2019, https://www.kaggle.com/cdc/mortality.

Danaei, G., Ding, E. L., Mozaffarian, D., Taylor, B., Rehm, J., Murray, C. J., et al. (2009). The preventable causes of death in the United States: comparative risk assessment of dietary, lifestyle, and metabolic risk factors. *PLoS Med*, 6(4), e1000058.

Danaei, G., Ding, E. L., Mozaffarian, D., Taylor, B., Rehm, J., Murray, C. J. L., et al. (2011). Correction: the preventable causes of death in the united states: comparative risk assessment of dietary, lifestyle, and metabolic risk factors. *PLos Medicine*, 8(1).

Deo, R., Norby, F. L., Katz, R., Sotoodehnia, N., Adabag, S., DeFilippi, C. R., et al. (2016). Development and validation of a sudden cardiac death prediction model for the general population. *Circulation*, 134(11), 806–816.

Mäkikallio, T. H., Huikuri, H. V., Mäkikallio, A., Sourander, L. B., Mitrani, R. D., Castellanos, A., et al. (2001). Prediction of sudden cardiac death by fractal analysis of heart rate variability in elderly subjects. *Journal of the American College of Cardiology*, 37(5), 1395–1402.

Mitushi, A., Parashar, P., Mathur, A., Utkarsh, K., Sinha, A., Suspicious activity detection in surveillance applications using slow-fast convolutional neural network, *International Conference on Advances in Data Computing, Communication and Security, Lecture Notes on Data Engineering and Communication Technologies*, Springer, pp. 1–10, 8–10, September 2021.

Mokdad, A. H., Marks, J. S., Stroup, D. F., Gerberding, J. L. (2004). Actual causes of death in the United States, 2000. *JAMA*, 291(10), 1238–1245.

Paras, C., Jain, S., Sinha, A., Sustainable approach for forest fire prediction, *Proceedings of International Conference on Futuristic Trends in Networks and Computing Technologies*, Springer, Singapore, pp. 456–469, 1–6, 2019.

Rowlandson, G. I. (2007). U.S. Patent No. 7,272,435. U.S. Patent and Trademark Office, Washington, DC.

Sinha, A., Puri, R., Balyan, U., Gupta, R., Verma, A. K., Sustainable time series model for vehicular traffic trends prediction in metropolitan network, *IEEE 6th International Conference on Signal Processing and Communication (ICSC 2020)*, pp. 74–79, 2020.

Sinha, A., Rathi, M., COVID-19 prediction using AI analytics for South Korea *Applied Intelligence*, pp. 1–19, 2021, doi: 10.1007/s10489-021-02352-z.

Waks, J. W., Sitlani, C. M., Soliman, E. Z., Kabir, M., Ghafoori, E., Biggs, M. L., et al. (2016). Global electric heterogeneity risk score for prediction of sudden cardiac death in the general population: the Atherosclerosis Risk in Communities (ARIC) and Cardiovascular Health (CHS) Studies. *Circulation*, 133(23), 2222–2234.

Xu, J., Kochanek, K. D., Murphy, S. L., Tejada-Vera, B. (2010). Deaths: final data for 2007. National vital statistics reports: from the Centers for Disease Control and Prevention, *National Center for Health Statistics, National Vital Statistics System*, 58(19), 1–19.

18

Modeling the Immune Response of B-Cell Receptor Using Petri Net for Tuberculosis

Gajendra Pratap Singh and Madhuri Jha
Jawaharlal Nehru University

CONTENTS

18.1 Introduction

A healthy, well-developed immune system is always required, as in our surroundings we cannot avoid the presence of billions of infectious microorganisms such as bacteria or viruses. Our immune system protects us from these infections whenever required. The word *immunology* is derived from *immunis* (*a* Latin term), meaning "exempt," which means they keep secure from infection. So, a healthy immune system is important to be protected by multiple antigens. A major challenge in the field of drug inventions is the highly complex programming and functioning of the human immune system. The immune system involves certain different cells, tissues, and organs whose contributions to fight with any antigen are different (Abbas et al., 2014; Janeway et al., 2001). Human immune cells have T cells and B cells, and both have different actions on antigens. This chapter deals with the study of the process of secretion of antibodies by the immune system that programs B cells to initiate the antibody response by binding with an outer antigen. Further, different downstream pathways of BCR-signaling act accordingly in different diseases (Loder et al., 1999; Zhou et al., 2007). So, discrete mathematical modeling that shows the stepwise progression of the system is important to get clear information (Gupta et al., 2018; Kumar et al., 2018; Singh et al., 2018). In this chapter, a widely used discrete model is being proposed and that is Petri net (PN).

DOI: 10.1201/9781003046431-18

Studying the system step by step can be proved as a better way to analyze human immunology. These days several computational and mathematical models are emerging rapidly to study biological networks (Raja et al., 2011; Singh et al., 2020a, b, c, d), and also in another field of sciences. One more discrete model PN graph is also being proved a useful tool to model such networks (Singh and Gupta, 2019; Cherdal and Mouline, 2018; Gupta et al., 2019a, b; Kansal et al., 2010, 2012; Singh and Verma, 2017). This tool provides a graphical and mathematical view of any system with several structural and behavioral properties. Its discrete behavior is useful as it can show the interrelationship between the states and reachability from one state to another involved in the system. *PN* has a huge application in almost every field of science, especially with networks and communication, as it can be related to a binary number system (Singh et al., 2013a, b; Singh and Kansal, 2016; Singh et al., 2020a, b, c, d). Several high-level PN variations help in modeling the concurrent and synchronized systems. Hence, PNs are a novel model in several fields of sciences. PN can also behave as a recommender system in several metabolic pathways or other technical use (Singh and Singh, 2019; Gupta et al., 2019a, b). In this chapter, the authors decide to model the BCR-signaling pathways and their nature and behavior in different deficiencies, especially the behavior that is independent of T-cell behavior. The authors also planned to model the mechanism of BCR pathways in tuberculosis (TB) and which signaling pathways are activated to fight Mtb bacterial infection. The disease TB is still contributing to the mortality rate in the world, which is because the Mtb bacteria have a drug-resistant behavior. The immune system plays a vital role in the cure of TB, which can help in the decrement of drug-resistant TB patients. Several mathematical models were implemented for the basic reaction during TB infection to the human body at an early stage. Earlier in the previous papers, the authors also modeled the molecular mechanism of the drugs involved in the First Line Drug treatment process in TB and concluded the mechanism behind the multidrug-resistant TB (MDR-TB) and extensive drug-resistant TB (XDR-TB) (Singh et al., 2020a, b, c, d). PN is also well capable of modeling the diagnostic path of any disease to analyze the shortest and longest paths of treatment. Previously the clinical diagnostic path followed for the disease TB was modeled with the PN graph and covered the sensitive aspects regarding the diagnosis of MDR-TB (Singh et al., 2020a, b, c, d).

18.2 Biological Background

The immune system consists of the B cell is an important part that has B lymphocytes, which are inside the white blood cells originating in the bone marrow of the human body. To bind the antigens, it has a specific type of receptors on its cell surface. Activation of B cell can be a T-cell-independent process or a T-cell-dependent process. Ninety-five percent of antigens are T-dependent and act differently from the T-independent. The T-dependent mechanism is followed by the generation of memory cells containing the antibodies, which will be called later at the time of the same infection. Figure 18.1 shows the B-cell receptor (BCR) consisting of regions like antigen-binding sites and various chains (Nemazee and Weigert, 2000; Ehlich et al., 1993).

In this chapter, B-cell activation independent of T cell is being modeled using a discrete model. The process in which plasma cells are generated inside the body is followed by the secretion of antibodies in the immune cell as well as in memory cells. Some antibodies further proceed to protect from infections by binding the microbes or the microbial toxins, and the antibodies in memory cells are to be called in the future at the time of the

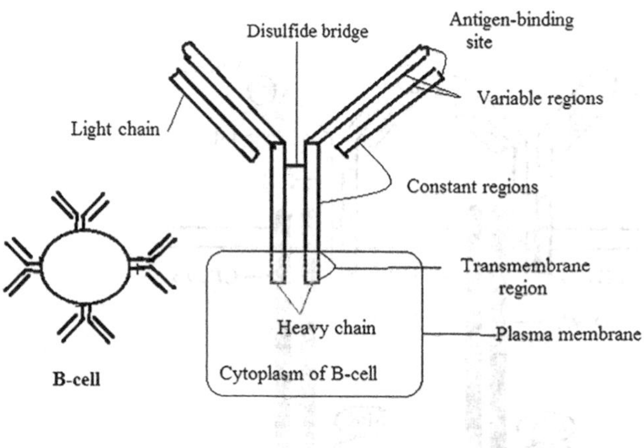

FIGURE 18.1
B-cell receptor (BCR).

same infection (Kenkre and Kahl, 2012). The BCR-signaling pathways are followed by four more downstream pathways and that are the *MAPK* walks/pathways, the pathways PI3K, NFAT, and NF-kβ, and BCR internalization (Szydłowski et al., 2014). In general, antigen-presenting cells (APCs) often activate B cells. The APCs capture antigens and show them on their cell surface. MAPK pathways are the sequence of proteins that transfer the signal received from the receptors on the cell surface to the DNA present in the nucleus of the cell. The PI3K pathway is important in the regulation of the cell cycle. NFAT pathways are the family of transcription factors that plays an important role in immune responses. The NF-kβ pathway is a family of transcription proteins that regulate the genetic information from DNA to RNA. This also plays a key role in immune responses. Activation and initialization of B cells by such membrane-related antigens require BCR-induced cytoskeleton reorganization. These downstream pathways followed in BCR signaling with the responsible enzymes in the functioning of BCR in the immune system are shown in Figure 18.2.

18.3 Methodology

In the present scenario modeling, any system graphically is being proven very useful to analyze the system systematically. One of the most used graphical tools is the PN, which is a specific type of bipartite directed graph that studies and analyzes the properties of the modeled system.

Mathematically, PN is a set of five components, $PN = \{ P, T, I, O, m_0 \}$

where P and T are the nonempty and finite sets of places and transitions, denoted by a circle, rectangle, or squares, respectively, I is the outflow mapping that depicts the number of the arc from any transition to output place, O is the inflow mapping that denotes the number of the arc from any place to output transition, and m_0 is represented preliminary marking represents the token present in a particular place. Graphically, tokens can be denoted by black dots putting in a place.

A transition "*t*" fires or triggers at any marking iff (if and only if) the number of the outgoing arc from all the places is less than or equal to the token present in that particular place.

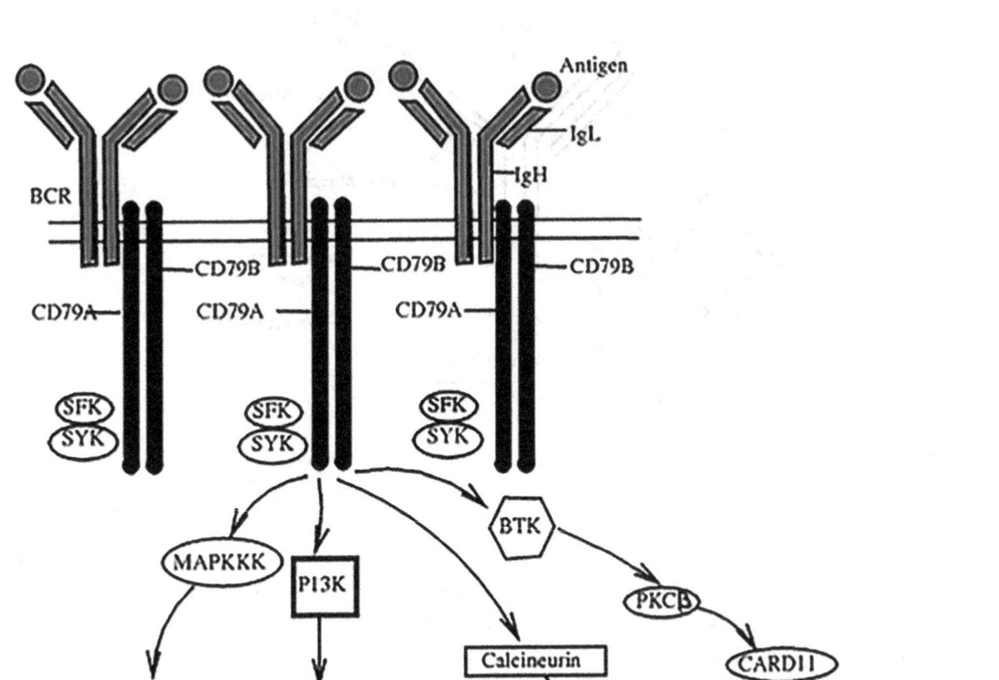

FIGURE 18.2
B-cell receptor signaling pathways independent of T cell.

Thus, any transition t will fire if,

$$I\,(p,t) \le m_0\,(p), \quad \forall\, p \cup P. \tag{18.1}$$

After firing of t the token moves as

$$m_1\,(p) = m_0\,(p) - I\,(p,t) + O\,(p,t) \tag{18.2}$$

This determines the next marking m_1 from m_0, and read as m_1 is reachable from m_0.

Once a transition fires the marking then tokens flowed from the input places and are deposited on the output places according to the next marking rule. When all these markings are grouped to form a graph, then it is called a reachability graph $(R\ (PN))$ of that PN.

An example of a simple PN is shown in Figure 18.3, in which the discrete value of tokens is allowed in places. The set of the firing of enabled places gives the reachability graph shown in Figure 18.3.

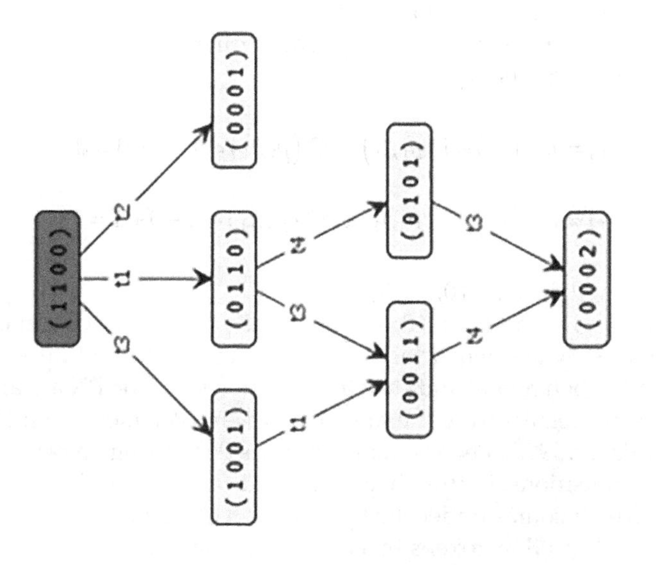

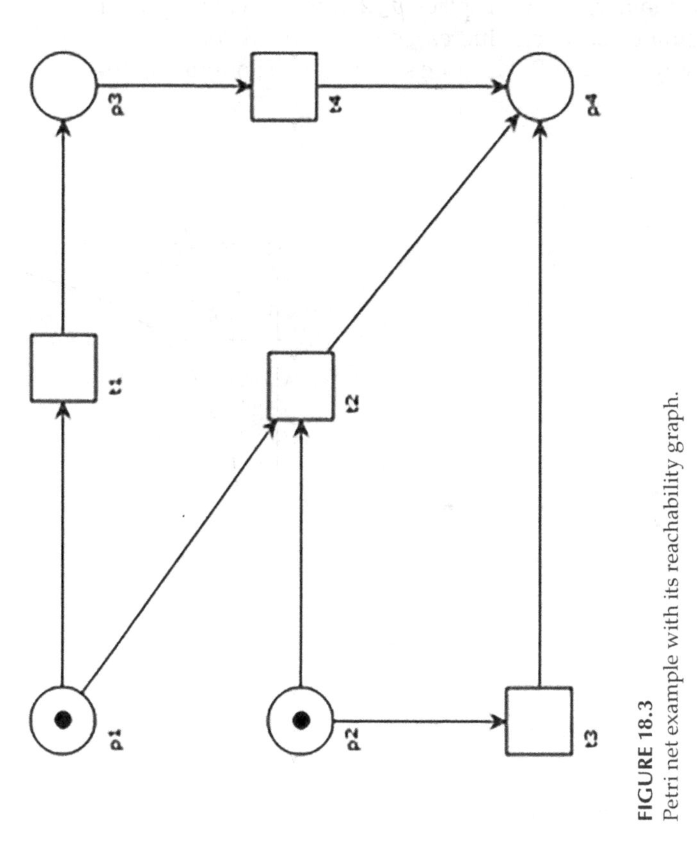

FIGURE 18.3
Petri net example with its reachability graph.

Initially, one token is taken on places p_1 and p_2, thus the initial marking is $m_0 = (1,1,0,0)$. At this stage, three transitions are enabled t_1, t_2, t_3, and can fire to change the marking. After firing these transitions next marking reaches according to the rule of shifting tokens. When t_1 fires, token shifts from p_1 to p_3.

$$m_1\left(p_1\right) = m_0\left(p_1\right) - I\left(p_1,t_1\right) + O\left(p_1,t_1\right) = 1 - 1 + 0 = 0 \tag{18.3}$$

$$m_1\left(p_3\right) = m_0\left(p_3\right) - I\left(p_3,t_1\right) + O\left(p_3,t_1\right) = 0 - 0 + 1 = 1 \tag{18.4}$$

Thus, the next marking will be $m_1 = (0,1,1,0)$.

Similarly, the firing occurs and the next marking reaches to get the reachability graph. The reachability graph is always beneficial to draw the connectivity of different stages in the process. Several other behavioral and structural properties of the PN are also very beneficial for drawing useful results from the modeled system (Peterson, 1981; Murata, 1989; Barkaoui and Pradat-Peyre, 1996). PN has many varieties depending on certain conditions applied to places and transitions. In this chapter, continuous place and continuous transition are being used, which comes under the hybrid PN. The tokens on continuous places can take a real value and can flow tokens in a continuous rate through continuous transitions (Alla and David, 1998). Figure 18.4 shows an example of the hybrid PN and how the tokens flow from one place to another place.

In Figure 18.3, the simulation result shows how the tokens flow from the places p_0 and p_1 and transfer to place p_2. With time the number of tokens on p_0 and p_1 decay by a fixed continuous rate and in place p_2 whose incoming token capacity is 2, so accordingly, the number of tokens increases with a fixed continuous rate. Modeling biological networks using a hybrid PN is more suitable as the number of tokens can take any positive real values.

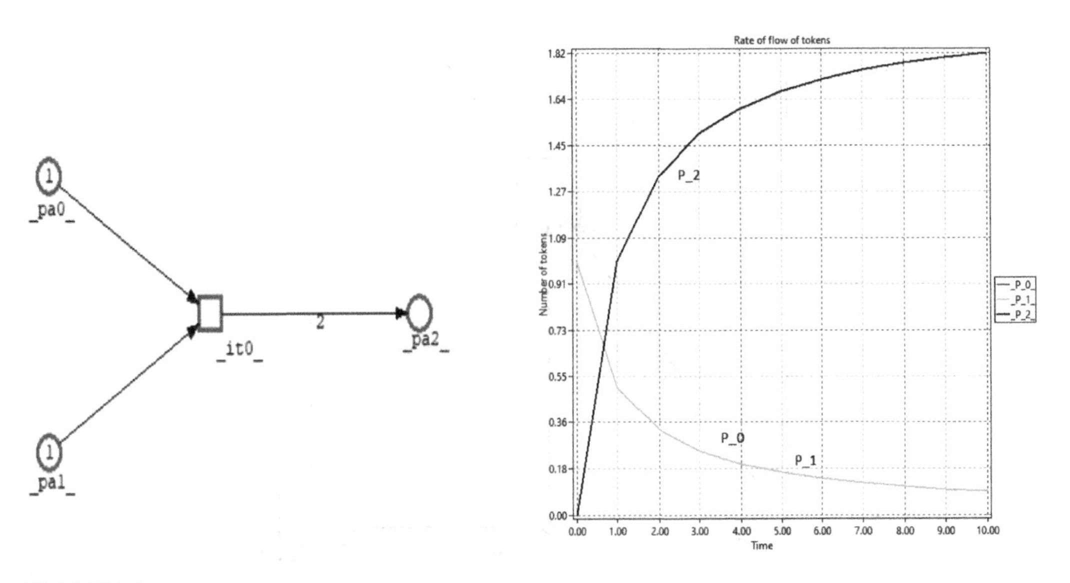

FIGURE 18.4
Example of a Petri net and its simulation.

18.4 Structural and Behavioral Properties of PN with Its Application in Modeling Biological Processes

Liveness: Any PN is called a live system if for m_0 if there always exists a sequence of firing to reach any marking from some other marking through an appropriate sequence of firing. Liveness refers to the process that is continuously happening and does not attain a deadlock stage.

Boundedness: If the count of tokens at each place of the PN Graph is finite at any stage, then the graph PN is said as Bounded PN. In a special case, if the token number never exceeds one then the PN is also called **1**-safe PN (Singh, 2013). Boundedness of any biological process infers that the formation or consumption of any adduct involved in the process is an infinite amount and can be controlled by the user.

Reversible: A PN is reversible if the initial marking can be obtained back from all the other reachable markings by some firing sequences. Reversibility of any net in biological process infers that the initial substrate that starts the reaction is also one of the intermediate or final products.

Reachability: In a PN graph, after getting the suitable firing states we can draw a reachable graph that is used to find whether a certain marking (state) is reachable from another marking. The graph tracing all these reachable states is called the PN reachability graph $R\ (PN)$. This helps to analyze whether certain outcomes are possible or not. In any biological process, it is used to find the ways of the formation of any particular substrate.

Siphon and Trap: Any subset of places $A \in P$ is called *Siphon* if

$$I(A,\ T) \subseteq O(A,\ PN) \tag{18.5}$$

i.e., any input transition of A is also an output transition of A.

On the other hand, a *trap* of a *PN* is a subset B of the places if

$$O(B,\ T) \subseteq I(B,\ T) \tag{18.6}$$

i.e., for each transition that takes something out of B also puts something back in. That means any output transition of B is also an input transition. Whenever a deadlock marking is reached, where no transition can fire, the set of empty places form a Siphon set.

Invariant: Two types of invariant are defined in PN: *p*-invariant and *t*-invariant.

p-invariant means the collection of places where the tokens remain unchanged in any set of firing. While *t*-invariant is the set of transitions that leave the marking of *PN* unchanged after firing. If $A_{m\times n}$ is the incidence matrix of *PN* defined by $A = \left(a_{i,j}\right)$, where $i = 1,\ 2,\ldots,p$ (transitions) and $j = 1,\ 2,\ldots..,q$ (Places), then the invariant may be obtained by $A^T M = 0$, for p-invariant, and $AM = 0$, for *t*-invariant. Here, M is a $\left(q \times 1\right)$ column vector that denotes the weighted sum of tokens in a PN.

18.5 PN Modeling of BCR-Signaling Pathways

Modeling any system by PN includes the selection of basic elements such as events and relationships within the pathways. Modeling by PN of the BCR-signaling pathways the

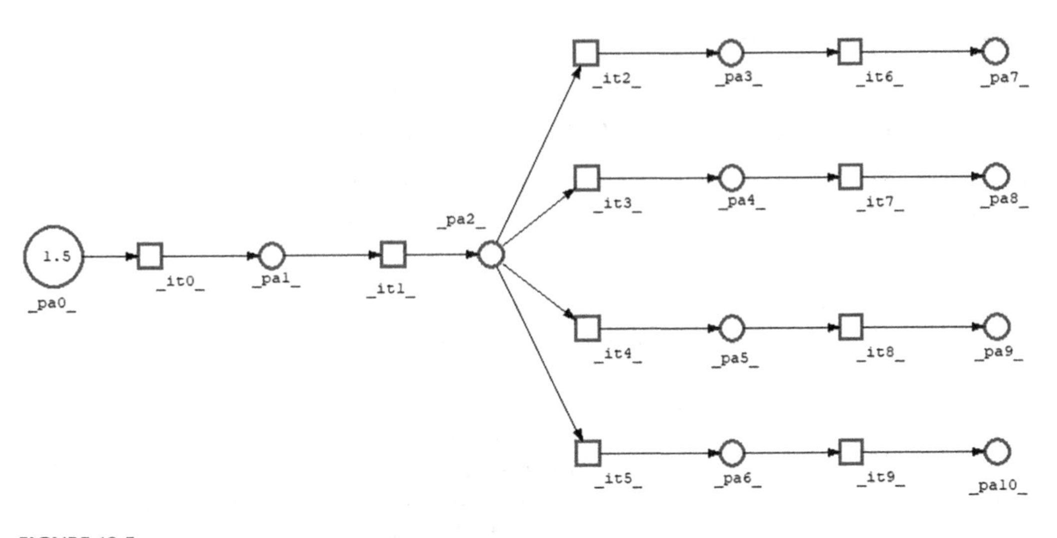

FIGURE 18.5
B-cell signaling pathway modeled using Petri net.

places covers the molecular activities or stable compounds, whereas the transitions cover the condition of the responses (proliferation, binding, etc.) or metabolic enzymes, and the arcs cover interrelationships between the functioning of the corresponding enzymes. The tokens show the presence of the respective compound and the tokens flow represents the involvement of a compound to another compound through some responsible enzymes. In this chapter, the hybrid PN is being used to model the BCR-signaling path as the tokens and its flow is not discrete rather is continuous. The PN model of BCR-signaling pathways is shown in Figure 18.5. Here in the model when an antigen comes in contact with B cell then BCR activates and with the active *CD79A, CD79B, SFK,* and *SYK* it starts making active *RAF* kinase, *AKT*, calcineurin, and active *BTK*. Table 18.1 summarizes transitions and place description taken in the model.

In TB, the first defender is always the immune system of the body. B cell has its role in defending *Mtb* bacteria by either generating antibodies or storing antibodies in memory cells. When *Mtb* bacteria encountered with macrophages then certain cytokines are

TABLE 18.1

Place and Transition Descriptions for Perti Net Model

Places	Description	Transitions	Description
pa_0	Antigens	it_0	Binding
pa_1	*Ig* (Immunoglobulins)	it_1	Activation
pa_2	BCR (B-cell receptor)	it_2	Receptor tyrosine kinase (*RTK*)
pa_3	*PI3K*	it_3	Tyrosine kinase activity
pa_4	Calcineurin	it_4	Mitogen phosphorylation
pa_5	*RAF*	it_5	*IL-1b* (Interleukin 1 beta)
pa_6	*BTK*	it_6	*PIP2, PIP3*
pa_7	Activation of *the PI3K* pathway	it_7	Dephosphorylation
pa_8	Activation of the *NFAT* pathway	it_8	Tyrosine kinase activity
pa_9	Activation of the *MAPK* pathway	it_9	*TNF-α*
pa_{10}	Activation of the *NF-kβ* pathway		

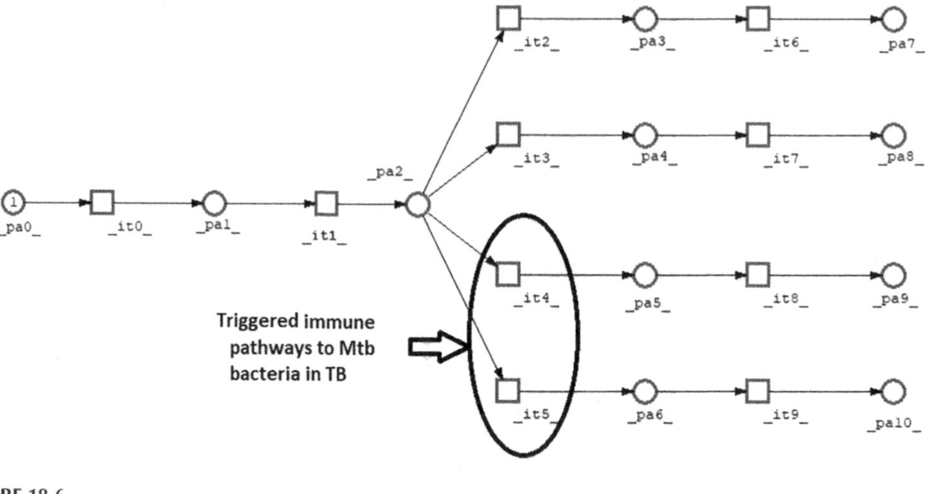

FIGURE 18.6
Petri net Immune model showing triggered immune pathways to *Mtb*.

generated in the response of the immune system. For host protective cytokines BCR regulates NF-kβ and MAPK signaling pathways. Several cytokines like *IL-1b* and *IL-6* are influential mediators acting as anti-TB immunity. Figure 18.6 shows the responsible pathways that got triggered while the body cells got infected by *Mtb* bacteria.

18.6 Results and Discussions

Checking the validity of the model of Figure 18.5, the real number say $m_0(pa_0) = 1.5$ is taken as a token on pa_0 that represents the availability of harmful outer antigens encountered inside the human body and is ready to bind with the BCR in the availability of immunoglobin receptor *IGH* and *IGR* through the transition it_0. The firing of the first transition that is it_0 will shift the tokens to pa_1, which is the receptor *Ig* that lies on the surface of B cell and is answerable for the further activation steps to recover from the effect of the antigen. Next, the place pa_2 is the BCR to bind with the antigen and can behave in four mains downstream. pa_3 represents the availability of *P13K* enzymes that are already present in the human body and activated by receptor tyrosine kinases (RTKs). Similarly, the transitions it_3, it_4, it_5 are enabled and lead to other signaling pathways, but the firing of these transitions depends on the BCR-associated antigen. The firing of these four transitions decides which pathway the immune system follows to minimize the worse effect of the antigens. Depending on the tyrosine kinase activity the immune cells produce the responsible antibodies. If it_2 will fire, then that depicts token shifts to place pa_3, which is the generation of *PI3K* protein. If it_3 will fire, then the token moves to pa_4 which means the activation of phosphatase calcineurin. In a similar manner, if it_4 fires, the token shifts to place pa_5, which means the generation of *RAF* kinase, which activates *the MAPK* pathway. Also, the firing of it_5 transfers to place pa_5 and *BTK* protein starts to activate and leads to NF-kβ pathway. Figure 18.7 shows the simulation result for the net model in Figure 18.5. The token on pa_0 starts decaying continuously and the rest of the other places start to get activated continuously as infection occurs in cells.

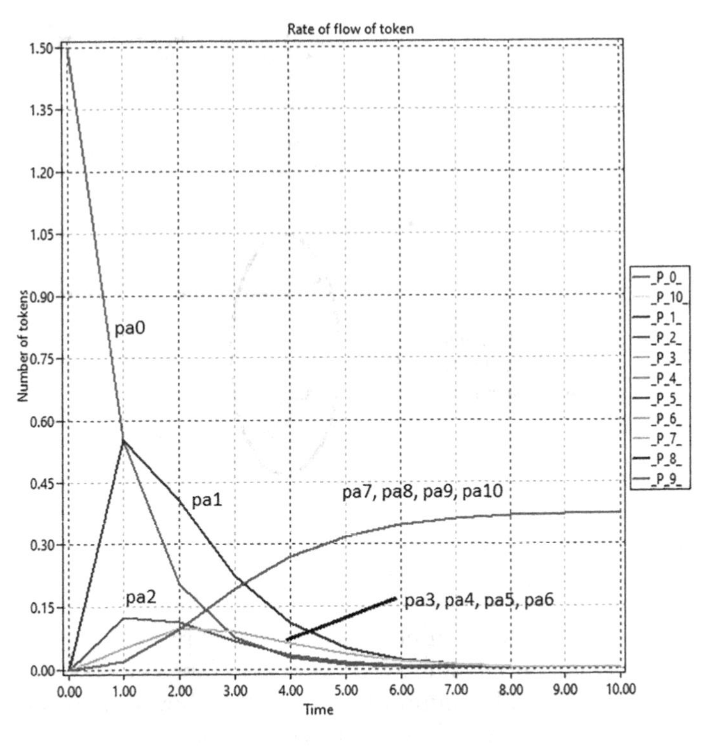

FIGURE 18.7
Simulation result of the rate of flow of tokens in the model of Figure 18.5.

18.7 Validation and Scope

Software Snoopy (Rohr et al., 2010) validated the model shown here and PIPE v4.3.0 (Bonet et al., 2007). Figure 18.8 shows the validation result of the structural properties of the net in Figure 18.5 through PIPE software.

It is concluded that it is a state machine so it's easy to track any changes in the net at any time. The net is bounded, which infers that the activation and formation of any

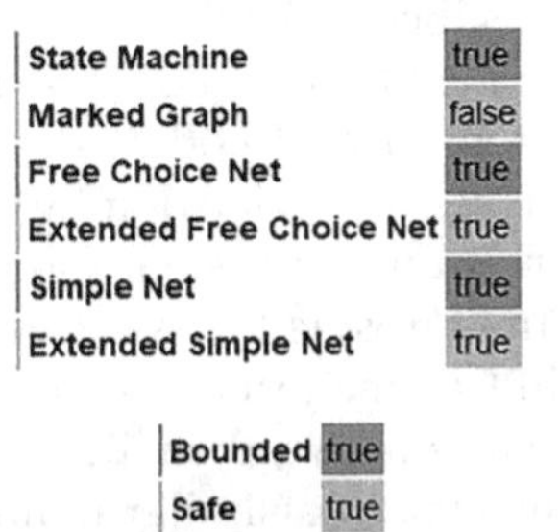

FIGURE 18.8
Validation of properties.

Petri net invariant analysis results

T-Invariants

T0	T1	T2	T3	T4	T5	T6	T7	T8	T9

The net is not covered by positive T-Invariants, therefore we do not know if it is bounded and live.

P-Invariants

P0	P1	P2	P3	P4	P5	P6	P7	P8	P9	P10
1	1	1	1	1	1	1	1	1	1	1

The net is covered by positive P-Invariants, therefore it is bounded.

P-Invariant equations

M(P0) + M(P1) + M(P2) + M(P3) + M(P4) + M(P5) + M(P6) + M(P7) + M(P8) + M(P9) + M(P10) = 1

Analysis time: 0.0s

FIGURE 18.9
The invariant property analysis of the modeled Petri net.

protein in the signaling pathways are limited at any stage. Also, the net has a deadlock stage as after the signaling pathways are activated the flow of tokens is limited up to a certain stage and there is no continuous process going on forever. The analysis of the interconnection of different states in this model can be performed if one knows about the path by which one state is directly reachable from another state. Again, in Figure 18.6, the two transitions it_4, and it_5 will fire after the firing of it_1 as when the antigen is *Mtb* bacteria at the place pa_0 the activation of the pathway in pa_9 and pa_{10} will prefer to stop TB bacteria from proliferating.

In Figure 18.9, the invariant analysis of the modeled net is shown to analyze the unchanging path in the system. Here, in this net, there is no *t*-invariant set, whereas the *p*-invariant set is the whole place setting of the net.

In Figure 18.10, the reachability/coverability graph of the immune model is shown. This graph shows the independence of the downstream pathways followed by BCRs.

18.8 Conclusion

In this chapter, modeling the function of the BCR-signaling path was discussed when it defends from the outer antigens attacking the human body. To fight with antigens certain memory immune cells are being called, which is according to the molecular mechanism of these downstream pathways and hence reducing the growth of infection. The four main downstream pathways followed by B cell are the *MAPK* pathway, *NF-kβ* pathway, *PI3K* pathways, and *NFAT* pathways, which work in their way in defending the responsible antigen in different diseases. In TB, *NFAT* pathways play an important role in susceptibility checking and also as the target detector for curing. Similarly, in curing cancer, the BCR follows the (*PI3K–AKT-mTOR*) pathway with the help of a *PI3K* inhibitor. In curing progressive tumors inside the human body, the followed pathway is the *NFAT* pathway. Also, in the process of drug designing, it includes excessive Ca^{2+} signaling. *MAP* kinase pathway is activated by the phosphorylation of tyrosine and threonine residues, which can

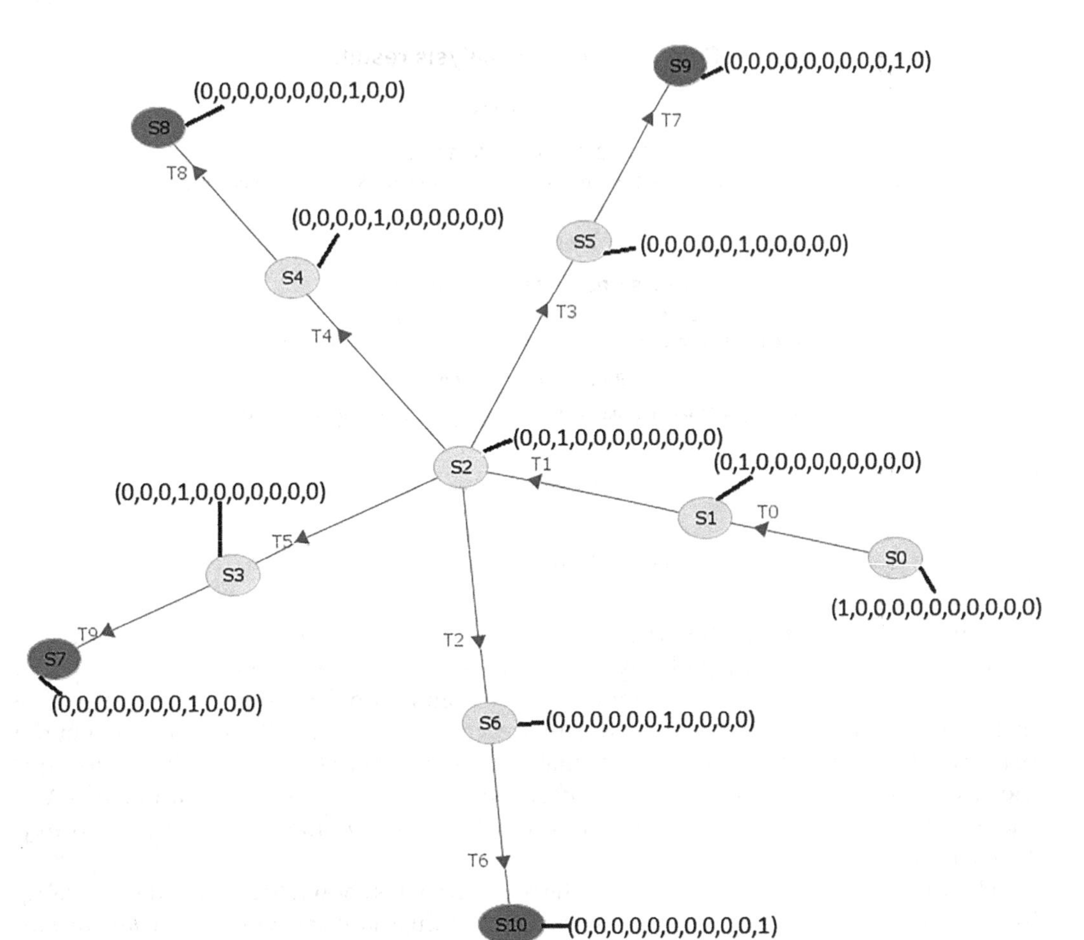

FIGURE 18.10
Coverability graph of the net in Figure 18.4.

reduce the lump of colon tumors. In the same way, various sorts of inflammations can be minimized by the activation of the *NF-kβ* pathway, which helps to degrade various inflammatory diseases. In the future, one can model this pathway differently to study its function in curing diseases. Modeling such pathways using PN methodology always helps to understand the mechanism behind the system by analyzing certain structural properties such as boundedness, liveness, finding the deadlock, and the reachability graph showing the interconnection of intermediate states of the system.

Acknowledgments

The first author of the manuscript thanks the funding agencies Science and Engineering Research Board, Govt. of India, India (project file number ECR/2017/003480/PMS), Department of Science and Technology Purse grant, University Grant Commission (Project id. 257 under University of Potential Excellence Scheme-II) and DBT, Govt. of India

(Project id BT/PR40251/BITS/137/11/2021) for providing financial help and instrumentation support.

Abbreviations

AKT(PKB), protein kinase B; CD79A, CD97B, heterodimers important in signaling component; NF-$\alpha\beta$, protein complex controls transcription of DNA; BTK, Bruton tyrosine kinase; SYK, spleen tyrosine kinase; SFK, SRC family kinase; IKK, inhibitor of kappa-B kinase; RAF, rapidly Accelerated fibrosarcoma; P13K, phosphoinositide 3-kinase; MAPK, microtubule-associated protein kinase; mTOR, mammalian target of rapamycin; PKC, protein kinase; IGL/IGH, immunoglobulin lambda/heavy locus; IGF1, insulin-like growth factor 1; PIP2, phosphatidylinositol-4,5-bisphosphate; PIP3, phosphatidylinositol-3,4,5-bisphosphate.

References

Abbas, A. K., Lichtman, A. H., & Pillai, S. (2014). *Cellular and Molecular Immunology E-Book*. Elsevier Health Sciences.

Alla, H., & David, R. (1998). A modelling and analysis tool for discrete events systems: continuous Petri net. *Performance Evaluation*, 33(3), 175–199.

Barkaoui, K., & Pradat-Peyre, J. F. (1996). On liveness and controlled siphons in Petri nets. In International *Conference on Application and Theory of Petri Nets*, 1091, 57–72.

Bonet, P., Lladó, C. M., Puijaner, R., & Knottenbelt, W. J. (2007). PIPE v2. 5: A Petri net tool for performance modelling. In Proc. 23rd Latin American Conference on Informatics (CLEI 2007).

Cherdal, S., & Mouline, S. (2018). Modeling and Simulation of Biochemical Processes Using Petri Nets. *Processes*, 6(8), 97.

Ehlich, A., Schaal, S., Gu, H., Kitamura, D., Müller, W., & Rajewsky, K. (1993). Immunoglobulin heavy and light chain genes rearrange independently at early stages of B cell development. *Cell*, 72(5), 695–704.

Gupta, S., Kumawat, S., & Singh, G.P. (2019a). Fuzzy Petri Net Representation of Fuzzy Production Propositions of a Rule-Based System. In International Conference on Advances in Computing and Data Sciences, 197–210.

Gupta, S., Singh, G. P., & Kumawat, S. (2018). Polyhydroxyalkanoates (PHAs) k Metabolic marg ka Petri net Nidarsan. *Ganitiya Visheshank (Aalekh Sidhant), Vigyan Garima Sindu*, CSTT, MHRD, Govt of India, ISSN: 2320-7736. 105, 10–18.

Gupta, S., Singh, G. P., & Kumawat, S. (2019b). Petri net recommender system to model metabolic pathway of polyhydroxyalkanoates. *International Journal of Knowledge and Systems Science (IJKSS)*, 10(2), 42–59.

Janeway, C., Travers, P., & Walport, M. (2001). Shlomchik. In Travers, P., Walport, M., Shlomchik, M. J., & Janeway, M. C. (Eds), *Immunobiology. The Immune System in Health and Disease*, 615–617.

Kansal, S., Acharya, M., & Singh, G. P. (2012). Boolean Petri nets. In Pawel Pawlewski (Ed), *Petri Nets-Manufacturing and Computer Science*, 381–406.

Kansal, S., Singh, G. P., & Acharya, M. (2010). On Petri nets generating all the binary n-vectors. *Scientiae Mathematicae Japonicae*, 71(2), 209–216.

Kenkre, V. P., & Kahl, B. S. (2012). The future of B-cell lymphoma therapy: the B-cell receptor and its downstream pathways. *Current Hematologic Malignancy Reports*, 7(3), 216–220.

Kumar, R., Singh, G. P., Pandey, S. K. & Shekhawat, V. J. (2018). Gair Charakiya Hydrocarbon: Sabhi Binary –Vectors ki Uttpatti. *Ganitiya Visheshank (Aalekh Sidhant), Vigyan Garima Sindu*, CSTT, MHRD, Govt of India, ISSN: 2320-7736. 105, 66–70.

Loder, F., Mutschler, B., Ray, R. J., Paige, C. J., Sideras, P., Torres, R., et al. (1999). B cell development in the spleen takes place in discrete steps and is determined by the quality of B cell receptor–derived signals. *Journal of Experimental Medicine*, 190(1), 75–90.

Murata, T. (1989). Petri nets: Properties, analysis, and applications. *Proceedings of the IEEE*, 77(4), 541–580.

Nemazee, D., & Weigert, M. (2000). Revising B cell receptors. *Journal of Experimental Medicine*, 191(11), 1813–1818.

Peterson, J. L. (1981). *Petri Net Theory and the Modeling of Systems*. Prentice-Hall.

Raja, V., Schilling, M., Böhm, M., Hahn, B., Kowarsch, A., Raue, A., et al. (2011). Dynamic mathematical modeling of IL13-induced signaling in Hodgkin and primary mediastinal B-cell lymphoma allows prediction of therapeutic targets. *Cancer Research*, 71(3), 693–704.

Rohr, C., Marwan, W., & Heiner, M. (2010). Snoopy—a unifying Petri net framework to investigate biomolecular networks. *Bioinformatics*, 26(7), 974–975.

Singh, G. P. (2013). A wheel 1-safe petri net generating all the {0, 1} n sequences. *International Journal of Computer Applications*, 84(16), 1–7.

Singh, G. P., Borah, A., & Ray, S. (2020a). A review paper on corona product of graphs. *Advances and Applications in Mathematical Sciences* 19(10), 1047–1054.

Singh, G. P., & Gupta, A. (2019). A Petri net analysis to study the effects of diabetes on cardiovascular diseases. *IEEE Xplore*, ISBN: 978-93-80544-36-6. (Accepted).

Singh, G. P., Jha, M., & Singh, M. (2020b). Petri net modeling of clinical diagnosis path in tuberculosis. In *Advances in Interdisciplinary Research in Engineering and Business Management*. Springer, Singapore, 401–412.

Singh, G. P., Jha, M., Singh, M., & Naina. (2020c). Modeling the mechanism pathways of first line drug in tuberculosis using Petri nets. *International Journal of System Assurance Engineering and Management*, 11(2), 313–324.

Singh, G. P., & Kansal, S. (2016). Basic results on crisp Boolean Petri nets. *Modern Mathematical Methods and High-Performance Computing in Science and Technology*, 83–88.

Singh, G. P., Kansal, S., & Acharya, M. (2013a). Construction of a crisp Boolean Petri net from a 1-safe Petri net. *International Journal of Computer Applications*, 73(17), 1–4.

Singh, G. P., Kansal, S., & Acharya, M. (2013b). Embedding an arbitrary 1-safe Petri Net into a Boolean Petri Net. *International Journal of Computer Applications*, 70(6).

Singh, G. P., & Singh, S. K. (2019). Petri net recommender system for generating of perfect binary tree. *International Journal of Knowledge and Systems Science (IJKSS)*, 10(2), 1–12.

Singh, G. P., Singh, S. K., & Jha, M. (2020d). Existence of forbidden digraphs for crisp Boolean Petri nets. *International Journal of Mathematical, Engineering and Management Sciences (IJMEMS)*, 5(1), 83–95.

Singh, G. P., & Verma, R. N. (2017). Petri net modelling for Biological Networks. *Vigyan Garima Sindu*, CSTT, MHRD, Govt of India, ISSN: 2320-7736. 102, 85–88.

Singh, R. K., Singh, G. P., & Kumar, V. (2018). Sattamak Taatparya Ki Samasya: Boolean Logic aur Strawvasan Logic ka Samikshatamak Aidhan. *Ganitiya Visheshank (Aalekh Sidhant), Vigyan Garima Sindu*, CSTT, MHRD, Govt of India, ISSN: 2320-7736. 105, 24–28.

Szydłowski, M., Jabłońska, E., & Juszczyński, P. (2014). FOXO1 transcription factor: a critical effector of the PI3K-AKT axis in B-cell development. *International Reviews of Immunology*, 33(2), 146–157.

Zhou, B., Wang, S., Mayr, C., Bartel, D. P., & Lodish, H. F. (2007). miR-150, a microRNA expressed in mature B and T cells, blocks early B cell development when expressed prematurely. *Proceedings of the National Academy of Sciences*, 104(17), 7080–7085.

19

Crop Prediction and the Sustainability of Farming

Megha Rathi, Palak Arora, Tanya Srivastava, Nitya Arora, Jigyasa Agarwal, and Abhinna Arjun

Jaypee Institute of Information Technology

CONTENTS

19.1 Introduction

The potential for artificial intelligence (AI) in agribusiness has developed exponentially as the innovation for this industry is persistently progressing. Food development started simultaneously with the advancement of humans, and it is a basic necessity for sustaining human life despite extraordinary advancements in economy, society, and living conditions. As of now, AI in horticulture is centered on observing and gauging ranch profitability. Today, the horticulture business stands outfitted with automatons and robots that are modified to accumulate information from creative fields to propel exactness agribusiness and amplify crop return and benefits to the farmers. Agribusiness is the essential wellspring of occupation for around 58% of India's population. The Indian food industry is prepared for a gigantic turn of events, gaining popularity globally and extending to supply of raw food worldwide which will eventually enhance the country's economy. India has enormous and various farming divisions and is currently among the three best worldwide producers of numerous harvests, including wheat, rice, pulses, cotton, fruits, and vegetables.

In India, a dominant part of the farmers are not able to the gather even ordinary yield because of a couple of reasons. With an expanding population and changing climate, farmers are confronted with settling on troublesome choices. The agricultural yield primarily depends on environmental conditions. Because of the development of new imaginative techniques and procedures, farmers are forced to cultivate artificial products or hybrid products that lead to unhealthy life. Currently, people are unaware of the correct practices that need to be implemented for the cultivation of crops. After analyzing the issues such as weather condition, e.g., drought, and temperature, rainfall, and right soil type, it was found

DOI: 10.1201/9781003046431-19

that there is no proper solution and technology to overcome the situations faced by the farmers. There are several ways to increase economic growth in the field of agriculture. There are numerous approaches to improve and amplify harvest yield and the nature of the harvests.

Predicting crop yield is an important agricultural problem. The capacity to conjecture last harvest yields beforehand and during a developing season is an amazing asset for farmers. If yield can be anticipated with more prominent exactness, there is the chance to prepare, at last, decreasing business hazard (Kantanantha et al., 2010). Determined harvests are additionally significant in controlling administration choices, for example, the use of manure (Raun et al., 2001). The spatial objectives of these yield desires play a significant part, as these allow the administrators to be custom-fitted to different fields inside a residence, or at the subfield level. Generally, ranchers and their specialists check their yield objectives dependent on experience and regular conditions, and afterward utilize this as a manual for developing board choices. Given that reap yield is compelled by the participation between the officials, soil, and atmosphere, this yield objective ought to fluctuate from season to season, yet in addition shift from area to area.

This examination looks at the utilization of machines to contemplate the forecast of rice yield. The current examination utilizes preset information from all the territories of India. Different climatic components are known to influence rice yields, for example, elevation, water quality, low selling price, fertilizer, area, creation were viewed as rice crop creation during the 2011–2015 season. Support Vector Regression (SVR) is applied to the current database. Result investigation was performed and ends were made about its adequacy in improving yield anticipating for prediction yield of rice crop.

19.2 Related Work

In a recent study, authors predicted the yield of rice crop using Support Vector Machine (SVM) in India considers a single state for only 4 years (1998–2002), with only two variables in rice crop (Gandhi et al., 2016b). It also applies only one AI model to anticipate the yield and does not apply any other models. SVM is utilized to make employments from a lot of labeled data with a mark. The information utilized contained precipitation, minimum temperature, etc., normal from the years 1998 to 2002 of Maharashtra only. It uses an SVM model and reaches an accuracy of 78.76%, which is very less than as compared to other models.

Kaur and Attwal (2017) studied the impact of temperature and precipitation on paddy yield by a priori calculation using data-mining apparatus (WEKA) for the investigation of temperature, precipitation, and paddy yield to predict the paddy yield. The dataset used consisted of precipitation, maximum temperature, and yield from the year 2001 to 2015 of Patiala and Ludhiana in Punjab. The research only considers two cities, Patiala and Ludhiana, with only two variables. It uses a small dataset. It also applies only one AI model to anticipate the yield and does not apply any other models. It uses an Apriori model and reaches an accuracy of 53.53%, which is very less than as compared to our model accuracy.

In another research work of crop yield prediction (Fan et al., 2015), big data techniques were utilized along with nearest neighbor and ARMA model to calculate the yield from the years 2009 to 2013. The dataset used consisted of air pressure, rainfall, temperature, humidity, wind speed, evaporation from the year 1955 to 2012 of China. The developed model's prediction accuracy was 93%. The research considers only environmental variables, which might have an effect on the crop yield and does not consider any other factor such as the previous production or fertilizer requirement.

In yet another novel contribution, machine learning models were applied to find out the minimum, maximum, and average crop production (Thombare, 2017). *k*-Means clustering along with the Apriori algorithm in association with Big Data Analytics is used to form clusters and find out the minimum, average, and maximum crop yield. The dataset used consisted of rainfall, state name, season, crop, area, and production from the years 2003 to 2014 of Indian states. The research considers basic parameters to find out the outcome for the crop in the state but does not provide any accuracy, which its model achieved or any other numerical parameter.

In the exploration work, creators proposed a model to foresee crop yield in which *k*-means grouping is utilized to set rules on the training dataset (Manjula and Djodiltachoumy, 2017). The dataset consisted of year, district, crop, well, area, tanks, production, and yield from the years 2002 to 2012 of the districts of Tamil Nadu. The research only considers the districts of Tamil Nadu with variables related to irrigation. It also applied a single AI model to anticipate the harvest and does not apply any other models. It uses a *k*-means clustering model and reaches an accuracy of 86%, which is very little as compared to our model accuracy.

The crop yield prediction model is developed in the research work (Dahikar and Rode, 2014) using artificial neural networks (ANNs). Predictive variables such as soil type, pH, nitrogen, phosphate, potassium, natural carbon, calcium, profundity, temperature, and precipitation were utilized to anticipate the right yield for cultivating in a specified area. Feed-forward back propagation ANN was used for training the network. On applying the model it was found that ANN is a useful and accurate tool to find out the crop yields in rural districts. It also concludes that regional soil parameter affects the most. Analysis was done in Matlab to make it more efficient.

Another significant contribution used the regional circulation model to assess the role of climate extremes of common winter and warm Mediterranean crop (Moriondo et al, 2011). Analysis was done by investigating the impact on sunflower and winter wheat. Direct impact from extreme hot stress was also kept to account. Winter and summer crops were demonstrated to be ideal for environmental change. The immediate impact of warmth stress has brought about a noteworthy decrease in yields while delivering sunflowers and the winter wheat crop shows an expansion in natural product yields. This model is determined with constrained information.

The creators built up a product called "Harvest Advisor" to consider the impact of environmental change on deals revenue (C4.5). The computation is utilized to decide the most significant climate boundaries in chosen organic product crops in chosen areas of Madhya Pradesh (Veenadhari et al., 2014). Giving a sign of a similar impact of various climatic synthetics, some agro-initiated natural product bearing info boundaries are not considered in this gadget, because, the utilization of these casings has changed with the current fields of solidarity. K-strategies utilized are utilized to quantify ecological contamination, a nearby neighbor *k* was fitted with everyday meteorological and other climate estimations, and different atmosphere changes are dispensed with utilizing SVMs (Choice tree count C4.5). The estimation was received and performed with Mo2.0 and VC++ 6.0 to make a horticultural land-scale framework for farming. Decision tree systems were applied to estimate future water prerequisites. The expectation accuracy of the created model shifted from 76% to 90% with the general forecast precision of 82% for the chosen crops and chosen districts.

In another study, enormous information with its huge size and unpredictability is progressively concerned, created, and utilized for all advances and exchanges (Huang et al., 2018) far off sensing, as presumably the greatest datum source, separates worldwide survey information and day-by-day scientific outcomes from satellite stages, human-made/unmanned airplane, and implicit structures. Far-sightedness is one of the most significant spine innovations in farming, which takes a gander at the decent variety of the field of nearby administration rather than a similar administration as in customary agribusiness. The way to rural affectability, with topographical and geological subtleties, is to deliver an assortment of room information for ensuing farming exercises. Rural remote detecting information, for example, standard remote hearing information, has all the highlights of enormous information. The location, handling, stockpiling, investigation, and representation of remote agrarian information are fundamental for farming achievement and exactness. This study takes a gander at remote sensor wellsprings of data, the most recent advances in remote detecting information the board innovation, and remote information knowledge tasks with exact horticultural administration. The five-dimensional satellite information the board framework (FLFL) is characterized and changed to make a three-by-three-meter remote information control framework (FLTL) exact farming data where sensors land on high-goal satellites, mechanical airplanes, illicit vehicles uninstalled, and underground structures. The structure of the FLTL is a framework for the administration and utilization of remote dialects that hear a lot of exact horticultural information and nearby ranch contemplates, mirroring the future mix of remote information the executives with local and homestead level applications.

Another study yield observation is significant for precise cultivating (Bendig et al., 2014), estimating the above biomass assists with observing plant vitality and foresee yield. In this examination, we compute new and dry biomass in a yearly grain feed trial of 18 g and two nitrogen (N)—utilizing plant obstruction (PH) from high return models (CSMs). With the biggest objective, most CSMs (1 cm/pixel) depend on red, green, blue (RGB) pictures taken in a little automated ethereal (UAV) vehicle. The mix of PH accounts of the signs got R2 of 0.92. The workplace of different diagnostics and treatments was tried during the "BiologischeBundesanstalt, Bundessortenamt und CHemischeIndustrie" (BBCH) Chapters 24–89. High collaborations between PH were found in CSMs with new biomass (R2=0.81) and dry biomass (R2=0.82). Five new kinds of estimations for biomass soil have been tried for cross-approval. Estimating biomass between N-medicines distinctive new biomass yielded better outcomes (R2=0.71). The fundamental confinement was the impact of putting the rest of the plants in stages that became later, delivering surprising plant statures. The strategy can utilize future applications by nonexperts, e.g., ranchers.

In another study, the facts confirm that the big has pulled in a ton of consideration from specialists in data science, arrangements, and choices made in government and business (Chen et al., 2014). As the pace of data development outperforms Moore's Law toward the start of this century, abuse makes difficult issues for individuals. Be that as it may, there are such a large number of and helpful qualities hidden in a lot of information. The new science worldview is conceived as the disclosure of logical information (DisD), otherwise called big data issues. An enormous number of parts and divisions, running from monetary and business exercises to policy management, from national security to logical examination in numerous zones, including big data issues. Then again, big data is basic to business profitability and the rise of advancement in logical examinations, which gives us numerous open doors for incredible improvement in numerous fields. There is no vulnerability that future contentions in business and advancement creation will unavoidably change into big data testing. Enormous data, then again, additionally presents numerous difficulties, for instance, inconvenience in getting data, data amassing, data examination, and survey data. This investigation hopes to show an increasingly basic look at big data, including the usage of big data, the incredible chances, and difficulties of data, just as the specialized and mechanical methodologies we right now use to address big data issues. We additionally talked about a couple of fundamental approaches to catch information floodplains, for instance, granular charts, distributed computing, PC roused bio and quantum processing.

A study foreseeing the enormous winter wheat reap, we have built up a harvest model and information assortment structures prove by the leaf-list (LAI) taken from Landsat TM and MODIS information on the WOFOST crop model (Huang et al., 2018). We evaluated LAI in seven natural classifications in two horticultural urban communities in China's Hebei province. To lessen cloud contamination, we have utilized Saltzky–Golay (S-G) to sift through MODIS LAI items to identify separated LAI. We at that point enlisted an LAI estimating field on LandsatTM in LandsatTM files to get multitemporal LAIs. We have built up a nonstraight way to deal with changing LAI by figuring the inconsistent sum between touchy sensor information and model adaptability. TM LAI and amended LAI information were consolidated into the WOFOST model to take into account an increasingly precise assessment of yield gauges. We have constructed a wide scope of information changeability estimation (4DVar) to imagine and show blunders during the significant development stage. We have executed complex elements of characteristic versatility—University of Arizona calculation to decrease the expense of 4DVar expenses among reasonable and adjusted LAI and to extend two significant WOFOST boundaries. At long last, we impersonated the winter wheat reap on a 1 km lattice of cells with at any rate half of their winter wheat consumption region utilizing improved WOFOST, and accumulated outcomes at a provincial level. The change rate fundamentally improved the precision of territorial wheat creation gauges contrasted and the prohibited outcomes and TM LAI results. Consequently, the synchronization execution relies to a great extent on the exactness of LAI recovery and estimation revision. Our exploration gives a framework for utilizing long-run information, soil estimation information, and harvest development model to improve crop yield gauges.

19.3 Methodology

Data preprocessing is a key feature in machine learning. Acquiring each parameter as the raw data set (Season, Area, Production, MSP, Yield) yearly records per means of each state from 2000 to 2015 from the Government of India resources was the first step for preprocessing then the water quality index was calculated by the minimum, maximum and average values from the city-wise data during the Kharif season and then data for the fertilizer (UREA) was acquired. All the data were then collected in one sheet using excel queries to find the total area and production under each state for each year, Microsoft Excel consists of columns: year, state, crop, season, area, production, water quality, rainfall, fertilizer, minimum support price, and yield.

Now, after getting the input, the data are appended in the created dataset without the corresponding value of yield, which will be predicted by our model. Here, data preprocessing would include encoding categorical data, parting the information into preparing and testing set, and finally feature scaling. In the encoding of categorical data, it will encode the categorical data into binary codes so that the algorithm can understand and distinguish between the values. Then the dataset is split into train and test sets and then scaled.

The row of input data from the processed dataset is now deleted so that the model can be applied and the data can be trained.

Farmers will be benefited from this analysis we will advise them about what factors they lack and what should they do to attain the maximum yield of the region at that time.

19.3.1 Dataset Description

The data were collected from different sources. All data utilized on the task were extricated from the open records of the Government of India. This was discharged in the year for the Kharif rice season. This was extracted for the year 2000–2015. Data were collected from https://data.gov.in/ and http://www.cpcbenvis.nic.in/. Table 19.1 describes dataset attributes.

Figures 19.1 and 19.2 present the snapshot of the initial dataset and dataset after preprocessing.

TABLE 19.1

Dataset Description

S. No.	Attribute	Description
1.	Year	Dataset was collected and stored for the years 2000–2015
2.	State name	All the parameters are collected for almost all the states of India
3.	Crop	Dataset is collected for rice crop
4.	Rainfall (mm)	The all-out climate estimate for the Khalif Season (June–November) for each state is determined from the nation's month to month precipitation for that year
5.	Area (Hectare)	The rice cultivating area for each state and each year was considered for this study
6.	Production (Tonnes)	The neighborhood rice utilization in every locale every year was utilized
7.	Yield (Tonnes/ Hectare)	Contingent upon the area and creation of each express every year, the yield is determined by research work
8.	Minimum support price (Unit is Rupees/Quintal)	Minimum support price was used for each year
9.	Fertilizer (Tonnes)	The fertilizer (UREA) required by each state from the years 2012 to 2015 was used
10.	Water quality (pH)	The water level of each state in the year was gathered and utilized

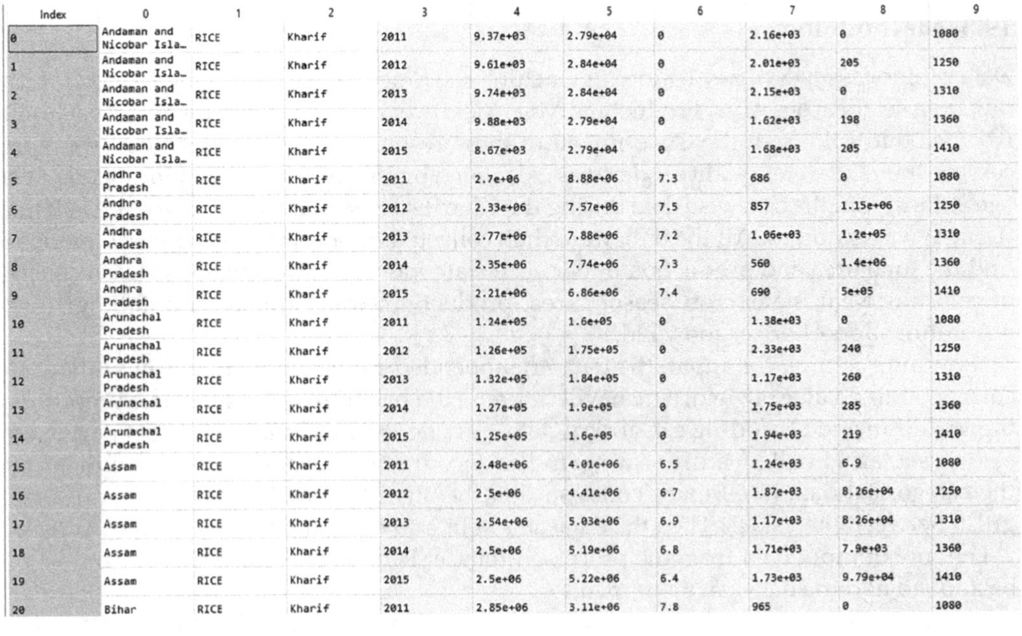

Index	0	1	2	3	4	5	6	7	8	9
0	Andaman and Nicobar Isla…	RICE	Kharif	2011	9.37e+03	2.79e+04	0	2.16e+03	0	1080
1	Andaman and Nicobar Isla…	RICE	Kharif	2012	9.61e+03	2.84e+04	0	2.01e+03	205	1250
2	Andaman and Nicobar Isla…	RICE	Kharif	2013	9.74e+03	2.84e+04	0	2.15e+03	0	1310
3	Andaman and Nicobar Isla…	RICE	Kharif	2014	9.88e+03	2.79e+04	0	1.62e+03	198	1360
4	Andaman and Nicobar Isla…	RICE	Kharif	2015	8.67e+03	2.79e+04	0	1.68e+03	205	1410
5	Andhra Pradesh	RICE	Kharif	2011	2.7e+06	8.88e+06	7.3	686	0	1080
6	Andhra Pradesh	RICE	Kharif	2012	2.33e+06	7.57e+06	7.5	857	1.15e+06	1250
7	Andhra Pradesh	RICE	Kharif	2013	2.77e+06	7.88e+06	7.2	1.06e+03	1.2e+05	1310
8	Andhra Pradesh	RICE	Kharif	2014	2.35e+06	7.74e+06	7.3	560	1.4e+06	1360
9	Andhra Pradesh	RICE	Kharif	2015	2.21e+06	6.86e+06	7.4	690	5e+05	1410
10	Arunachal Pradesh	RICE	Kharif	2011	1.24e+05	1.6e+05	0	1.38e+03	0	1080
11	Arunachal Pradesh	RICE	Kharif	2012	1.26e+05	1.75e+05	0	2.33e+03	240	1250
12	Arunachal Pradesh	RICE	Kharif	2013	1.32e+05	1.84e+05	0	1.17e+03	260	1310
13	Arunachal Pradesh	RICE	Kharif	2014	1.27e+05	1.9e+05	0	1.75e+03	285	1360
14	Arunachal Pradesh	RICE	Kharif	2015	1.25e+05	1.6e+05	0	1.94e+03	219	1410
15	Assam	RICE	Kharif	2011	2.48e+06	4.01e+06	6.5	1.24e+03	6.9	1080
16	Assam	RICE	Kharif	2012	2.5e+06	4.41e+06	6.7	1.87e+03	8.26e+04	1250
17	Assam	RICE	Kharif	2013	2.54e+06	5.03e+06	6.9	1.17e+03	8.26e+04	1310
18	Assam	RICE	Kharif	2014	2.5e+06	5.19e+06	6.8	1.71e+03	7.9e+03	1360
19	Assam	RICE	Kharif	2015	2.5e+06	5.22e+06	6.4	1.73e+03	9.79e+04	1410
20	Bihar	RICE	Kharif	2011	2.85e+06	3.11e+06	7.8	965	0	1080

FIGURE 19.1
Initial dataset.

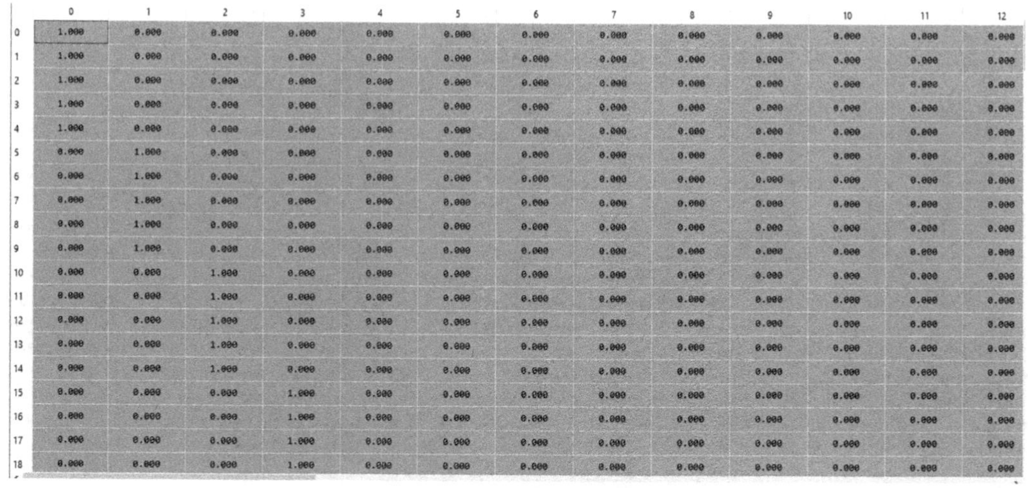

	0	1	2	3	4	5	6	7	8	9	10	11	12
0	1.000	0.000	0.000	0.000	0.000	0.000	0.000	0.000	0.000	0.000	0.000	0.000	0.000
1	1.000	0.000	0.000	0.000	0.000	0.000	0.000	0.000	0.000	0.000	0.000	0.000	0.000
2	1.000	0.000	0.000	0.000	0.000	0.000	0.000	0.000	0.000	0.000	0.000	0.000	0.000
3	1.000	0.000	0.000	0.000	0.000	0.000	0.000	0.000	0.000	0.000	0.000	0.000	0.000
4	1.000	0.000	0.000	0.000	0.000	0.000	0.000	0.000	0.000	0.000	0.000	0.000	0.000
5	0.000	1.000	0.000	0.000	0.000	0.000	0.000	0.000	0.000	0.000	0.000	0.000	0.000
6	0.000	1.000	0.000	0.000	0.000	0.000	0.000	0.000	0.000	0.000	0.000	0.000	0.000
7	0.000	1.000	0.000	0.000	0.000	0.000	0.000	0.000	0.000	0.000	0.000	0.000	0.000
8	0.000	1.000	0.000	0.000	0.000	0.000	0.000	0.000	0.000	0.000	0.000	0.000	0.000
9	0.000	1.000	0.000	0.000	0.000	0.000	0.000	0.000	0.000	0.000	0.000	0.000	0.000
10	0.000	0.000	1.000	0.000	0.000	0.000	0.000	0.000	0.000	0.000	0.000	0.000	0.000
11	0.000	0.000	1.000	0.000	0.000	0.000	0.000	0.000	0.000	0.000	0.000	0.000	0.000
12	0.000	0.000	1.000	0.000	0.000	0.000	0.000	0.000	0.000	0.000	0.000	0.000	0.000
13	0.000	0.000	1.000	0.000	0.000	0.000	0.000	0.000	0.000	0.000	0.000	0.000	0.000
14	0.000	0.000	1.000	0.000	0.000	0.000	0.000	0.000	0.000	0.000	0.000	0.000	0.000
15	0.000	0.000	0.000	1.000	0.000	0.000	0.000	0.000	0.000	0.000	0.000	0.000	0.000
16	0.000	0.000	0.000	1.000	0.000	0.000	0.000	0.000	0.000	0.000	0.000	0.000	0.000
17	0.000	0.000	0.000	1.000	0.000	0.000	0.000	0.000	0.000	0.000	0.000	0.000	0.000
18	0.000	0.000	0.000	1.000	0.000	0.000	0.000	0.000	0.000	0.000	0.000	0.000	0.000

FIGURE 19.2
Encoded dataset.

19.3.2 Techniques Used

The SVM can be utilized as a reworking technique, keeping all the primary highlights indicating the calculation (high marx). SVR utilizes guidelines that are not available to SVM in stages, with only a couple of examinations. As an issue of need first, considering the yield is a genuine number thing being what they are, it is hard to see current information, with incredible potential. On account of backsliding, the resilience line (epsilon) is set to SVM, which may as of now have been mentioned for this issue. Isolated preparation and test information appear in Figures 19.3 and 19.4.

Notwithstanding, the fundamental thought continues as before: to limit blunder, to modify the hyper plane that extends the line, recalling that piece of the mistake is okay.

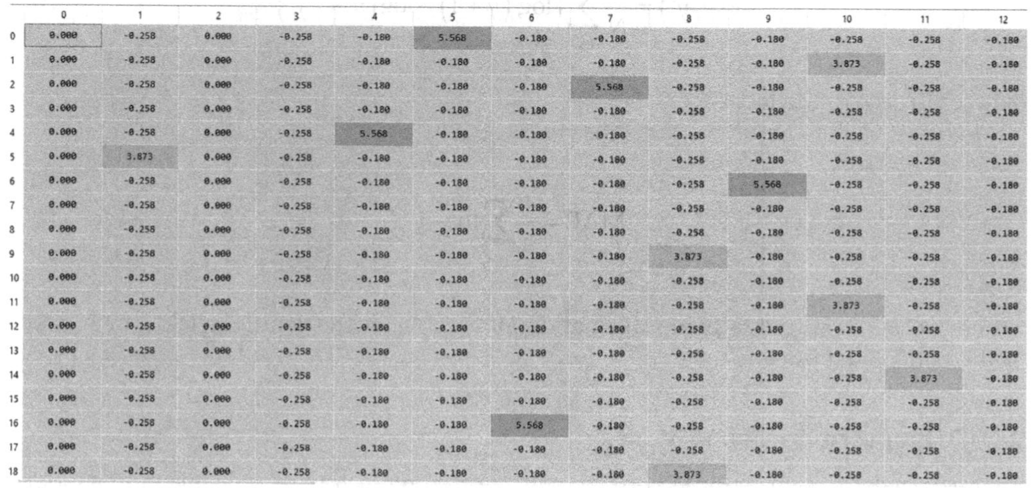

	0	1	2	3	4	5	6	7	8	9	10	11	12
0	0.000	-0.258	0.000	-0.258	-0.180	5.568	-0.180	-0.180	-0.258	-0.180	-0.258	-0.258	-0.180
1	0.000	-0.258	0.000	-0.258	-0.180	-0.180	-0.180	-0.180	-0.258	-0.180	3.873	-0.258	-0.180
2	0.000	-0.258	0.000	-0.258	-0.180	-0.180	-0.180	5.568	-0.258	-0.180	-0.258	-0.258	-0.180
3	0.000	-0.258	0.000	-0.258	-0.180	-0.180	-0.180	-0.180	-0.258	-0.180	-0.258	-0.258	-0.180
4	0.000	-0.258	0.000	-0.258	5.568	-0.180	-0.180	-0.180	-0.258	-0.180	-0.258	-0.258	-0.180
5	0.000	3.873	0.000	-0.258	-0.180	-0.180	-0.180	-0.180	-0.258	-0.180	-0.258	-0.258	-0.180
6	0.000	-0.258	0.000	-0.258	-0.180	-0.180	-0.180	-0.180	-0.258	5.568	-0.258	-0.258	-0.180
7	0.000	-0.258	0.000	-0.258	-0.180	-0.180	-0.180	-0.180	-0.258	-0.180	-0.258	-0.258	-0.180
8	0.000	-0.258	0.000	-0.258	-0.180	-0.180	-0.180	-0.180	-0.258	-0.180	-0.258	-0.258	-0.180
9	0.000	-0.258	0.000	-0.258	-0.180	-0.180	-0.180	-0.180	3.873	-0.180	-0.258	-0.258	-0.180
10	0.000	-0.258	0.000	-0.258	-0.180	-0.180	-0.180	-0.180	-0.258	-0.180	-0.258	-0.258	-0.180
11	0.000	-0.258	0.000	-0.258	-0.180	-0.180	-0.180	-0.180	-0.258	-0.180	3.873	-0.258	-0.180
12	0.000	-0.258	0.000	-0.258	-0.180	-0.180	-0.180	-0.180	-0.258	-0.180	-0.258	-0.258	-0.180
13	0.000	-0.258	0.000	-0.258	-0.180	-0.180	-0.180	-0.180	-0.258	-0.180	-0.258	-0.258	-0.180
14	0.000	-0.258	0.000	-0.258	-0.180	-0.180	-0.180	-0.180	-0.258	-0.180	-0.258	3.873	-0.180
15	0.000	-0.258	0.000	-0.258	-0.180	-0.180	-0.180	-0.180	-0.258	-0.180	-0.258	-0.258	-0.180
16	0.000	-0.258	0.000	-0.258	-0.180	-0.180	5.568	-0.180	-0.258	-0.180	-0.258	-0.258	-0.180
17	0.000	-0.258	0.000	-0.258	-0.180	-0.180	-0.180	-0.180	-0.258	-0.180	-0.258	-0.258	-0.180
18	0.000	-0.258	0.000	-0.258	-0.180	-0.180	-0.180	-0.180	3.873	-0.180	-0.258	-0.258	-0.180

FIGURE 19.3
Scaled trained data.

	0	1	2	3	4	5	6	7	8	9	10	11	12
0	-0.205	-0.157	-0.205	-0.157	-0.183	-0.183	-0.183	-0.183	-0.157	-0.183	-0.157	-0.157	-0.183
1	-0.205	-0.157	-0.205	-0.157	-0.183	-0.183	-0.183	-0.183	-0.157	-0.183	-0.157	-0.157	-0.183
2	-0.205	-0.157	-0.205	-0.157	5.477	-0.183	-0.183	-0.183	-0.157	-0.183	-0.157	-0.157	-0.183
3	-0.205	-0.157	-0.205	-0.157	-0.183	-0.183	-0.183	-0.183	-0.157	-0.183	-0.157	-0.157	-0.183
4	-0.205	-0.157	-0.205	-0.157	-0.183	-0.183	-0.183	-0.183	-0.157	-0.183	-0.157	-0.157	-0.183
5	-0.205	-0.157	-0.205	-0.157	-0.183	-0.183	-0.183	-0.183	-0.157	-0.183	-0.157	-0.157	-0.183
6	-0.205	-0.157	-0.205	-0.157	-0.183	-0.183	-0.183	-0.183	-0.157	-0.183	-0.157	-0.157	-0.183
7	-0.205	-0.157	-0.205	-0.157	-0.183	5.477	-0.183	-0.183	-0.157	-0.183	-0.157	-0.157	-0.183
8	-0.205	-0.157	-0.205	-0.157	-0.183	-0.183	-0.183	-0.183	-0.157	-0.183	-0.157	-0.157	-0.183
9	-0.205	-0.157	-0.205	-0.157	-0.183	-0.183	-0.183	-0.183	-0.157	-0.183	-0.157	-0.157	-0.183
10	-0.205	-0.157	-0.205	-0.157	-0.183	-0.183	-0.183	-0.183	-0.157	-0.183	-0.157	-0.157	-0.183
11	-0.205	-0.157	-0.205	6.351	-0.183	-0.183	-0.183	-0.183	-0.157	-0.183	-0.157	-0.157	-0.183
12	-0.205	-0.157	-0.205	-0.157	-0.183	-0.183	-0.183	-0.183	-0.157	-0.183	-0.157	-0.157	5.477
13	-0.205	-0.157	-0.205	-0.157	-0.183	-0.183	5.477	-0.183	-0.157	-0.183	-0.157	-0.157	-0.183
14	-0.205	-0.157	-0.205	-0.157	-0.183	-0.183	-0.183	-0.183	-0.157	-0.183	-0.157	-0.157	-0.183
15	-0.205	-0.157	-0.205	-0.157	-0.183	-0.183	-0.183	-0.183	-0.157	-0.183	-0.157	-0.157	5.477
16	4.879	-0.157	-0.205	-0.157	-0.183	-0.183	-0.183	-0.183	-0.157	-0.183	-0.157	-0.157	-0.183
17	-0.205	-0.157	-0.205	-0.157	-0.183	-0.183	-0.183	-0.183	-0.157	-0.183	-0.157	-0.157	-0.183
18	-0.205	-0.157	-0.205	-0.157	-0.183	-0.183	-0.183	-0.183	6.351	-0.183	-0.157	-0.157	-0.183

FIGURE 19.4
Scaled test data.

In the regression model, accuracy is judged according to the R-square value calculated as follows:

$$R_{adj^2} = 1 - \frac{\mathrm{Var}(e_i)/(n-k-1)}{\mathrm{Var}(y_i)/(n-1)} \tag{19.1}$$

where "n" is the number of cases used to fit the model and "k" is the number of indicator factors in the model.

Other factors include median squared logarithmic error calculated as:

$$L(y, y') = \frac{1}{N}\sum_{i=0}^{N}\left(\log\,(y+1) - \log(y'+1)^2\right) \tag{19.2}$$

where y' is the predicted value.

Median Absolute Error is defined as:

$$\mathrm{MAE} = \frac{1}{n}\sum_{i=1}^{n}\left(|y - y'|\right) \tag{19.3}$$

where "n" is the all-out tally of information focuses, "y" is the genuine yield worth, and y' is the anticipated esteem.

19.3.3 Proposed Prediction Model

In this research, we first trained our dataset, containing factors responsible for the yield of Kharif crop, i.e., rice for 2011–2015 for overall India using SVR because we analyzed that on this kind of dataset, model of SVR works the best. It predicted the yield of the crop for a particular input given by the user containing urea, rainfall, water quality, minimum support price, area, and production.

It also calculates the fault in the input parameters by giving the difference between the ideal conditions and your input. Here, ideal conditions correspond to the maximum yield within a state. For example, one entered parameters corresponding to "Madhya Pradesh" (suppose), then the code first predicts the yield and then searches for the maximum yield and the corresponding parameters in that state's entries. This row of parameters and yield is compared to your input, thereby giving the gap between predicted and maximum yield along with the factors contributing to the same.

19.4 Results and Analysis

In this section, we have provided detailed description of outcomes obtained in the wake of applying the SVR procedure to the rice organic product database of a considerable number of regions of India. The various boundaries used to anticipate are fertilizer, water quality, precipitation, area, and creation. The calculation has discovered a precision of 78.12%. The model errors results mean an absolute error of 0.23, a median error of 0.35. Figures 19.5–19.8 show the water quality, rainfall, fertilizers, and MSP v/s yield representation.

The various factors that were considered for the acquired outputs include yield versus water quality as represented in Figure 19.5.

Another factor that was considered for the procured yields incorporates yield versus rainfall as spoke to in Figure 19.6.

Another factor that was considered for the acquired outputs includes yield versus fertilizer as represented in Figure 19.7.

Another factor that was considered for the procured yields incorporates yield versus MSP as spoke to in Figure 19.8.

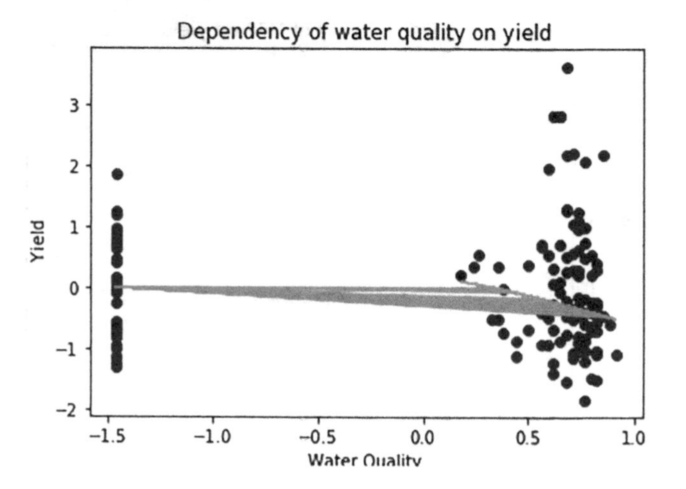

FIGURE 19.5
Yield vs water quality.

FIGURE 19.6
Yield vs rainfall.

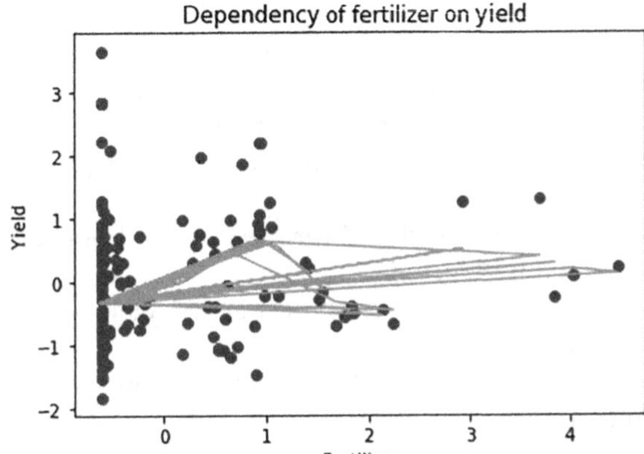
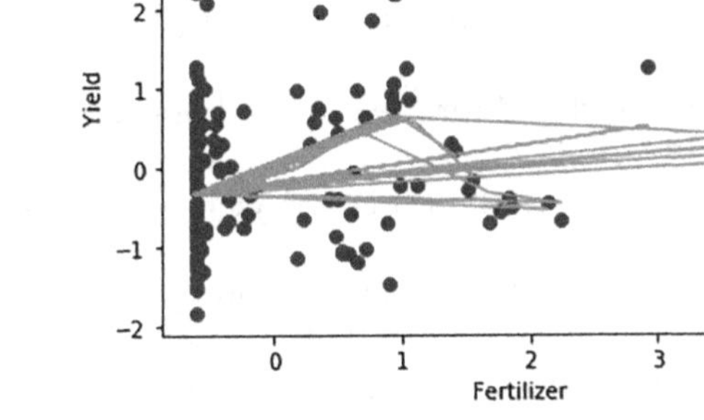

FIGURE 19.7
Yield vs fertilizer.

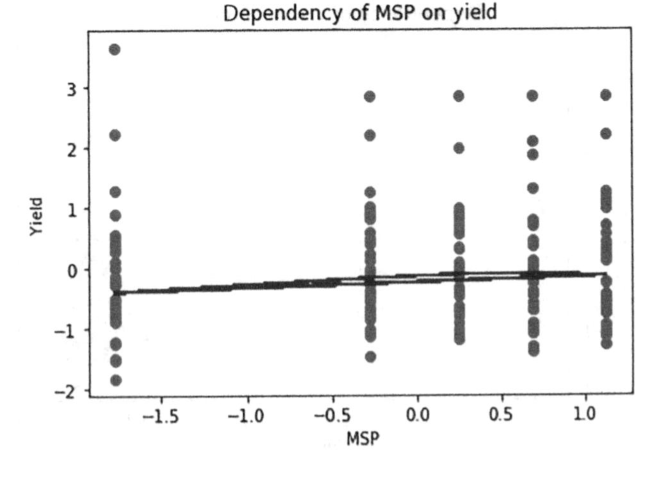

FIGURE 19.8
Yield vs MSP.

TABLE 19.2

Tabulated Result

S. No.	Model Parameter	Value
1.	Accuracy	78.12%
2.	Mean absolute error	0.23
3.	Median error	0.35

19.5 Conclusion

As of late, there have been comparable endeavors to foresee the rice crop (Table 19.2). Making explicit harvest yield demonstrating models utilizing information and technology. Technologies can bolster ranchers and different accomplices can improve limits by national food/transportation what's more, food security. Rice is one of the hugest results of India. It is built up everywhere throughout the nation and contributes over 40% of the all-out grain creation of Singh and Mustard (2012). Given the noteworthiness of rice in world food security, any improvements in choosing the yield of rice under various climatic conditions and planting will be profitable. This test indicated the normal yield of rice utilizing a solitary AI framework, SVM. Test outcomes have demonstrated that distinctive classifiers, for instance, Naïve Bayes, BayesNet, and Multilayer Perceptron have been fundamentally improved in accomplishing the most elevated exactness, affectability, and closeness contrasted with the SMO class with less accuracy, affectability, and clearness determined before with similar determinations record Gandhi et al. (2016a,b) It will be viewed as that the different dividers utilized in the current database and have subtleties before proposing further improvement of the rice conjecture gauge.

References

Bendig, J., Bolten, A., Bennertz, S., Broscheit, J., Eichfuss, S., Bareth, G. (2014, October). Estimating biomass of barley using crop surface models (CSMs) derived from UAV-based RGB imaging *Remote Sensing*, 6, pp. 10395–10412

Chen, Y., Lin, Z., Zhao, X., Wang, G., Gu, Y. (2014, March). Deep learning-based classification of hyperspectral dat. *IEEE Journal of Selected Topics in Applied Erath Observations and Remote Sensing*, 7, pp. 2094–2107.

Dahikar, S. S., Rode, S. V. (2014). Agricultural crop yield prediction using an artificial neural network approach. *International Journal of Innovative Research in Electrical, Electronics, Instrumentation and Control Engineering*, 2(1), 683–686.

Fan, W., Chong, C., Xiaoling, G., Hua, Y., Juyun, W. (2015, December). Prediction of crop yield using big data. In *2015 8th International Symposium on Computational Intelligence and Design (ISCID)* (Vol. 1, pp. 255–260). IEEE, Hangzhou, China.

Gandhi, N., Armstrong, L. J., Petkar, O. (2016a, September). Predicting rice crop yield using Bayesian networks. In *2016 International Conference on Advances in Computing, Communications, and Informatics (ICACCI)* (pp. 795–799). IEEE, Jaipur, India.

Gandhi, N., Armstrong, L. J., Petkar, O., Tripathy, A. K. (2016b, July). Rice crop yield prediction in India using support vector machines. In *2016 13th International Joint Conference on Computer Science and Software Engineering (JCSSE)* (pp. 1–5). IEEE, USA.

Huang, Y., Zhong-Xinchen, Tao, Y., Xiang-Zhi, H., Xing-Fa, G. (2018, January). Agricultural remote sensing big data: Management and applications. *Journal of Integrative Agriculture*, 17(9), 1915–1931.

Kantanantha, N., Serban, N., Griffin, P. (2010). Yield and price forecasting for stochastic crop decision planning. *Journal of Agricultural, Biological, and Environmental Statistics*, 15(3), 362–380.

Kaur, K., Attwal, K. S. (2017, January). Effect of temperature and rainfall on paddy yield using data mining. In *2017 7th International Conference on Cloud Computing, Data Science & Engineering-Confluence* (pp. 506–511). IEEE, Noida, India.

Manjula, E., & Djodiltachoumy, S. (2017). A model for prediction of crop yield. *International Journal of Computational Intelligence and Informatics*, 6(4), 298–305.

Moriondo, M., Giannakopoulos, C., & Bindi, M. (2011). Climate change impact assessment: The role of climate extremes in crop yield simulation. *Climatic Change*, 104(3–4), 679–701.

Raun, W. R., Solie, J. B., Johnson, G. V., Stone, M. L., Lukina, E. V., Thomason, W. E., & Schepers, J. S. (2001). In-season prediction of potential grain yield in winter wheat using canopy reflectance. *Agronomy Journal*, 93(1), 131–138.

Singh, S., & Mustard, A. (2012). India grain and feed annual. *Global Agricultural Information Network (GAIN) Report no IN2026*. USDA-Foreign Agricultural Services.

Thombare, R., Bhosale, S., Dhemey, P., & Chaudhari, A. (2017). Crop yield prediction using big data analytics.

Veenadhari, S., Misra, B., & Singh, C. D. (2014, January). Machine learning approach for forecasting crop yield based on climatic parameters. In *2014 International Conference on Computer Communication and Informatics* (pp. 1–5). IEEE, Bandung, Indonesia.

20

Personalized Heart Disease Framework for Health Sustainability

Megha Rathi, Amogh Sanjeev Gupta, Ayush Raj, Rizul Singh, and Vishrut Sacheti

Jaypee Institute of Information Technology

CONTENTS

20.1 Introduction

Determination of the exact cause of heart disease is a crucial task in the domain of data mining and machine learning. Globally heart diseases account for the leading cause of mortality in terms of disease mortality rate. There is a need for specific and accurate data-mining models, which can accurately find out initial symptoms or factors contributing to heart disease. The paramount objective of the proposed research work is to decide the best model in the setting of coronary illness prediction and the development of a recommendation system. The proposed model is validated on different evaluation criteria, anticipates the odds of coronary illness, and recommends the appropriate diet and exercise based on the details entered by the user.

20.1.1 Basics

World Health Organization shows Ischemic heart disease and stroke to be the main causes for deaths in the world, where the combined deaths were 15 million in 2015. Similarly, the statistics from the 2017 report of the American Heart Association showed cardiovascular

DOI: 10.1201/9781003046431-20

disease (CVD) death count approximately around 800,000 in the United States, or one out of every three deaths. The majority of CVD deaths were caused by coronary heart disease (CHD), followed by stroke and heart failure. Globally, 31% of deaths were caused by CVD. An estimation carried out relating to the cost of therapy for CVD would be $1,044 billion by 2030. One of 20 deaths in the United States is from strokes. As of 2017, death rates caused by strokes have declined. Stroke has become the leading cause of long-term disability in the United States.

Since the last few years, machine learning has been very impactful in the healthcare industry. Some of its application areas include Medical Imaging, Diagnosis, and Recommendation Engine for patients, Drug Discovery, etc. Apart from these, a few other application areas in healthcare include Scaled-Up Medical Data Collection, Drug Side Effects, Treatment Personalization, and Robotic Surgery. Similarly, the scope of machine learning in the prediction of diseases has been broadening significantly, and heart diseases are one of them. Machine-learning techniques improve accuracy in cardiovascular risk prediction and recommendation, which helps in avoiding patients who don't require treatment at an initial stage rather than mistreating them, which would put a physical and financial burden on the patients.

In the proposed work, we compared and analyzed four machine-learning models for the prediction of heart disease. Cleveland, Hungarian, and Switzerland Heart Disease Datasets were utilized in the study. All three datasets were related to heart disease. All these datasets were picked from the UCI Machine Learning Repository. The machine-learning techniques that we used for working on them were: Logistic Regression, Decision Trees, Support Vector Machine (SVM), and Neural Networks (NNs). The most significant technique was SVM, with the highest accuracy. Thus, it is the basic algorithm behind the whole recommendation system. The recommendation system is designed using the Shiny package and developed using Item-Based Collaborative Filtering. The main goal of this project was to determine which model worked the best from the four, based on different evaluation criteria, predict the chances of heart disease, and recommend the appropriate diet and exercise based on the details entered by the user.

20.1.2 Background Study

Listed are the four models that are used to calculate the accuracy and find out the best suitable one for our purpose.

20.1.2.1 Decision Tree

A decision tree is a flowchart-like structure or graphical representation of all the possible solutions to a decision in which each decision is based on some conditions, and the outcome of the test is communicated to each branch, and the class name is determined with each leaf hub. A decision tree and the stably associated results graph are utilized as a perceptible and explanatory option help device, where the conventional peculiarity of contending alternatives is determined (Mythili et al., 2013).

Total three vital nodes are present in the decision tree:

- End node
- Decision node
- Chance node

20.1.2.2 Logistic Regression

On the off chance that the reliant variable is dichotomous (parallel), strategic relapse is the reasonable relapse examination to perform. Like all break down of relapse, a prescient examination is a strategic relapse. Analysts made the calculated capacity, additionally called the sigmoid capacity, to outline the qualities of populace development in nature, developing swiftly and amplifying at the ecological conveying limitation. It is an S-formed bend that can take any obvious esteemed sum and guide it to a provocation somewhere in between zero and one, however on no account to those limits precisely.

$$1/\left(1+e^{\wedge -}\text{value}\right) \tag{20.1}$$

Where e is the base of the natural logarithm and value is the numerical value to be transformed (Mujawar and Devale, 2015).

20.1.2.3 Neural Network

NN is the interconnection of various neurons. The NN works on the same principle as that of the brain and can recognize different structures or patterns.

NN has immense applications running from infection expectation to confront acknowledgment to question location and so on. There are two sorts of neural systems: one is an organic neural system comprising real natural neurons, while the other is a counterfeit neural system used to fathom human-made reasoning (AI) issues (Manogaran et al., 2018).

20.1.2.4 Support Vector Machine

SVM is a regulated learning strategy that dissects data utilized for arrangement and relapse evaluation. A huge amount of arranging information is given, named possessing a spot accompanied with both requests; by then SVM makes a framework that hands over new instances to one target label or the other, creating it as a nonprobabilistic technique for classification. SVM framework is a diagram of the models as space is pinpointed, organized by an undeniable opening as expansive as basic to confine the individual request models. New cases are then masterminded into a similar room, anticipating that they have a spot with a class subject to which class they exactly belong (Sharma et al., 2016).

20.1.2.5 Item-Based Collaborative Filtering

This method is utilized to fabricate the proposed framework for recommendations. In the calculation, the similitude between various things in the dataset are determined utilizing one of a few comparability estimations, and afterward, this closeness esteems are utilized to anticipate client thing pair appraisals that are absent in the dataset. First, the similarity is found between all pairs of items, and hence, the model-building stage is executed. There can be many types of this similarity function—correlation between scores or cosine of those scores. Similarity features can use standardized scores (correcting, for example, for the average rating of each user) (Szummer et al., 2010).

20.2 Literature Survey

Numerous works linked to disease prediction and recommendation systems have been performed in medical centers using various data-oriented and machine-learning techniques.

In a recent study (Sharma et al., 2008), procedures were executed that can anticipate constant sickness by mining the information contained in chronicled wellbeing records using a Decision Tree, Naïve Bayes, SVM, and Artificial Neural Network (ANN). A counterfeit neural model (ANN), regularly just called a "neural system" (NN), is a numerical model dependent on a natural neural system. At the end of the day, it is an imitation of an organic neural framework. In this work, the heart ailment expectation system has been created. All the 15 properties are available in the dataset out of which 13 traits were utilized for forecast, yet this exploration work joined two additional qualities, for example, heftiness and smoking for effective analysis of coronary illness. A relative report is performed on classifiers to quantify the better execution at a precise rate. At first, missing values were recognized in the dataset and they were supplanted with proper values. Further, different information-mining methods have been broken down on the coronary illness database. Disarray lattice is acquired for every classifier. From this test, SVM gives the most elevated exactness rate, though for diabetes Naïve Bayes gives the most elevated precision.

In another significant study (Anooj, 2012) authors developed a decision support system for the risk prediction of heart patients. Proposed framework comprised of two phase, in first phase fuzzy rules are generated and in second phase decision system is developed. Earlier, PC was utilized to fabricate data-oriented medical strategies using a sincerely strong framework, which employs data from clinical authorities and moves these data into PC figurines genuinely. This method is dreary and genuinely contingent on clinical authorities' appraisals, which may be enthusiastic. To manage this issue, AI methodology has been made to obtain data normally from models or unrefined data. Here, a large standard medical decision steady system is displayed for the investigation of coronary ailment thus achieving data from medical information of a patient. The proposed system for heart illness patients includes two aspects: self-operating technique for the time of weighted soft rule and working up a fuzzy rule-based decision. In the chief stage, we have used the mining system, characteristic assurance, and credit weightage technique to secure the weighted soft norms. By then, the cushioned structure is worked according to the weighted soft rules and picked properties.

In a study (Mythili et al., 2013), a model was created for early anticipation of cardiovascular maladies, which further guided in settling on choices to the way of life adapted to high-hazard sufferer and thus declined their intricacies. The examination has endeavored to target the potential cause of coronary illness just as precisely foreseeing the general hazard utilizing homogenous information-mining methods. The late examination has dove into amalgamating these procedures utilizing approaches, for example, half and half information-mining calculations. The authors provided a learning framework to contrast the exactness of appertaining rules with the individual after-effects of the help vector machine, decision trees, and calculated deterioration on the Cardiac Disease dataset to introduce a precise architecture of foreseeing coronary illness.

In yet another novel work (Parthiban and Subramanian, 2008), another methodology dependent on induction framework (CANFIS) was introduced for the assumption of heart-related illness. The proposed model joined the neural system's versatile abilities and the rational subjective methodologies, which is then engulfed with computation to explore the proximity of the malady. The exhibitions of the CANFIS model were assessed with regard to preparing exhibitions and grouping correctness's and the results exhibited that the proposed CANFIS model has phenomenal potential in predicting the coronary ailment. Changes were executed by flipping a piece at a double locus, with the goal that a "0" piece was changed over to a "1," or a "1" piece was changed over to a "0." In this chapter, for the enhancement of the CANFIS model, GA utilized the sequential strategy for parallel sort, roulette-wheel in the choice administrator, tow-point hybrid in the hybrid administrator, and limit in the transformation administrator. Programmed assurance of the chromosome length used to ideally hunt is one of the most significant capacities of NeuroSolution programming. Subsequently, all the chromosomes were naturally set in this product with the goal that they consisted of a number of information neurons and participation capacities, learning rates, and energy.

The investigation (Pattekari and Parveen, 2012) externalized the assessment to create an advanced model using data-oriented methods. It is executed as an online system module in which the customer addresses the pre-set queries. It recoups covered information from the set aside database and complexities the customer regard and arrange the educational file. It can answer complex requests for diagnosing coronary ailment and thus help social protection experts to choose keen clinical decisions, which standard decision genuinely steady systems cannot. By giving feasible prescriptions, also helps with reducing treatment costs. The system can discover and focus disguised data related to ailments (coronary scene, harmful development, and diabetes) from a recorded coronary sickness data set. It can respond to difficult requests for diagnosing sickness and in like manner help human administrations pros to choose canny clinical decisions which traditional decision sincerely strong systems can't. By giving amazing medications, furthermore helps with reducing treatment costs. To improve the portrayal and effortlessness of interpretation, it shows the results in plain and PDF structures.

In a research work (Krishnaiah et al., 2016), system is built up such that it expresses the medicinal services exchanged normally when the clinical determination is finished ordinarily by specialist's information. PC-assisted support module plays a significant assignment in the clinical area. Data mining gives strategy, innovation to change these huge amounts of information into valuable data for dynamic purposes. By utilizing intelligent methods, it takes light effort for the forecast of the ailment with precise exactness. The expanding work on coronary illness foreseeing framework has happened noteworthy to classes the examination results and gives a blueprint of the current coronary illness expectation methods in every classification. Data-mining instruments can respond to exchange addresses that are routinely being used a lot of time abrogating to choose. In this work numerous researches are surveyed, which depicts a computational method of mining and is further utilized for the expectation of coronary illness. Also, it is seen that techniques based on fuzzy logic increment the exactness of the coronary illness expectation framework. Thus, mostly utilized strategies for heart disease prediction are summed up in this work.

The author in the study (Xing et al., 2007) stated that the forecast of endurance of CHD has been a difficult exploration issue in clinical society. This study aims at creating data-mining calculations for foreseeing the endurance of CHD patients dependent on 100 cases. We complete a clinical perception and a 3-month study to incorporate 100 CHD cases. The data of each case is achieved through development. In light of the information, we utilized three mainstream data-mining calculations to build up the forecast models utilizing the 50 cases. The outcomes demonstrated that the SVM is the best indicator with 92% exactness, then neural systems came out to be the second with 90.0 precision, and the decision tree was third with 89.6% exactness. A similar investigation of numerous expectation models for the endurance of CHD patients alongside a ten-overlap cross-approval gave us a piece of knowledge into the relative forecast capacity of various information.

In a recent survey (Dwivedi, 2018), various information-mining calculations are contrasted and investigated all together with discovering the best calculation for illness identification. Machine-learning models such as Tree-based models, NN-based approaches, Nearest Neighbor, Regression techniques, and SVM have been thought about and it has been discovered that logistic regression beat different calculations as far as exactness and other measurable boundaries.

A shrewd proposal framework is created in the examination (Lafta & AL-Araji 2011) that utilizes an imaginative time arrangement forecast calculation to give suggestions to coronary illness patients in the tele-wellbeing condition. In light of the investigation of every patient's clinical tests in records, the framework furnishes the patient with choice help for the need of clinical tests. It can additionally be improved to anticipate the capacity of the calculation to upgrade the accuracy and outstanding task at hand sparing execution. In the recent study (Sinha and Rathi, 2021), analysis on COVID-19 data is done for uncovering the factors that lead to death due to COVID-19.

20.3 Methodology

First, the data are analyzed for the missing values and those values are replaced by the attribute's mean/median/mode accordingly. After preprocessing, a 7:3 split is done on data for train/test. After this, the following two systems are developed for the working of the proposed model. The first system works on heart disease prediction, which depends on details of the user while the second one recommends the diet and exercise based on the predicted outcome of heart disease probability and details entered by the user with regard to other disease records.

20.3.1 Predictive Model

The following four algorithms have been implemented at first. Datasets have been trained individually for all algorithms. All of them were tested after that.

- SVM
- Decision tree
- Logistic regression
- NN

The algorithms were trained using the University of California Irvine data set; 7:3 split is done on data for train/test. In addition, some steps have been taken to improve the accuracy of the algorithms. These steps include preprocessing of the dataset.

Based on the statistical calculations SVM outperform other techniques and produce accuracy approximately equals to 88%. It was along these lines picked for the primary application. The primary application is a Web application that acknowledges the client's various boundaries as information and ascertains the result. The outcome will be shown alongside the forecast exactness.

UCI repository is used for the comparison of different classification methods. The dataset contains 76 attributes of which 13 attributes are utilized for this analysis as depicted (Table 20.1).

TABLE 20.1

Dataset Details

No.	Area	Details	Type	Interval/Values
1.	Age	Patient's age	Continuous	0–100 years
2.	Gender	Patient's gender	Discrete	0 for Female & 1 for Male
3.	Chest Ache	Chest ache type	Discrete	1-Normal angina 2-Aberrantl angina 3-Non-angina 4-asymptotic
4.	BP in the course of rest	for the time of rest	Discrete	mm Hg
5.	Cholesterol	Serum Cholesterol	Continuous	mg per dL
6.	Fasting blood sugar	Blood sugar before meal if >120 mg/dL	Continuous	0 for False & 1 for True
7.	Electro-cardiogram	Results for resting ECG	Discrete	0-Normal, 1-STT Wave
8.	Max heartbeat	Max heart rate	Discrete	Beats/min
9.	Exercise infused angina	If pain has been infused by workout	Continuous	1-Positive, 0-negative
10.	Earliest peak	ST-segment depression infused by workout	Continuous	0–4
11.	The slope of peak exercise	Peak slope workout ST segment	Discrete	1 for raised slope, 2 for unvarying slope&3 for lower slope
12.	Ca	Fluoroscopy colored holder	Continuous	0–3
13.	Thal		Discrete	3 for standard, 6 for decided flaw & 7 for reversible flaw
14.	Num	Cardiovascular illness diagnoses (result)	Discrete	0, 1 (0: <50% 1: >50%)

Inputs: Data set, user data
Outputs: Result, accuracy, recommendations
Algorithm

1. Start
2. Input the details
    i. Check the format of details
    ii. Process the details
    iii. Ignores spaces as delimiters
3. Train dataset
4. Prediction of the result using SVM algorithm
5. Display result
6. End

Figure 20.1 depicts the overall working of the algorithm pictorially.

20.3.2 Recommendation System

A Web-based application is designed using the Shiny package of RStudio. The recommender recommends a suitable diet (i.e., foods) and exercises which one should take to lower the chances of risk.

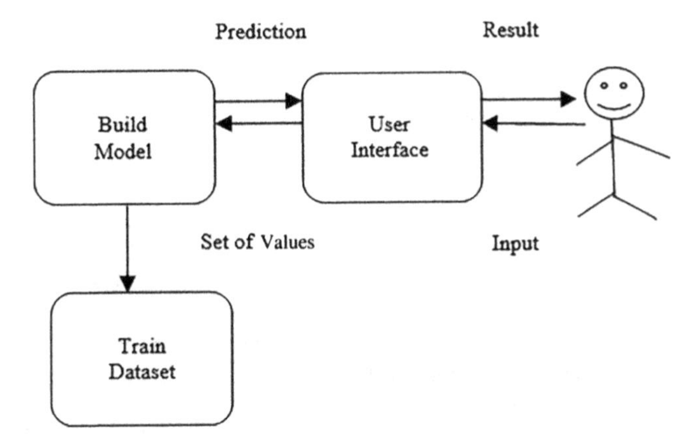

FIGURE 20.1
Architecture diagram.

Thing-based shared sifting is utilized as the suggestion framework calculation. Thing-based agreeable separation for proposals is a model-based calculation. In the calculation, the similitudes between various things in the dataset are determined utilizing cosine closeness estimations, and afterward, this comparability esteems are utilized to anticipate client thing pair appraisals that are absent in the dataset. To begin with, the likeness is found between all sets of things and, thus, the model structure stage is executed. As in client frameworks, similitude capacities can utilize standardized appraisals (rectifying, for example, for every client's normal rating).

Second, the framework executes a proposal stage. The most comparable thing to a client as of now evaluated things is utilized, to produce suggestions. This is generally straight relapse or the weighted total. This sort of exhortation is like "people who emphatically rate thing *X*, similar to you, likewise will in general firmly rate thing *Y*, and you have not yet rate thing *Y*, so you should give it a shot."

Collaborative filtering of items is more efficient than cooperative filtering between user and user. Moreover, its less dynamic model was less frequently calculated and stored in a narrower matrix, so the output of the item–item scheme was better than the user–user scheme (Table 20.2).

```
Pseudo-Code for Recommendation System
function recommend (a,d,t) {
datanew = data(-id,-items)
if(a,d,t==0) i,j,k=0 else 1
if(svmPrediction==NotDisease) h=0 else h=1
tes = (h,i,j,k)
function Cosinefun (X,Y)
{
return ((X.Y) / (|X|*|Y|))
}
similarity = matrix(rows=84,cols=2)
for(i=1:rows)
{
similarity[i,] = Cosinefun(tes,datanew[i,])
}
similarity[,1] = (1:84)
```

TABLE 20.2

Recommendation Dataset

Id	Items	Heart	Diabetes	Asthma	Thyroid
1	Acerola	Yes/1	Yes/1	Yes/1	Yes/1
2	Apple	Yes/1	Yes/1	Yes/1	Yes/1
3	Apricots	Yes/1	Yes/1	Yes/1	Yes/1
4	Avocado	Yes/1	Yes/1	Yes/1	Yes/1
5	Banana	Yes/1	No/0	Yes/1	Yes/1
6	Blackcurrant	Yes/1	No/0	Yes/1	Yes/1
7	Breadfruit	Yes/1	Yes/1	Yes/1	No/0
8	Eggplant	Yes/1	Yes/1	Yes/1	No/0
9	Carambola	Yes/1	Yes/1	Yes/1	No/0
10	Padangusthasana	Yes/1	Yes/1	No/0	No/0
11	Janu Sirsasana	Yes/1	Yes/1	Yes/1	No/0
12	Supta Padangusthasana	Yes/1	No/0	No/0	Yes/1
13	Setu Bandha Sarvangasana	Yes/1	No/0	Yes/1	Yes/1
14	Dhanurasana	Yes/1	No/0	No/0	Yes/1
15	Bhujangasana	Yes/1	Yes/1	Yes/1	Yes/1
16	Paschimottanasana	Yes/1	No/0	Yes/1	No/0
17	Sarvangasana	Yes/1	Yes/1	No/0	Yes/1

```
similarity[order(similarity[,2])]
index = similarity[1:8,1]
for(k = 1:cols(index))
{
for (l = 1:rows(data))
{
if(index[1,k]==data.id[l])
{
print(data.items[l])
}}}}
```

20.4 Results

Datasets have been trained individually for all algorithms. All of them were tested after that. Based on different criteria, the most efficient algorithm was to be selected. We found out that the SVM algorithm was the most efficient out of the four with an accuracy of 87.64%. Decision Tree, Logistic Regression, and NN had an accuracy of 83.14%, 84.52%, and 79.78%, respectively. Thus, SVM algorithm was further implemented in the form of a Web application using a better user interface. R studio's Shiny Web tool has been used for this. This would help end-users get a preliminary forecast of their heart condition. Since heart disease is a major killer in India and around the world, the application of promising technology such as machine learning to initial heart disease prediction will have a profound impact on society (Figure 20.2).

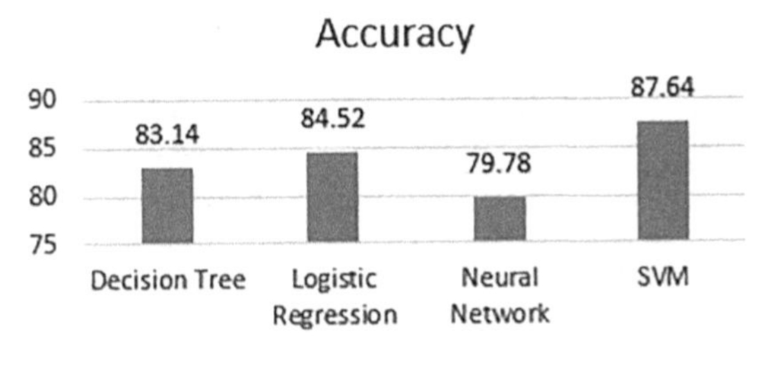

FIGURE 20.2
Accuracy graph.

The accuracy achieved by SVM is 87.64% and the precision matrix is as summarized in Table 20.3.

The accuracy achieved by logistic regression is 84.52% and the confusion matrix is summarized in Table 20.4.

The accuracy achieved by decision trees is 83.14% and the confusion matrix is summarized in Table 20.5.

The accuracy achieved by NN is 79.78% and confusion matrix is summarized in Table 20.6 (Figure 20.3).

The evaluation criteria carried out was comparing the accuracy in the two scenarios; one, considering all the variables, and second, considering significant variables. The significant variables obtained by logistic regression are shown in Figures 20.4–20.7. The other evaluation criteria carried out was generating the ROC curve for all variables and the significant variables, respectively.

TABLE 20.3
SVM Model Precision Matrix

Forecasted/Real	0	1
0	48	7
1	4	30

TABLE 20.4
Logistic Regression Precision Matrix

Forecasted/Real	0	1
0	41	6
1	7	37

TABLE 20.5
Decision Tree Precision Matrix

Forecasted/Real	0	1
0	49	12
1	3	25

TABLE 20.6
Neural Network Precision Matrix

Forecasted/Real	0	1
0	45	12
1	7	30

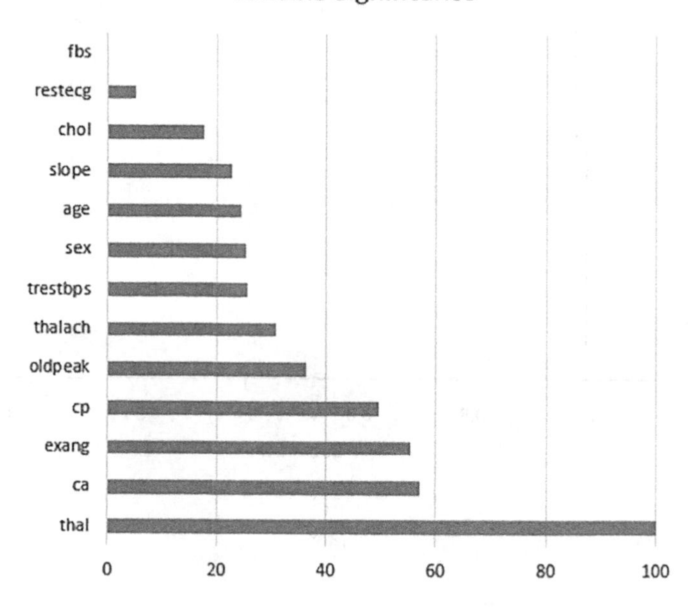

FIGURE 20.3
Significant variable.

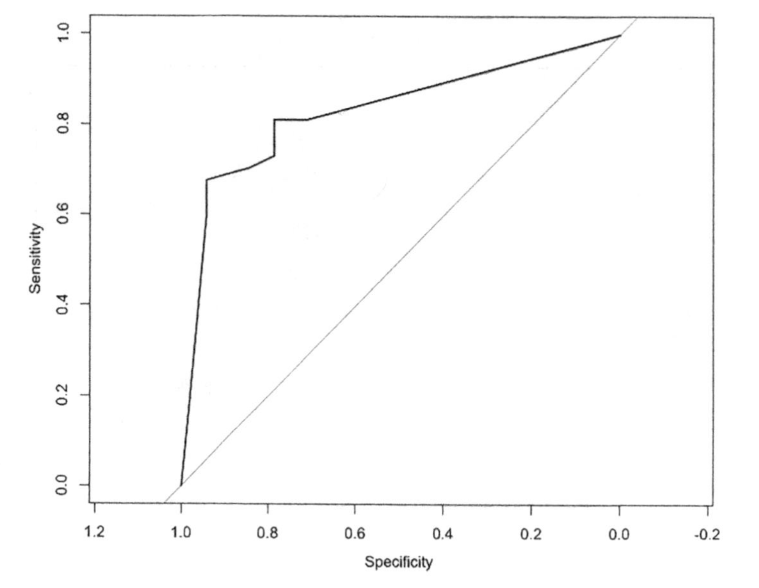

FIGURE 20.4
SVM–ROC curve.

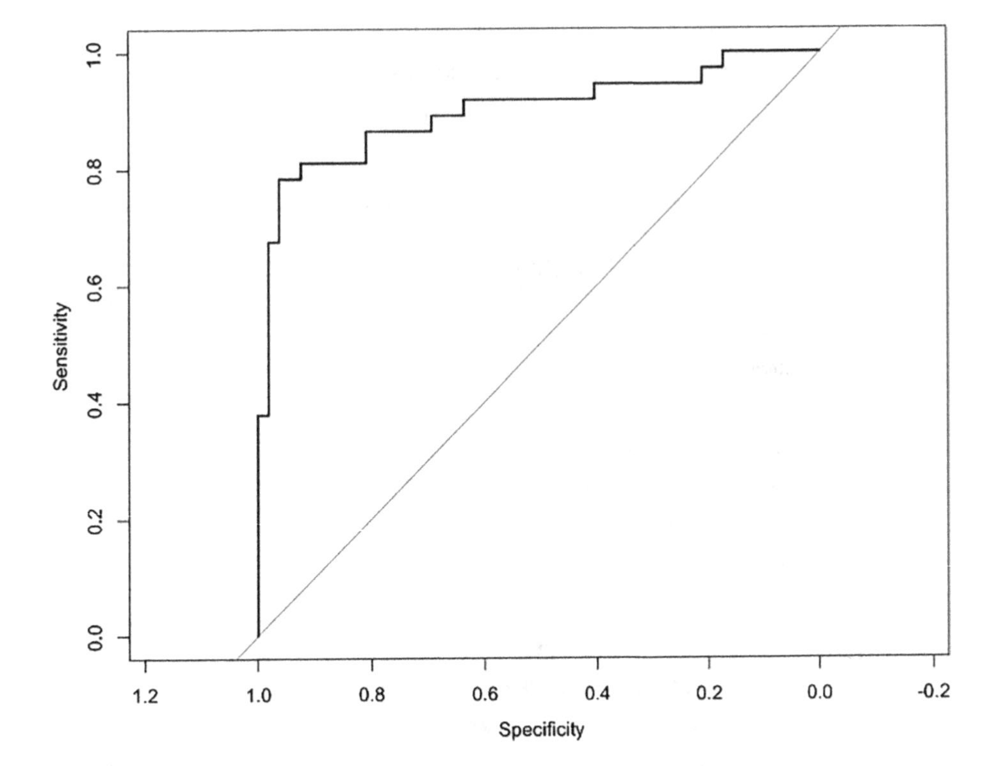

FIGURE 20.5
Decision tree ROC curve.

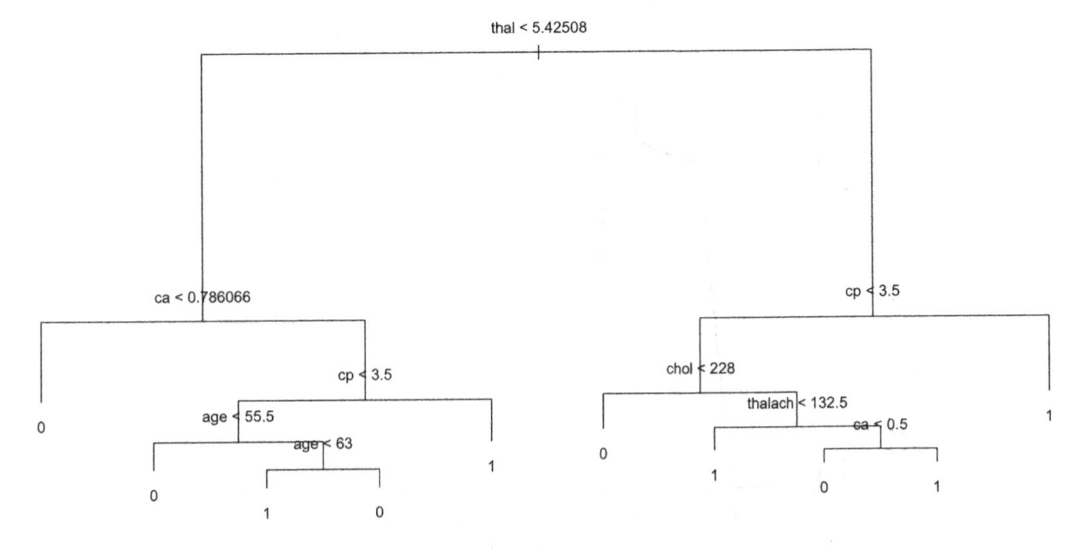

FIGURE 20.6
Pruned tree.

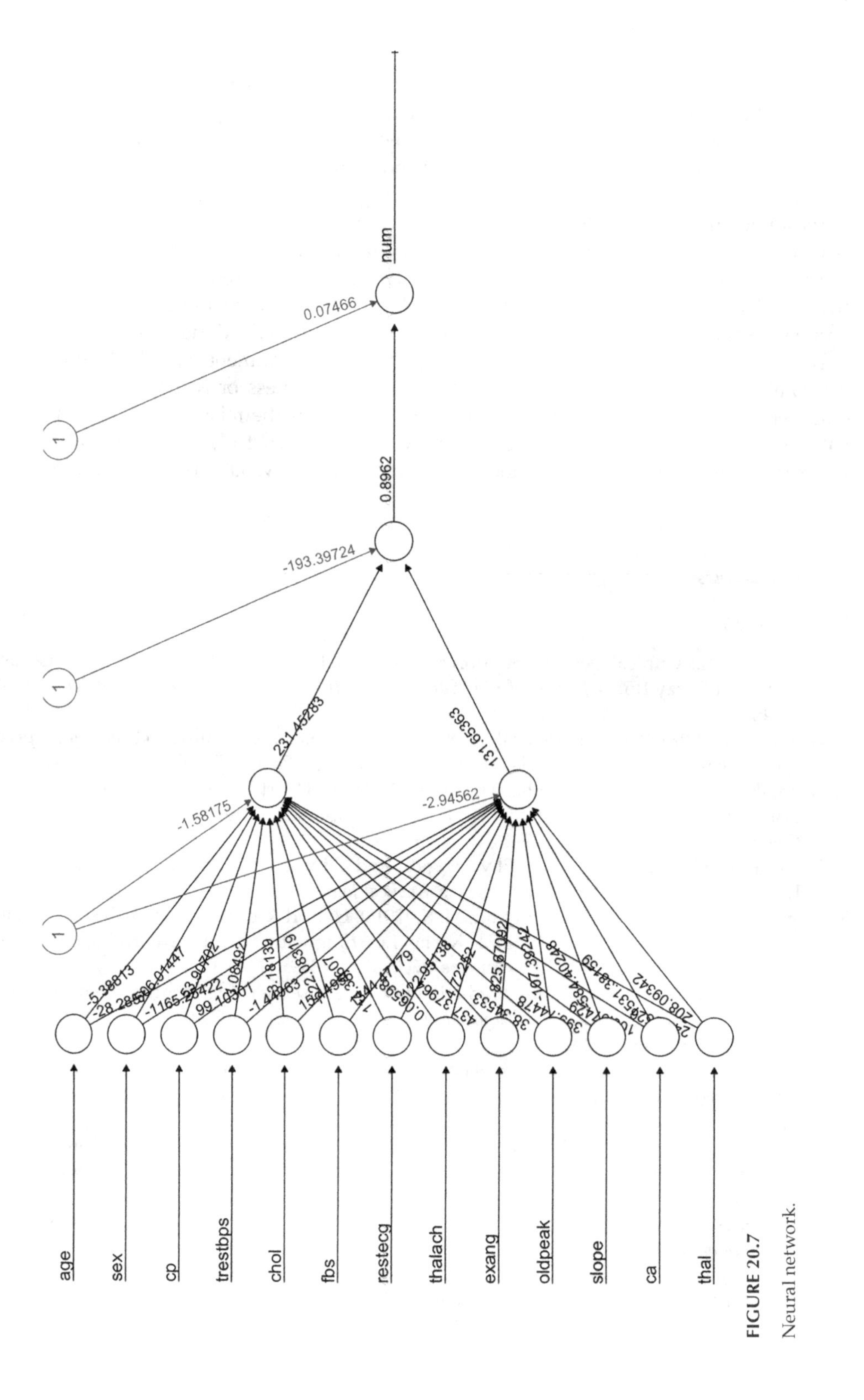

FIGURE 20.7
Neural network.

20.5 Conclusion

This exploration chapter is chiefly engaged to foresee coronary illness probability and suggest diet and exercise dependent on the patient's wellbeing profile, utilizing information mining or AI approach to improve the exactness or accuracy of the infection location master framework. This study additionally shows the related work investigation of various methodologies, for example, NN, logistic regression, SVM, and decision tree, and it is reasoned that SVM performs best when contrasted with the other existing strategies. Because of the study, the proposed calculation is planned to utilize SVM calculation and the trial result shows the precision of 87.64%. Thing-based shared separating is utilized as the calculation for the suggested framework. It suggests suitable nourishments one should admission and activities one should rehearse to bring down the danger of coronary illness.

This exploration chapter is primarily engaged to foresee ailment plausibility utilizing the information mining or AI approach to improve the exactness or accuracy of the infection discovery master framework. This chapter likewise shows the related work investigation of various methodologies, for example, NN, SVM, Naïve Bayes, fully Convolutional Network, K-Nearest Neighbors, and so forth, and it is presumed that SVM outperforms other algorithms.

References

Anooj, P. K. (2012). Clinical Decision Support System: Risk Level Prediction of Heart Disease Using Weighted Fuzzy Rules. *Journal of King Saud University-Computer and Information Sciences*, 24(1), 27–40.

Dwivedi, A. K. (2018). Performance evaluation of different machine learning techniques for prediction of heart disease. *Neural Computing and Applications*, *29*(10), 685–693.

Krishnaiah, V., Narsimha, G., & Subhash Chandra, N. (2016). Heart Disease Prediction System using Data Mining Techniques and Intelligent Fuzzy Approach: A Review, *International Journal of Computer Applications*, 136(2), 43–51.

Lafta, S. H., & AL-Araji, S. M. (2011). Physiological and Clinical Importance of Calcium Score and Risk Factors in Coronary Artery Disease. *Medical Journal of Babylon*, *8*(4).

Manogaran, G., Varatharajan, R., & Priyan, M. K. (2018). Hybrid Recommendation System for Heart Disease Diagnosis Based on Multiple Kernel Learning with Adaptive Neuro-Fuzzy Inference System. *Multimedia Tools and Applications*, 77(4), 4379–4399.

Mujawar, S. H., & Devale, P. R. (2015). Prediction of Heart Disease using Modified K-means and by using Naïve Bayes. *International Journal of Innovative research in Computer and Communication Engineering*, 3, 10265–10273.

Mythili, T., Mukherji, D., Padalia, N., & Naidu, A. (2013). A Heart Disease Prediction Model Using SVM-Decision Trees-Logistic Regression (SDL). *International Journal of Computer Applications*, 68(16).

Parthiban, L., & Subramanian, R. (2008). Intelligent Heart Disease Prediction System Using CANFIS and Genetic Algorithm. *International Journal of Biological, Biomedical and Medical Sciences*, 3(3), 157–160.

Pattekari, S. A., & Parveen, A. (2012). Prediction System for Heart Disease Using Naïve Bayes. *International Journal of Advanced Computer and Mathematical Sciences*, 3(3), 290–294.

Sharma, B., Agarwal, S., Karbowitz, S., Gumpeni, R., Lee, P. C., & Paul, S. (2008). A man with end-stage renal disease and dyspnea. *The Journal of Respiratory Diseases*, *29*(8), 302.

Sharma, P., Saxena, K., & Sharma, R. (2016). Heart Disease Prediction System Evaluation Using C4. 5 Rules and Partial Tree. In *Computational Intelligence in Data Mining*, (Vol. 2 pp. 285–294). Springer, New Delhi.

Sinha, A., & Rathi, M. COVID-19 Prediction Using AI Analytics for South Korea. *Applied Intelligence*, pp. 1–19, 2021, doi: 10.1007/s10489-021-02352-z.

Szummer, K., Lundman, P., Jacobson, S. H., Lindbäck, J., Stenestrand, U., Wallentin, L., & Jernberg, T. (2010). Cockcroft-Gault Is Better than the Modification of Diet in Renal Disease Study Formula at Predicting Outcome after Myocardial Infarction: Data from the Swedish Web-System for Enhancement and Development of Evidence-Based Care in Heart Disease Evaluated According to Recommended Therapies (SWEDEHEART). *American Heart Journal*, 159(6), 979–986.

Xing, Y., Wang, J., & Zhao, Z. (2007, November). Combination Data Mining Methods with New Medical Data to Predicting the Outcome of Coronary Heart Disease. In *2007 International Conference on Convergence Information Technology (ICCIT 2007)* (pp. 868–872). IEEE.

21

Sports Analytics for Classifying Player Actions in Basketball Games

Ayush Gupta, Ekam Singh Chahal, Ankur Haritosh, Adwitiya Sinha, and Satish Chandra

Jaypee Institute of Information Technology

CONTENTS

21.1 Introduction

Computer vision is one of the most promising fields in computer science. The application of computer vision is visible in everyday life. Retail, driving, healthcare, agriculture, finance, manufacturing are the areas that have gained a lot of attention from researchers. But one of the untapped fields is sports analytics. Sports analytics amalgamates many AI subproblems, such as human-pose estimation, image reconstruction, object detection, and tracking.

Today sports has become an enormous industry with an estimated value of US$80 billion. Games, such as football, basketball, hockey, and baseball, have gained a lot of interest from people around the world. Because of the mass appeal sports has gathered in recent times, stakes are very high in every game. Each team wants to gain an upper hand over their opponent. Gaining an advantage over an opponent is no longer only about having better stamina or an enhanced physique or positioning of players on the field (Li et al. 2018).

DOI: 10.1201/9781003046431-21

Sometimes an edge is achievable by observing the minute pattern in the player's movements during the match (Ludl et al. 2019). That is the reason today that numerous football and basketball clubs have a separate analytics division. Recording the minute details leads to a better algorithm. We do not think these games will become solely about the algorithm's ability to crunch data and provide results. Humans will continue to play sports, and skill, talent, and enthusiasm will always be paramount. The uncertainty makes sports exciting, the unpredictable moves that no machine can ever tell. In our work, we have analyzed basketball match videos to identify player actions.

As far as the results are concerned, the convolutional 3D network (C3D model) proved to be more suitable for this type of task by offering an average accuracy on the test of 98%. On the other hand, the fully connected neural network (FCNN) model trained on pose estimations to classify the actions of players using their joint coordinates obtained via OpenPose has achieved an accuracy of 58% on the test data, making it less suitable from the point of view of accurate results than the C3D approach. Finally, the Hybrid model, which integrated the outputs of the two models, resulted in minor gains, bringing the total accuracy to a maximum of 99%. Thus, preparing data for testing videos using court detection and player tracking has excellent results, despite not using any trained model for it as we did for other tasks.

21.2 Related Work

We have presented research work related to pose estimation and player action classification in a video by different authors in this section.

A much basic and highly efficient approach for learning spatial–temporal features using a three-dimensional convolutional network, also known as 3D convnets, when trained on large-scale datasets containing videos is proposed in Tran et al. (2015). Their approach achieved a state-of-the-art accuracy on four different tasks: action recognition, action similarity labeling, and scene and object detection. They discovered that 3D convnets are much more relevant for any video-related jobs than 2D convnets because they simultaneously learn appearance and motion for a given video input. By experimenting with the depth parameter in kernel size, they discovered that using smaller and homogeneous receptive fields (kernel size) of $3\times3\times3$ throughout the network produces the best performing architecture for 3D convnets for video-related tasks. Their proposed model has a frame rate of 313 compared to other current models with a frame rate of 4, but it is less accurate when trained on larger datasets.

A novel model for the detection of 2D poses of multiple people in a single image was proposed by Cao et al. (2017). The proposed model beats other state-of-the-art models for real-time estimation in almost every field, i.e., accuracy, accurate positive detections, and run-time complexity. The model's originality takes a bottom-up technique for pose estimation. It uses a new representation of images in the middle layers of the network called part affinity fields (PAFs), a nonparametric representation. These PAFs make the whole process much easier to learn how to associate different body parts with a person in the given image and make real-time pose estimation feasible at the same time.

A system to track the movements of ten players using basketball game videos was proposed in the study by Cheshire et al. (2013). First, a player's position is tracked and then mapped onto a basketball court image. A custom dataset of several YouTube videos was employed, with only those selected showing all-important court lines. The modules of their system include Court Detection, Individual Detection, Color Classification, Player Tracking, and Mapping. The plan was built using hog detectors and color segmentation from the OpenCV library. For every frame, the boxes of players and their positions are stored. In addition, player positions during various game circumstances are taken into account for tracking. Hence, in a basketball arena, individual players can be identified and mapped.

Redmon et al. (2018) highlighted improvements made to earlier versions of the Yolo model, an object detection model. The locations of the objects are detected by predicting their bounding boxes and also indicating the object's class using the Yolo model. Yolov3 employs the Darknet-53 architecture to extract features from input photographs containing things to be detected and additionally providing the model with the bounding box coordinates of all objects in the frame. First, three anchors are estimated using the k-means clustering algorithm from the training dataset. Each anchor contains the height and width of the bounding box on three different scales. Therefore, a total of nine values are estimated. The author treated object detection as a logistic regression problem. The model predicts five values, i.e., center coordinates x and y, width and height offset of the bounding box, and the confidence score that the bounding box contains an object. Also, the probability scores of the things belonging to each class are calculated and can be calculated for three different scales. Finally, the type of an object is determined by calculating a conditional probability based on the confidence score. The bounding box with the value of this conditional probability higher than a threshold is selected and displayed as output. The author used the COCO Dataset to determine the model's performance and compared the results to other models. Although the model achieves a relatively low processing time per frame, its mAP value is lower than other models.

Real-time detection and tracking of basketball players in basketball match videos are proposed by Acuna (2017) using the Yolov2 object detection model and the SORT object tracking framework. The proposed model uses a deep learning-based method. First, using the Yolov2 object detection model, player detection is achieved. The Yolov2 model's architecture adjusted the number of filters to training it on the NCAA Basketball Dataset. To track players, they used the information from the previous and current frames. Each object's inter-frame displacement was approximated. Each player state was determined by a vector that contains the horizontal and vertical center pixel of the player, scale, and aspect ratio of the player bounding box detected by the Yolov2 model. A cost matrix is estimated using IOU as the distance to correctly associate players between current and previous frames. Players having IOU less than a threshold are not considered. Finally, precision–recall curves are estimated using the two criteria to evaluate the performance of the model. Based on the first criteria, i.e., $IOU > 0.5$ AP, achieved by the model is 0.89, and based on the second criteria, i.e., $IOU > 0.7$, achieved by the model is 0.63 AP. The tracking part of the model lost the tracking of the player during the cases when two or more players are occluded at the same place. These occurrences resulted in a significant failure in tracking a section of the model that can be improved.

A model for pose estimation, specifically for crowded places, was proposed by Golda et al. (2019). A mixture of two datasets was utilized for training the model: real-time dataset CrowdPose and artificially produced JTA. Crowd Pose contains 20K images and 80K human poses with 14 labeled vital points. At the same time, JTA was created by extracting videos from the video game GTA V. A combination of single-person pose estimator using transposed convolution and ResNet50 network as a backbone was adapted for the model. This combination is simple and performs better than other models. In the first step, they detect every object, i.e., the human in the picture. Then every object's pose was determined by mapping key points. Similarity (OKS) of CrowdPose was used for the evaluation metric Object Key Points. The main disadvantage of this measure is that it accurately depicts crowded photos taken from a shallow viewing angle but fails to reflect the crowdedness of images taken from a higher viewing angle. Relying on self-made data is also not sufficient, and there is a need for more real-time data of human crowded spaces to bridge the gap. Their model resulted in more key points and legitimate estimates for occluded ones.

In another contribution, an architecture was developed for using convolutional networks in typical pose machines to estimate a pose of a person in real-time by Wei et al. (2016). In standard pose machines, belief maps were produced using simple classifiers, but these classifiers were replaced with convolutional networks by the paper's authors. Convolutional networks assist the model in realizing the spatial relationships between different body parts by increasing the size of receptive fields, which is directly connected with model accuracy. Each level generates a belief map that passes through the additional network and then serves as an input to the next stage; this cycle continues until the final output layer. Each step refines the likelihood of the body component being in that specific position. The authors of the research also used intermediate supervision to avoid the vanishing gradient problem during backpropagation. As a result, the proposed model achieved a state-of-the-art accuracy on different pose estimation datasets.

Shuai et al. (2019) proposed an action classification model for videos based on 2D coordinates derived from an instantaneous pose estimation model. These 2D coordinates of body key points are prominent features that can aid in classifying actions. Joint coordinates of a person in "*f*" different number of frames were derived, and coordinates were accumulated over "*f*" number of frames to construct an accumulated matrix, which acts as an input to the action classification model. A simple CNN neural network consisting of only three convolutional layers, a pooling, and a classification layer for classifying the actions of a person, was utilized as the model architecture (Ren et al. 2015). The proposed architecture outperformed many other state-of-the-art action classification models on the KTH and UCF50 datasets.

Cai et al. (2006) proposed a multitarget tracking system to track hockey players. This system is built using a particle-filtering approach. A rectification method was used that calculates the relationship between the coordinates of a hockey player and video frame that at the same time enhances player dynamics as well as handles variability of camera motion. Particle filter distribution tracks were added, with boosting detections from the global nearest neighbor data-association algorithm. The linear optimization model can add the boosting detections to current tracks optimally. While ongoing mutual occlusion, the mean-shift algorithm was used to stabilize the trajectories of robust tracking targets. Their results conclude that the algorithm can track various targets and maintain their identities without any effect of camera motion, mutual occlusion, or background clutter.

Another approach to identify and track basketball players in a single feed video of a basketball match was proposed in the research by Sangüesa et al. (2019). The suggested method analyzed various players in different frames to find a similarity match, allowing for the tracking of the same player. The study used video frames as the model input, then detected all the players inside the court (by utilizing a court segmentation algorithm) and estimated their poses using a pretrained state-of-the-art model such as OpenPose. Later, they compared the detected players using three different methods. They compared distances between their respective bounding boxes, pose similarities, and deep features derived using a pretrained VGG-19 model on imagenet. The study demonstrated that using basic and traditional computer vision techniques improves the overall performance of the tracker.

To the best of our knowledge, there is no research on categorizing player actions using a model that uses both the player's visual and posture coordinates. In this work, we present a Hybrid model that uses both a player's visual features and the coordinates of a single frame to identify its action label throughout the whole video.

21.3 Methodology

Figure 21.1 represents the flowchart of our proposed method. First, OpenPose extracts 18 human points from an input video. It does so for each player in each frame. The data generated are utilized by court segmentation and player tracking algorithms. We then derive a patch for the court in court segmentation and use the player's foot coordinates lying on

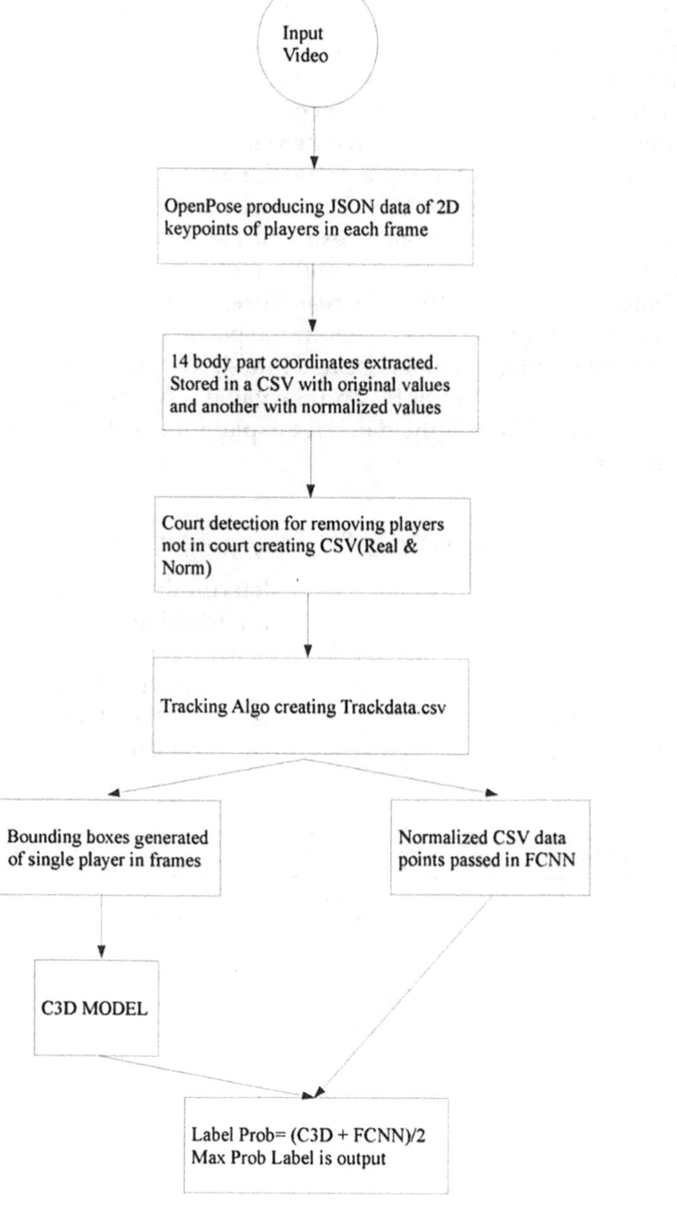

FIGURE 21.1
Flowchart of the proposed method.

that zone for further processing by player tracking. The coordinates obtained from court segmentation are sent to player tracking, where it assigns a unique id to each tracked player. It then provides tracked player data to C3D and FCNN, which use it to identify player action by comparing each player's picture across multiple tracked frames.

21.3.1 Dataset

We trained models using a labeled action classification dataset (described in Section 21.3). Dataset can be downloaded and accessed online at the given link https://github.com/simonefrancia/SpaceJam (accessed on 1/03/2020). The whole dataset consists of 332 different datasets having names—clip dataset and joints dataset, each containing a total of 37,085 samples, and together with these datasets, we also have a dictionary named "annotations.json," where the key is some ID and their corresponding action label ([0, 9]).

Clip dataset consists of videos centered on only a single basketball player, each having a duration of 3–4 seconds, performing a particular action for an interval of a total of 16 RGB frames stored as .mp4 extension. This dataset was created using HD basketball game videos, by carefully removing anomalous examples and padding each frame to a particular dimension of height 176 pixels (y-axis) and width 128 pixels (x-axis). Videos used had a frame rate between 25 and 30 fps. Whereas, corresponding joints dataset contains information on coordinates (x, y) of the player in the particular frame. These coordinates were derived using CPM (convolutional pose machines). The output of a CPM is a dictionary of 14 joints where the key is the joint ID, and the value is (x, y) coordinate.

Tables 21.1 and 21.2 quantify the data and explain the features/characteristics of different individual classes.

21.3.2 Use of OpenPose for Getting Player's Coordinates

We have utilized the OpenPose Library for determining the players' coordinates in the video, which serves its main purpose in player tracking. OpenPose has trained on the COCO, COCO+foot, and MPII dataset. MPII dataset consists of training data with 3,844 samples and testing data with 1,758 image samples having 14 distinct parts of the body. The Foot dataset contained around 15,000 images. OpenPose for every player in each frame extracts 18 points of the human body form, which has utilized only 14 points, as the rest of

TABLE 21.1

Dataset and Label Information

Label	Action	Number of Samples
0	Block	996
1	Pass	1,070
2	Run	5,924
3	Dribble	3,490
4	Shoot	426
5	Ball in hand	2,362
6	Defense	3,866
7	Pick	712
8	No action	6,490
9	Walk	11,749
Total		37,085

TABLE 21.2

Description of Labels

Action	Features/Characteristics
Block	The player intercepts the opponent player
Pass	Exchanging of the ball within the same team
Run	Players only run without performing any other action
Dribble	The player keeps the ball in the whole time frame
Shoot	The player taking a shot
Ball in hand	Possession of ball and performing no action
Defense	The player maintains a folded position on his or her legs in a defensive position
Pick	The player without a ball makes a block on an opponent to free a partner, who may or may not have the ball
No action	No action (player is ideal)
Walk	Player simply walks

the 4 points are redundant and of no benefit for our task. It can detect Face, Foot, and Hand along with detecting overall body points. Thus, available in two models, Body Training and Whole Body Training. Additionally, the results are available for 2D and 3D images/videos.

In their method, they passed an input frame through a CNN model. The backbone of the CNN is VGG-19, with the first ten layers used through transfer learning. It produces 2D confidence maps of the location of body parts. Their approach requires a nonparametric representation, which is also called part affinity fields (PAFs), for mapping body parts and players in a frame. PAFs are bottom-up representations of association scores, which store the location and limb orientations in a frame. These confidence maps, along with PAFs, are parsed to form 2D key points. The points extracted can be saved in JSON format. To train for Foot points, they have used the Human Foot Keypoint Dataset.

After extracting 18 body points of a particular individual in a given frame and using only 14 of them, we may describe temporal connections between past and present frames using these points for each unique individual. Using the previous frame points for a particular player, we can track its position in the current frame over a 2D plane. We tracked 14 points with the past frame for each player to create a possible mapping with the closest identified player from previous frames. This mapping continues until the end of the video for all players. Hence, this results in the tracking of each player over the entire video length (Figure 21.2).

21.3.3 Court Segmentation to Optimize Number of Detected Players

In court segmentation, we aim to segment the basketball court area as a continuous patch of white, and the rest of the background in black color. Then we try to minimize the number of players detected using OpenPose by choosing those players whose feet are on the segmented court.

For implementation, first, we converted the color space of the image from RGB to HSV. This color space represents an image using hue (which determines the dominant color), saturation (which determines the shade in color), and value (which determines the brightness in color). The reason why we convert the color space is that HSV color space is more robust toward minor changes in external lighting (like shadows, dullness, and so on), but these changes result in significant differences in RGB values. After converting to HSV color space, we manually selected the HSV values for the court. This value is then subtracted and added by some constants to produce a range of HSV color space that represent

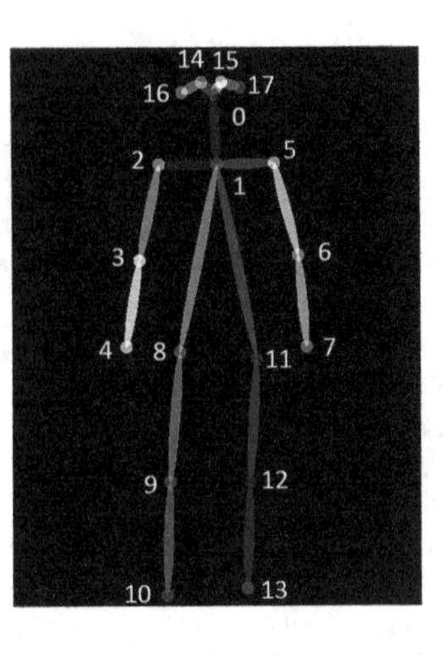

FIGURE 21.2
18 Body parts estimation points.

minor changes (shades, brightness) in the court. Then a mask is generated of the court. A pixel value of more than 0 represents the court. This mask consists of dark patches that occur because of the presence of players, different colored areas of the court, lines present in the court, etc. Dilation is employed to fill these dark patches and achieve an approximate continuous patch of court.

The resultant court segment (patch) generated is then used to filter out the persons detected by OpenPose. In this, we take the foot coordinates of all the people found by OpenPose. If the foot's coordinate lies on the court segment, then that person is selected as a player, otherwise, we discard it.

21.3.4 Player Tracking

Automatic tracking of objects in a sequence of images is one of the most challenging ongoing researches in the computer vision community. Object tracking is a problem of tracking the same object as it moves in a series of frames of a video (Karpathy et al. 2014). The player is considered an object in this method. Player tracking has several challenges. Because of low-resolution images, face recognition is not possible here, even for humans to recognize players. We could use face recognition only if multiple cameras are available throughout the field. Also, it is difficult to identify the jersey numbers of players. Continuous orientation and pose changes of players along with deformed jerseys are some of the issues. It's tough to get a frontal view of jersey numbers or players' faces with one camera. Apart from these, occlusion creates problems in detection when some players are closer to each other. They are harder to identify, even harder to track in the mixup. There are certain approaches to address such issues while tracking the player position (Ballan et al. 2007; Lu et al. 2011; Shitrit et al. 2011; Ye et al. 2005). In our method, we have approached this problem of the player uniquely and devised an algorithm to solve this problem which is presented below.

Player Tracking is a principal issue that demands the estimation of distinct human-pose values and combining them momentarily across various video frames. Player tracking cannot be done without first identifying the players per frame in a given video. We need to identify every single player having a unique identity across all frames. Once all the players' positions are extracted, we can move forward toward our tracking algorithm. To extract these points is a difficult task because of the problems discussed earlier.

We have utilized the OpenPose Library in this method. Also, we used the players detected by court detection for tracking. In the zeroth frame of the video, each player was assigned an id (from 1 to the number of players). We processed the current and previous frames in successive iterations and mapped them from the previous frame to the current frame. If there are no players detected in a particular frame then this frame is not processed further.

We used the players' pose coordinates to map the same player between frames (estimated using OpenPose). The intuition behind using pose coordinates for tracking was that pose coordinates for players between frames will not change much. First, we constructed a distance matrix for all the detected players in frame i with all the players in frame $i+1$. For this, a row of the matrix represents details of the previous frame, and columns represent the details of the current frame. For each row, we calculated the minimum distance. Then, a dictionary was used for mapping row numbers to column numbers containing the minimum value only if it was less than 15.0. For computing distance between two players, the coordinates for neck, shoulder, elbow, hip, and knee are used, and then the corresponding coordinate is only selected if it was detected for both the players. By using this mapping, the relevant ids of the previous frame's players are assigned to the current frame's players. The new ids are assigned to the players in the current frame, which have no mapping with any previous frame's players. Then, using the player ids, another dataset is created. All players tracked for 16 frames in a row are chosen and added to this new dataset. This dataset's rows contain the range of frame numbers (0–15). Each row contains the information of a specific player. This information includes the player's id and positional index in these 16 frames. Then, this information is passed to the C3D algorithm for further processing.

21.3.5 Proposed Model Consisting of C3D Model and FCNN

This section will explain the different deep learning models used for the problem along with their architectures. We used two separate models for the problem: a variant of C3D and a simple FCNN, and created a Hybrid model that uses both of these models to classify player actions in a basketball video.

21.3.5.1 C3D

In this section, we will elaborate on details of the C3D model used. C3D is a short abbreviation for a 3D convolution network and uses 3D convolutional and 3D polling layers with a kernel size of the depth of 3. 3D convnets are networks that use a space–time learning approach, which will be useful for action classification for a particular player throughout frames. It has been discovered experimentally that these models learn appearance features in the initial layers and temporal features in the later stages of the network. These types of networks are preferred over 2D convnets because of their ability to learn space-temporal features, whereas 2D convnets are only capable of learning spatial features (space features) as temporal information is lost in them. For our problem, we have fine-tuned the C3D model pretrained on the Sports1M dataset. In the process. of fine-tuning, we have removed the last classification layers in the original network and added our classification layers with the number of output neurons of 10 instead of 487. Sports1M dataset is one of the most prominent datasets used for sports video classification; it contains about 1.1 million videos of 2 seconds each. In this dataset, each particular video belongs to one of the 487 classes of different sports. Authors of the C3D model have trained their model on this dataset with (16, 112, 112) random crop sizes, SGD as the optimizer, batch size of 30, and initial learning

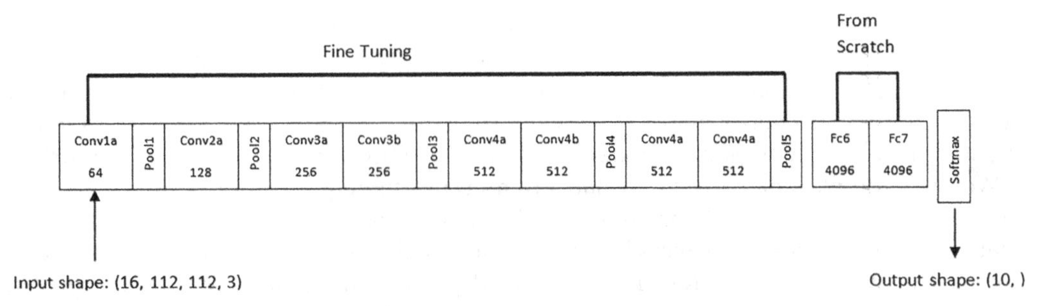

FIGURE 21.3
C3D model proposed architecture.

rate of 0.003 with learning rate scheduler, which divides the learning rate by a factor of 2 after every 150 thousand iterations. To initialize the model, these pretrained weights were used and the resultant features for a specific image were derived (Figure 21.3).

The clips dataset was used for model fine-tuning, with each frame cropped to a shape of (112, 112, 3). Each video in the clips dataset was divided into 16 frames, which aided in converting each video into an input vector of shape (16, 112, 112, 3) for the model's input. This model was trained over 20 epochs using a batch size of 12 and ADAM as the optimizer.

21.3.5.2 FCNN

The second model used for action classification was a simple FCNN, which takes as input 14 (x, y) coordinates derived using CPM for 16 frames. Joints dataset was used for accomplishing training on this network. The joints dataset has 16 different lists (marking different frames) of identified joints for a particular video sample. As a result, we have 16×14 $[x, y]$ shaped vectors, which equal 448 features as an input for a video. For a video, the input vector is like Equation 21.1:

$$\text{Input vector} = \left[X_i\ Y_i\right], \quad i \in \left[0, 223\right] \tag{21.1}$$

All the (x, y) coordinates are normalized between [0, 1], and the value of (–1, –1) is entered if a specific joint is not present or not recognized by the CPM model (Figure 21.4).

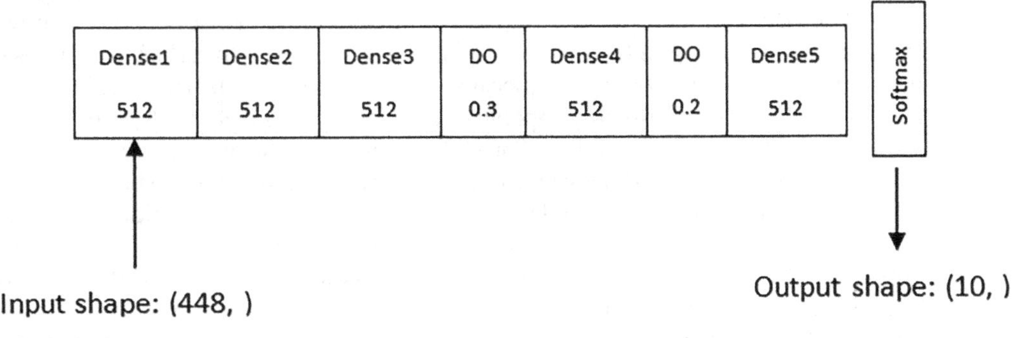

FIGURE 21.4
FCNN proposed architecture.

This model is trained for 50 epochs with a batch size of 12. The learning rate of 1e-3 is used for the first 20 epochs. For the remaining 30 epochs, the learning rate of 1e-3 is used. ADAM is used as the optimizer.

21.3.5.3 Hybrid Model

For using both models in the Hybrid model, we have utilized an average scoring function determined using hyper as library (for hyperparameter tuning), which gives 89% weightage to C3D model predictions and 11% to FCNN model predictions. Results for both of these are combined by adding their respective softmax probabilities using Equation 21.2:

$$\text{Label Probability} = 0.871 * C3D + (1 - 0.871) * \text{FCNN} \tag{21.2}$$

Label with maximum probability is the predicted label for the individual basketball player.

21.4 Results

This section presents the results obtained at different steps in our methodology.

21.4.1 OpenPose Results

Figure 21.5 depicts different body points obtained by the OpenPose for a particular player. The dots represent the joints identified by OpenPose.

FIGURE 21.5
Result of OpenPose on test image.

21.4.2 Court Segmentation Results

Figures 21.6 and 21.7 show the results generated by using the court segmentation algorithm. The figures depict the number of players identified by the OpenPose. Then, the court segmentation method reduces this count of detected players by only considering the players who are present on the patch of the court. This algorithm was tested on a video consisting of 210 frames. The persons detected by OpenPose in these frames were 8,949, implying an average of 42 persons being detected in each frame. After filtering out we obtained 1,264 total players which indicate an average of six players per frame. The filtering reduces the data and saves processing time for tracking algorithms by reducing the number of players to be tracked.

21.4.3 Player Tracking Results

Figures 21.8–2.10 show sample results of multiple players tracked over four consecutive frames. The id on the player's bounding box represents this tracking.

21.4.4 Hybrid Model Results

Accuracy is utilized to evaluate the performance of the Hybrid model, C3D model, and FCNN model. The Hybrid model considered the visual features of the players and their joint coordinates to perform the classification. The Hybrid model achieved 95% accuracy, with a near follow-up by C3D Model (Table 21.3). However, the accuracy achieved in the case of the FCNN model is around 58%, hence it is not suitable for the classification purpose. The video samples showing the actions of players with the Hybrid model are depicted in Figure 21.11.

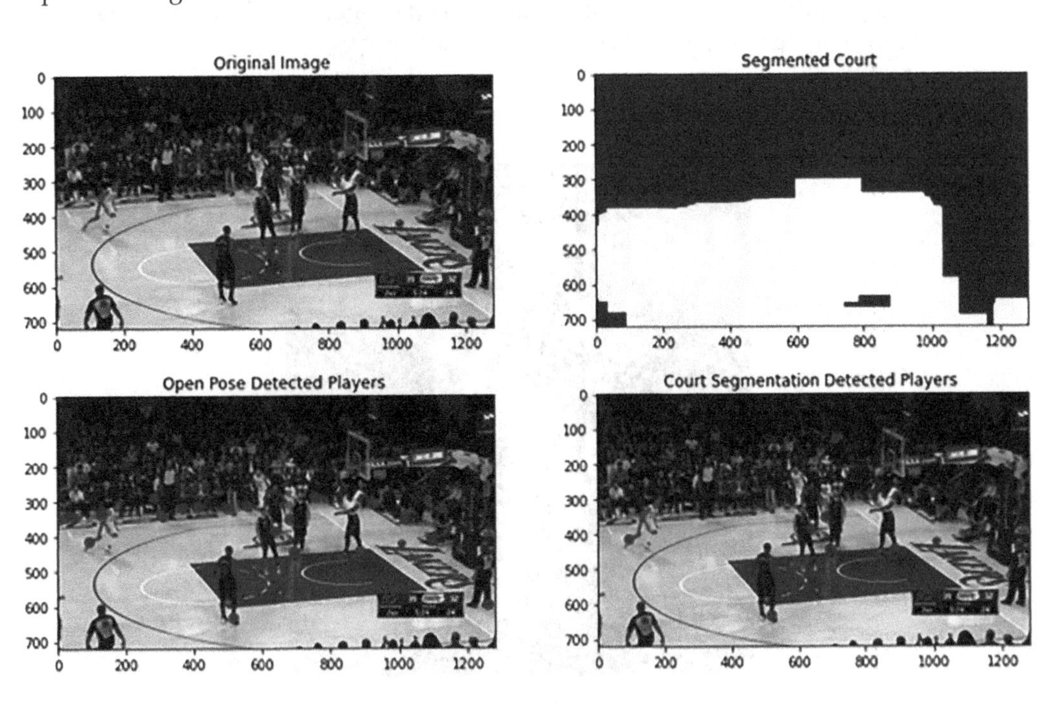

FIGURE 21.6
Court segmentation result sample 1.

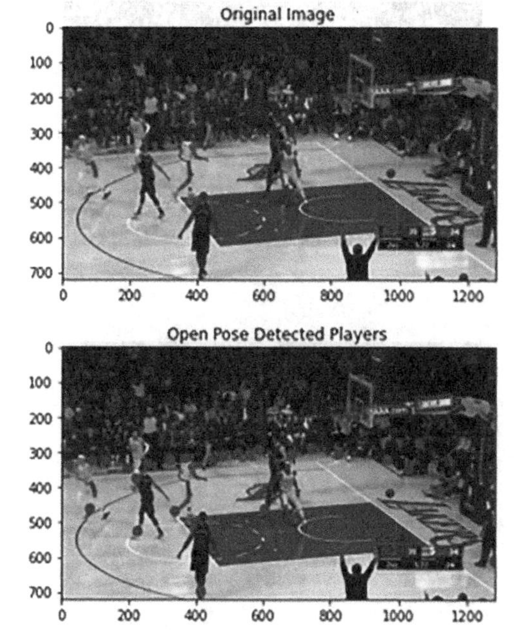

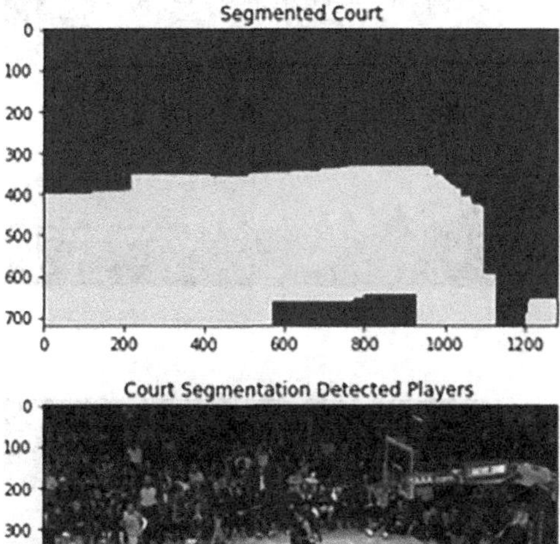

FIGURE 21.7
Court segmentation result sample 2.

FIGURE 21.8
Player tracking result sample 1.

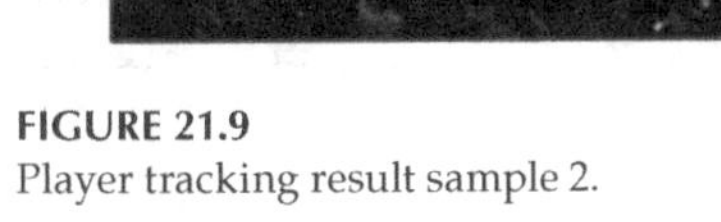

FIGURE 21.9
Player tracking result sample 2.

FIGURE 21.10
Player tracking result sample 3.

TABLE 21.3
Result Statistics

S. No.	Precision	Recall	F1-Score	Support
0	0.95	0.93	0.94	90
1	0.96	0.90	0.93	98
2	0.95	0.96	0.95	582
3	0.96	0.98	0.97	369
4	1.00	0.92	0.96	39
5	0.97	0.96	0.96	254
6	0.95	0.96	0.95	390
7	0.97	0.95	0.96	79
8	0.95	0.94	0.95	659
9	0.94	0.95	0.94	1,149
Accuracy	0.95	0.95	0.95	3,709
Macro avg.	0.96	0.94	0.95	3,709
Weighted avg.	0.95	0.95	0.95	3,709

FIGURE 21.11
Multiplayers result.

21.5 Conclusion

The classification of player actions in basketball game videos using a deep neural network has proven to be a good approach for the defined problem, yielding good, albeit lower, results when compared to state-of-the-art models on action recognition. For posture classification, the convolutional 3D model is discovered suitable. The FCNN model trained on pose estimations to classify the actions of players using their joint coordinates obtained using OpenPose has achieved an accuracy of 58% on the test data, making it less suitable from the viewpoint of accuracy results than the C3D approach. Finally, the Hybrid model where the output of the two models is combined to produce an accuracy of 95%.

References

Acuna, D. "Towards real-time detection and tracking of basketball players using deep neural networks," In *Proceedings of the 31st International Conference on Neural Information Processing Systems (NIPS)*, pp. 1–6, 2017.

Arbués-Sangüesa, A., Ballester, C., & Haro, G. Single-camera basketball tracker through pose and semantic feature fusion. arXiv preprint arXiv:1906.02042, 2019

Ballan, L., Bertini, M., Bimbo, A. D., and Nunziati, W. "Soccer players identification based on visual local feature," In Proceedings of the 6th ACM international conference on Image and video retrieval (pp. 258–265), 2007, July.

Cai, Y., Freitas, N. D., and Little, J. J. "Robust visual tracking for multiple targets," In *European Conference on Computer Vision*, pp. 107–118. Springer, Berlin, Heidelberg, 2006.

Cao, Z., Simon, T., Wei, S., and Sheikh, Y., "Realtime multi-person 2D pose estimation using part affinity fields," In *Conference on Computer Vision and Pattern Recognition(CVPR)*, pp. 7291–7299, 2017.

Cheshire, E., Halasz, C., and Perin, J. K. "Player tracking and analysis of basketball plays," In *European Conference of Computer Vision*, 2013.

Golda, T., Kalb, T., Schumann, A., and Beyerer, J. "Human pose estimation for real-world crowded scenarios," In *16th IEEE International Conference on Advanced Video and Signal Based Surveillance (AVSS)*, 2019.

Karpathy, A., Toderici, G., Shetty, S., Leung, T., Sukthankar, R., and Fei-Fei, L. "Large-scale video classification with convolutional neural networks," In *2014 IEEE Conference on Computer Vision and Pattern Recognition, Columbus, OH*, pp. 1725–1732, 2014, doi: 10.1109/CVPR.2014.223.

Li, X., and Chuah, M. C. "ReHAR: robust and efficient human activity recognition," In *2018 IEEE Winter Conference on Applications of Computer Vision (WACV), Lake Tahoe, NV*, pp. 362–371, 2018, doi: 10.1109/WACV.2018.00046.

Lu, W. L., Ting, J. A., Murphy, K. P., and Little, J. J. "Identifying players in broadcast sports videos using conditional random fields," In *CVPR*, pp. 3249–3256, 2011.

Ludl, D., Gulde, T., and Curio, C. "Simple yet efficient real-time pose-based action recognition," In 2019 IEEE Intelligent Transportation Systems Conference (ITSC) (pp. 581–588), 2019, October.

Redmon, J., and Farhadi, A. "YOLOv3: an incremental improvement." ArXiv:1804.02767 [Cs], 2018. arXiv.org, http://arxiv.org/abs/1804.02767.

Ren, S., He, K., Girshick, R., and Sun, J. "Faster R-CNN: towards real-time object detection with region proposal networks." In *Advances in Neural Information Processing Systems*, pp. 91–99, 2015.

Shitrit, H. B., Berclaz, J., Fleuret, F., & Fua, P. Tracking multiple people under global appearance constraints. In 2011 IEEE International conference on computer vision (pp. 137–144)., 2011, November.

Siyi, S., Kavitha, M., Junichi, M., and Takio, K. "Action classification based on 2D coordinates obtained by real-time pose estimation," In *International Workshop on Frontiers of Computer Vision (IW-FCV)*, 2019.

Tran, D., Bourdev, L., Fergus, R., Torresani, L., and Paluri, M. "Learning spatiotemporal features with 3D convolutional networks," In Proceedings of the IEEE international conference on computer vision (pp. 4489–4497), 2015.

Wei, S., Ramakrishna, V., Kanade, T., and Sheikh, Y. "Convolutional pose machines," In *IEEE Conference on Computer Vision and Pattern Recognition (CVPR)*, pp. 4724–4732, 2016.

Ye, Q., Huang, Q., Jiang, S., Liu, Y., and Gao, W. "Jersey number detection in sports video for athlete identification,". In Visual Communications and Image Processing 2005 (Vol. 5960, pp. 1599–1606). SPIE, 2005, June.

Index

Note: **Bold** page numbers refer to tables; *italic* page numbers refer to figures.

9780367495220